OXFORD MEDICAL

A Guide to Effective Care in Pregnancy and Childbirth

A Guide to
Effective Care in Pregnancy and Childbirth

Second Edition

Murray W. Enkin
Professor Emeritus, Departments of Obstetrics and Gynaecology, Clinical
Epidemiology and Biostatistics, McMaster University,
Hamilton, Ontario, Canada

Marc J. N. C. Keirse
Professor of Obstetrics and Gynaecology, The Flinders University of
South Australia, Flinders Medical Centre, Adelaide, Australia

Mary J. Renfrew
Professor of Midwifery Studies, Centre for Reproduction Growth and
Development, Research School of Medicine, University of Leeds, Leeds, UK

James P. Neilson
Professor of Obstetrics and Gynaecology, University of Liverpool,
Liverpool Women's Hospital, Liverpool, UK

with the editorial assistance of
Eleanor Enkin

Oxford   New York   Tokyo   Toronto
OXFORD UNIVERSITY PRESS

Oxford University Press, Great Clarendon Street, Oxford OX2 6DP

Oxford New York

*Athens Auckland Bangkok Bogota Bombay Buenos Aires
Calcutta Cape Town Dar es Salaam Delhi Florence Hong Kong
Istanbul Karachi Kuala Lumpur Madras Madrid Melbourne
Mexico City Nairobi Paris Singapore Taipei Tokyo Toronto*

*and associated companies in
Berlin Ibadan*

Oxford is a trade mark of Oxford University Press

*Published in the United States
by Oxford University Press Inc., New York*

*First published 1995
Reprinted 1996 (twice)*

A catalogue record for this book is available from the British Library

Library of Congress Data available on request

*ISBN 0 19 262326 5 Hbk
ISBN 0 19 262324 9 Pbk*

*Printed and bound in Great Britain by
Biddles Ltd, Guildford and King's Lynn*

*Dedicated to the women who participated
in the controlled trials
on which so
much of the evidence in this
book is based.*

Preface

Work towards *A guide to effective care in pregnancy and childbirth* started almost twenty years ago when Iain Chalmers began to assemble a register of controlled trials in perinatal medicine. He was joined by the members of the present editorial team over the ensuing years, and together we began the systematic review of the effects of care during pregnancy and childbirth which forms the basis for this book. We first established explicit criteria to identify studies which were likely to provide the best evidence for evaluating care. As well as using the MEDLINE database to locate studies that met our criteria, we organized a systematic hand search of over 60 key journals, from the 1950 issues onwards. We wrote to over 40 000 obstetricians and paediatricians in 18 countries in an attempt to identify unpublished studies; we also asked the authors of many published studies for relevant unpublished data. Having assembled the raw material for our review, we and our colleagues began a systematic synthesis of all these data, using methods to reduce bias and random error. The result of this work was two major publications: *Effective care in pregnancy and childbirth*, a 1500 page book in two large volumes, and a regularly updated electronic database of reviews. The latter is now incorporated in *The Cochrane database of systematic reviews* and is available on disc as *The Cochrane pregnancy and childbirth database*.

We prepared *A guide to effective care in pregnancy and childbirth* to make the conclusions of these larger publications more readily available, in a more portable format, to all who are involved in the care of childbearing women. The first edition of *A guide to effective care in pregnancy and childbirth* appeared in 1989. It received critical acclaim because its conclusions were based on sound evidence, and it has been widely used by policy-makers, obstetricians, midwives, childbirth educators, and childbearing women themselves as a reference source for evaluating care policies and practices. However, it has become out of date. While new evidence has been steadily accumulating, and has been regularly incorporated into the systematic reviews published in successive disc issues of *The Cochrane pregnancy and childbirth database*, the printed word stood still. It is time for a new edition.

This second edition of *A guide to effective care in pregnancy and childbirth* has been extensively revised and rewritten, incorporating the results of the important evidence about the effects of care during pregnancy and childbirth that has accumulated since the first edition was published. References to the primary sources of the information it contains can be found in *Effective care in pregnancy and childbirth* and in the systematic reviews cited from *The Cochrane pregnancy and childbirth database*. Each of these publications fills a specific niche. We hope that they will help those providing and receiving care during pregnancy and childbirth.

Murray Enkin, Marc Keirse, Mary Renfrew, Jim Neilson

Acknowledgements

We owe particular thanks to the authors of each chapter of *Effective care in pregnancy and childbirth*, and to those who prepared the systematic reviews for the *Cochrane pregnancy and childbirth database*. All of this book is derived from, and extensively quotes, their contributions. We are well aware and appreciative of the countless hours that they spent in reviewing and analysing the data on which those contributions were based. The credit for the review and analysis is theirs; the blame for any errors that may have crept into our condensation of their work rests with us.

We are also grateful to those who have read and given valuable comments and suggestions on portions of this book: in particular, Hilda Bastian, Eva Bild, Beverley Chalmers, Iain Chalmers, Patricia Crowley, Caroline Crowther, Lelia Duley, Diana Elbourne, Jane Enkin, Bill Fraser, Jo Garcia, Adrian Grant, Gill Gyte, Mary Hannah, Jean Hay-Smith, Andrew Herxheimer, Ellen Hodnett, Justus Hofmeyr, Charlotte Howell, David Jewell, Richard Johanson, Frank Johnstone, Karyn Kaufman, Michael Kramer, Sandra Lang, Judith Lumley, Cheryl Nikodem, Andy Oxman, Walter Prendiville, Chris Redman, Penny Simkin, Fiona Smaill, Jon Tyson, Steve Walkinshaw, Gavin Young, and Alvin Zipursky.

The Research and Development Division of the Department of Health (England) provided the necessary support for the Cochrane Pregnancy and Childbirth Group, and additional funding from the North East Thames Regional Health Authority helped to expedite the preparation of the work. Oxford University Press was helpful at all times. The World Health Organization provided the initial funding for the register of controlled trials in perinatal medicine on which much of this work is based, and the Rockefeller Foundation gave us the opportunity to work on the first edition at their study centre in Bellagio. McMaster University, Leiden University, the University of Liverpool, and the Department of Health (England) allowed the editors the time necessary to work on this project. Support for individual reviewers was received from the British Heart Foundation; College of Physicians and Surgeons, Ireland; Department of Health (England); Leiden University; McGill University; McMaster University; Monash University; National Health Service R&D Programme, England; Overseas Development

Administration, UK; Trinity College, Dublin; University of Adelaide; University of Bristol; University of Keele; University of Liverpool; University of Toronto; University of Utah; University of Vermont; University of Zimbabwe; University of the Witwatersrand; Université Laval; and West Midlands (England) Regional Health Authority.

Finally, we thank Jini Hetherington who kept us on course and Lizi Holmes who helped in so many ways to keep this project moving through the innumerable drafts of the manuscript. This task would have been impossible without them.

Murray Enkin, Marc Keirse, Mary Renfrew, Jim Neilson

Contents

BASIC CARE

SCREENING

PREGNANCY PROBLEMS

CHILDBIRTH

PROBLEMS DURING CHILDBIRTH

TECHNIQUES OF INDUCTION AND OPERATIVE DELIVERY

CARE AFTER CHILDBIRTH

SYNOPSIS

BASIC CARE

1 Effective care in pregnancy and childbirth

The care that women and babies receive during pregnancy and childbirth should be effective. Although no one is likely to disagree with this in principle, there is still a great deal of disagreement as to what constitutes effective care. The disagreement includes differences of opinion both about what the objectives of care should be and about the best means of achieving them.

The objectives of care, or the relative emphases placed on particular objectives, depend on what individuals or communities think is most important. Priorities may range from women's enjoyment of the experience of childbirth to reducing perinatal mortality by a fraction of a percentage point, regardless of the human and economic resources expended on the effort. This diversity has resulted in widely differing recommendations for care during pregnancy and childbirth.

Differing views about the objectives of care also help to explain the disparate indicators used to assess the effects of care. Some people rate women's satisfaction with the care that they have received as the most important measure of its effectiveness. Others seem to regard indirect measures of the baby's well-being, such as fetal heart rate tracings or estimates of the acid–base status of umbilical cord blood, as more important.

These differences of opinion show themselves in the dramatic variations in the forms of care practised from country to country, from community to community, from institution to institution, and from one caregiver to another. Innumerable examples come to mind: the variety of methods used to assess the risk status of the mother and the well-being of the fetus; the place, if any, for routine iron or vitamin supplements during pregnancy; the utility of routine examination of the cervix at each antenatal visit; the value of routine ultrasound visualization of the fetus for all pregnancies; the need for bed-rest for women with an uncomplicated twin pregnancy; the value of corticosteroids for fetal lung maturation; the appropriate indications for the use of forceps, vacuum extraction, or caesarean section. The list could be endless.

A number of other factors, in addition to priorities and evidence of effectiveness, may help to explain these variations in practice. Some relate to differences in the needs or circumstances of childbearing women and their babies. Others reflect differences in culture, tradition, status, and fashion. Still others result from differences in resources, including personnel, hospital beds, and equipment. There may be differences in the need to provide opportunities for clinicians in training to gain experience, in the extent to which malpractice litigation is feared, and in the way that caregivers are paid. Commercial pressures from pharmaceutical companies, equipment manufacturers, and others may also influence practice.

The focus of this *Guide to effective care in pregnancy and childbirth* relates to none of these. Rather, it addresses the uncertainty felt by those who provide or receive care about the effectiveness and safety of many of the elements of care given during pregnancy and childbirth.

Sometimes the belief that one form of care is better than another is based on informal impressions derived from personal experience or previous teaching. Such impressions are sometimes right and sometimes wrong. The impression that women are less likely to sustain injury during delivery with the vacuum extractor than with forceps has been confirmed by the results of formal studies mounted to investigate this possibility. Other beliefs, such as the widely held conviction that diethylstilboestrol could prevent miscarriages and fetal death, have been refuted by the results of properly controlled studies. All too often this confirmation or refutation happens only after irreparable damage has already been done. Unless the validity of informal impressions about the effects of care is assessed by formal evaluation, effective forms of care will not be recognized as such and will not be brought into use as promptly as possible. Ineffective or harmful forms of care will not be detected efficiently and may do harm on a wider scale than necessary.

In the next chapter we outline the rationale, materials, and methods that have been used to arrive at the conclusions presented in subsequent chapters in this book.

This chapter is derived from Chapter 1 in EFFECTIVE CARE IN PREGNANCY AND CHILDBIRTH. References to primary sources and more complete data for the statements made in this chapter can be found in the source chapter.

2 Evaluating care in pregnancy and childbirth

1 Introduction

The number of formal studies that attempt to address uncertainties about the effects of aspects of care during pregnancy and childbirth is overwhelming. However, not all of these studies provide reliable information. If judgements about the effects of care based on formal studies are to be valid, careful consideration must be given to the strengths and weaknesses of the methods used by the investigators. In this way a rational basis can be established for selecting those studies that are most likely to provide useful evidence.

Formal studies can be arranged in a hierarchy that reflects the likelihood that biases (systematic errors) will result in misleading conclusions. In addition, it is possible to estimate the extent to which the play of chance (random errors) may be misleading.

2 Minimizing systematic errors (biases)

Sometimes past experience provides a sufficient basis for making a valid assessment of the effects of care. This will only be the case when these effects are dramatically different from what would have been expected on the basis of past experience. For example, case reports have shown that prostaglandin administration may be life-saving when used to treat otherwise uncontrollable haemorrhage because the uterus fails to contract after birth. Dramatic effects such as this are rare. Usually, one is trying to detect more modest differences (which are nevertheless important) in the effects of alternative forms of care. Single case reports and case series without formal comparison groups (controls) cannot provide a secure basis for making valid judgements about the effects of care. Such studies are subject to a variety of biases that may either mask real differences between alternative forms of care or suggest that differences exist when, in fact, they do not.

Bias may also affect studies that do have formal comparison groups. In the sections that follow we discuss two of the most important sources of bias. The first of these occurs during the selection process that leads people to receive one form of care over another; the second can result when those providing, receiving, or evaluating care know which one of two or more alternative forms of care has been received.

2.1 *Minimizing bias in the selection of controls*

Uncontrolled observations of events after a particular form of care has been given usually do not answer what might have happened if a different form of care had been provided. Judgements about the effects of care require comparisons of what happens to people who have received one form of care with what happens to the 'controls' who have received an alternative form of care (or no care at all).

Differences between the experiences of people who have received alternative forms of care can be due to either differential effects of the different forms of care or differences in the pretreatment characteristics (prognoses) of the people who received these forms of care. How much of the differences in outcome can be attributed to the effects of the treatment thus depends on the extent to which the people who received the different forms of care are comparable in every other respect that matters. The challenge is to select comparison groups that are comparable in every important respect.

2.1.1 *Studies using historical controls*   One approach to the selection of controls involves making comparisons between people who have received a recently introduced form of care with similar people who received a different form of care in the past. The use of such 'historical controls' sometimes leads to valid inferences, but at other times it can be seriously misleading. For example, studies with historical controls suggested that administration of diethylstilboestrol during pregnancy dramatically decreased the risk of miscarriage and stillbirth, thus giving rise to one of the best known examples of harmful treatment.

Even when outcome following current treatment appears to differ substantially from outcome after earlier forms of care, the difference may simply be a reflection of changes in other, undocumented, factors that have modified the outcome over time. Without concurrent comparisons between alternative forms of care there is no way of knowing which of the studies using historical controls provide reliable data about the effects of care and which do not. The most useful role for comparisons using historical controls may therefore be as 'screening tests' for promising new forms of care, which can then be assessed in properly controlled prospective experimental studies.

2.1.2 *Case–control studies* The underlying principle of a case–control study is straightforward: groups of people who have and who have not experienced a particular outcome are assembled. Then the frequencies with which each group has received the form of care in question are compared. This approach is particularly valuable when the postulated outcome of care is rare or when it cannot be ascertained for some months or years after the particular care has been received. For example, when cases of cerebral palsy were compared with controls, no difference in the frequency of substandard care during labour and delivery was detected, thus casting further doubt on the widespread belief that the quality of intrapartum care is an important factor in the aetiology of cerebral palsy.

Although case–control studies may sometimes offer the only practicable research strategy for evaluating some of the postulated effects of care during pregnancy and childbirth, they are subject to a variety of biases that restrict their value. Some of these biases may be eliminated by careful matching of cases and controls, but it is never possible to know how successful such measures to reduce selection and other biases have been. No amount of matching based on information about known confounding factors can ever eliminate the effect of unrecognized confounding factors. Thus conclusions about causes and effects based upon case–control studies are often insecure.

Like the results of studies using historical controls, the results of case–control studies are sometimes supported and sometimes not supported by those of studies that are less subject to bias. In many instances, however, there are simply no unbiased comparisons available for judging the validity of inferences based on the results of case–control studies. In these circumstances, consistent findings from a number of well-designed case–control studies may provide the best evidence that is ever likely to be available.

2.1.3 *Studies using non-randomized concurrent controls* A common approach to controlled evaluation involves concurrent comparison of two or more groups of individuals who have received different forms of care. Before making causal inferences about the effects of care on the basis of such studies, one must be convinced that 'like has been compared with like'.

There are a number of ways in which bias can affect comparisons between non-randomized concurrent groups (cohorts) receiving different forms of care. For example, comparisons have been made between infants with very low birthweight delivered by caesarean section and other such babies delivered vaginally. In most reports of these comparisons, infants delivered by caesarean section were more likely to survive than those delivered vaginally. Some people have concluded from these observations that caesarean section is a safer method of delivery for babies with very low birthweight. This conclusion would not be justified unless the two groups of babies could be shown to be at comparable prior risk of death and morbidity. Caesarean section is less likely to be used to deliver babies whose chances of survival are judged to be minimal anyway; vaginal delivery is more likely to have occurred when labour has been precipitate, in itself a risk factor for poor outcome. These and other factors of prognostic importance can introduce bias in the non-randomized comparisons of these two methods of delivery. As in studies using historical controls and in case–control studies, the conclusions drawn from non-randomized cohort studies may be invalid because important, but unknown, selection biases have not been controlled adequately.

2.1.4 *Studies using randomized controls*   There is only one certain way to overcome the bias that results from people at different prior risk selectively receiving one of the alternative forms of care that are compared. This is to conduct a prospective experiment in which chance (randomization) is used to determine which of the alternative forms of care a particular woman or baby should receive. Randomization not only controls selection biases from factors known to be important; it is the *only* known way to control for *unknown* selection biases.

Randomization does not guarantee, nor does it need to guarantee, that the comparison groups will be exactly matched in respect of all characteristics of prognostic importance. What randomization does guarantee is that the members of the comparison groups will be selected by chance rather than by any biased form of selection. The statistical test procedures used to compare the outcomes in the two groups take into account the probabilities that chance imbalances may affect the study results.

The logic underlying the use of randomized controls in prospective experiments to create comparable groups of people for comparing alternative forms of care has great force once it is clearly perceived. Indeed, the randomized controlled trial has become widely accepted as the methodological 'gold standard' for comparing alternative forms of care.

The fact that a formal comparison of two or more alternative forms of care is reported to be a randomized controlled trial is not a guarantee that selection bias has been eliminated. Unless adequate precautions are taken, potential participants in a 'randomized' comparison may be selectively recruited into the study depending on their or their caregivers' prior knowledge of the group to which they have been allocated. Futhermore, they may be selectively 'withdrawn' from the study either before or after formal entry.

The selection bias that results from tampering with the make-up of the randomized groups in these ways is sometimes a more important determinant of the differences in outcome than the effects of the forms of care that are compared; comparisons of the groups can then be misleading. For these reasons it is important to assess the likelihood of selection bias in studies purporting to be randomized comparisons of alternative forms of care when assessing whether the results should be used to guide practice.

2.2 Minimizing other biases

The second major source of bias results when those receiving, providing, or evaluating care know which of the alternative forms of care has been received. This bias can be reduced and sometimes eliminated by 'masking' or 'blinding' (keeping those administering or receiving care unaware of the particular form of care that is being used).

Masking is worthwhile when one or more of the forms of care being compared is likely to have psychologically mediated effects on the outcomes of interest. The expectation that a form of care will have certain effects may result in a self-fulfilling prophecy. This phenomenon is known as the 'placebo effect' (literally, 'I will please') when the effects are pleasant or beneficial in some other way. When the effects are unpleasant or unwanted, it is referred to as 'symptom suggestion'.

Another reason for trying to keep caregivers unaware of the forms of care that are compared is to reduce the extent to which they may adjust the remainder of their care in the light of this knowledge. This 'co-intervention' may make the results of the comparison more difficult to interpret.

Lastly, knowledge of which form of care has been received can affect people's perception of the outcomes. This can occur, for example, if the people assessing the outcome of treatment consciously or unconsciously believe that one of the forms of care is superior to the other; they may tend to record the outcomes in ways that confirm their expectations.

Protection against these observer biases is not a problem when the outcome in question is unambiguous (death is the obvious example). Observer biases among those assessing less unambiguous outcomes of care, for example neonatal jaundice, can be reduced, and sometimes abolished, by masking what care has been received. When it is either not practicable or not possible to mask the identity of the alternative forms or care, observer bias may be eliminated by having the outcomes assessed by 'independent' observers who are not aware of the treatment allocation.

3 Minimizing random errors (the play of chance)

Even after successful control of selection biases and other biases that can distort comparisons of alternative forms of care, the results of such comparisons may still be misleading because of *random errors*. Unlike systematic errors, random errors result from the play of chance and they are reduced by increasing the size of the sample studied.

Tests of statistical significance are used to assess the likelihood that the observed differences between alternative forms of care may simply be a reflection of random errors (chance). These tests are used to prevent people inferring that a real difference exists when it does not. Unfortunately, differences between alternative forms of care that are not statistically significant tend to be dismissed as simply reflecting the play of chance. Often, this conclusion is not warranted. Failure to detect a difference does not mean that a difference does not exist.

Random errors will be reduced as samples yielding larger numbers of the outcomes of interest are studied. This can be achieved both by conducting larger trials than has been usual in the past and by incorporating all the available data from broadly similar trials within a particular field of enquiry in systematic reviews (meta-analyses).

Estimating the range within which the true differential effects of alternative forms of care is likely to lie, by calculating a confidence interval for the statistically estimated differences in the outcome of care, provides further protection against being misled by random error.

4 Applying the results of research

Even if one is reasonably certain that systematic and random errors have been adequately controlled in a particular study or in an overview of similar studies, questions may remain about the extent to which this research evidence forms a valid basis for guiding care of individual women or babies. This 'external validity' of research may be compromised either because there are crucial differences between participants in these studies and people receiving care in other contexts, or because there are differences in the nature of the care given within the studies and that provided in usual clinical practice.

Sometimes the limited applicability of the study or overview results may be relatively clear. The results of trials assessing the value of vitamin D supplementation in women deprived of sunlight, for example, may well not be relevant to other populations. More usually, it is not possible to conclude with any confidence whether the results of controlled trials are or are not applicable in practice. It would be rare, however, for a particular form of care to have opposite effects in specific subcategories of individuals. Any real differences in the effects of care between participants in controlled trials and apparently similar people seen in the context of everyday clinical practice are more likely to be differences in the magnitude, rather than the direction, of the effects. Judgements to guide practice must then be made more in terms of whether the size of the effect is sufficient to warrant changes in usual clinical practice. These judgements will often involve social and economic dimensions.

5 Conclusions

The consequences of being misled by systematic errors (biases) and random errors (the play of chance) are that some women and babies will be denied effective care during pregnancy and childbirth, while others will receive care that is ineffective or actually harmful. In the hierarchy of evidence that we have adopted for arriving at the conclusions presented in this book, the results of studies comparing groups of people who happened, for one reason or another, to have received one or other form of care (observational evidence) have been distinguished from evidence derived from comparisons of people who have been randomized, in prospectively planned trials, to alternative forms of care (experimental evidence). The distinction between these two kinds of evidence is crucial. Uncontrolled case series and studies using non-randomized controls so frequently lead to biased estimates of the effects of different forms of care that they should usually be seen as 'screening tests' for identifying forms of care that *may* be valuable, rather than be used as a basis for guiding clinical practice.

Although small well-designed randomized trials may offer protection against the possibility of being misled by bias (systematic error), too often they provide little protection against being misled by the play of chance (random error). In the analyses on which this book is based, this problem has been addressed, whenever possible, by grouping similar studies together in systematic reviews (meta-analyses). Our objectives in this book have been to use the strongest available evidence, and to bring that evidence together in a way that will help those who wish to give or receive effective care in pregnancy and childbirth.

This chapter is derived from the chapters by Iain Chalmers (1), Iain Chalmers, Jini Hetherington, Diana Elbourne, Marc J. N. C. Keirse, and Murray Enkin (2), Jane Robinson (4), and Miranda Mugford and Michael F. Drummond (5) in EFFECTIVE CARE IN PREGNANCY AND CHILDBIRTH. References to primary sources and more complete data for the statements made in this chapter can be found in the source chapters.

3 Social, financial, and psychological support during pregnancy and childbirth

1 Introduction   2 Social and financial support
3 Access to care   4 Psychological support   5 Caregivers
6 Conclusions

1 Introduction

Outcomes of pregnancy and childbirth depend to a large extent on the social policies and health care organization of the country in which the woman lives. Her health, her use of and response to health services, and her ability to follow the advice that she is offered are affected by her own social circumstances and by the wider social, financial, and health care policies.

Public and private concerns meet in many aspects of maternity care. Antenatal advice about rest or admission to hospital during pregnancy, for example, often does not take into account a woman's circumstances. The woman must weigh the benefits that she may gain from following the advice against its financial and social costs. Soundly based dietary advice may be ineffective if women are unable to follow it because of cost, or because it fails to take into account the constraints of cultural or family customs regarding food.

There is an association between a woman's social situation and both her health and her utilization of health services. This can be modified to only a limited extent by policies of social and financial support for childbearing families. In complex societies, fiscal, economic, social, and other policies all interact.

2 Social and financial support

Most industrialized countries provide direct financial aid to childbearing families, although the amount of aid women receive varies dramatically among countries. In addition to maternity benefits, there are other aspects of the welfare and taxation systems to be considered, such as family or tax allowances. In The Netherlands, for example, provision for 'maternity aides' is an integral part of the maternity system. These specially trained women provide support for up to seven days postpartum for the substantial proportion of women who are at home for some or all of this period.

Most industrialized countries also have legislation intended to protect the fetus, newborn, and mother from the general and specific harmful effects of work, to protect employment by enabling parents to keep jobs while caring for children, and to provide income maintenance for parents during breaks in employment. Many countries have laws that restrict the type of work that pregnant women can do. Contact with low temperatures, lead, ionizing radiation, and other hazards may be controlled by law. Pregnant women may be barred from night work or long working hours. In some countries, employers may not be allowed to employ a woman in the period just before or just after delivery.

There has been considerable debate about this type of legislation because, while it has laudable aims, it can restrict women's employment opportunities and result in them losing earnings during the childbearing period. Legislation of this nature, in the absence of adequate unemployment benefits or alternative work, may lead some women to conceal their pregnancies and to avoid seeking care. One way of avoiding some of this effect is to have laws that protect women from dismissal on the grounds of pregnancy and that offer alternative work or compensation to women if their usual jobs are thought to be dangerous.

The other main area of legislation concerns leave, reinstatement, and income maintenance during maternity or parental leave. Most industrialized countries allow all employed women to have paid leave around the time of birth. For legislation to be effective in protecting parents and infants from stress and hardship, the level of income replacement during leave must be adequate. In some countries the maternity allowance is the same as, or close to, the woman's usual earnings, while in others it is fixed at a lower rate. If benefits are too low, women will be more likely to work during their period of maternity leave.

There is no evidence at present which would allow one to determine the optimal timing or length of leave. Some types and aspects of work seem more likely than others to compromise health during pregnancy. Different women may have different requirements, and more flexible leave arrangements are needed to allow some pregnant women to take time off earlier in pregnancy.

3 Access to care

The best care will not be effective if it is not available to those who need it. The cost of obtaining care can be a major impediment to access. For low-income people there is a close association between the lack of support for medical costs and low uptake of medical services. In many countries, teenagers, immigrants, and socially marginal women may delay seeking care because of feeling ill at ease in conventional care settings. They are made uncomfortable by their difficulty in communication with staff, the frequent impossibility of following the advice that they are given, and, often, the reactions of caregivers. Partly because of problems of access and communication, women from lower social classes, as well as women from ethnic minorities, tend to be less well informed about the progress of pregnancy and birth, about potential problems, and about preventive and curative care. In giving greater priority to the preventive aspects of care than to the alleviation of symptoms, the current system of antenatal care is more adapted to the usual behaviour of middle- and upper-class women.

4 Psychological support

The interests of mothers are sometimes forgotten by those who profess an interest in promoting maternal and child health. The social, psychological, and physical problems experienced by pregnant women are often substantial. Those providing care must be sufficiently aware of them. Social and psychological support should be an integral element of all the care provided for pregnant women.

Social support during pregnancy may be particularly important for socially disadvantaged mothers. Such support has been shown to reduce the likelihood of a number of adverse outcomes for the baby, including suspected child abuse/neglect, severe diaper (nappy) rash, chronic otitis media, hospitalization, and delayed well-child immunizations. Supported mothers are less likely to become pregnant again during the next 18 months.

During recent years most attention has been given to trials of support during pregnancy for women who are at increased risk of adverse outcomes (whether or not they are socially disadvantaged). Data from eleven trials, mostly of excellent quality, involving over 8000 women in nine countries on five continents have been reported. The results have been disappointing. Psychosocial support interventions for at-risk pregnant women are not associated with improvements in any medical outcomes for the current pregnancy. Specifically, when compared with women receiving standard care, women at high risk who receive enhanced support during pregnancy experience similar rates of stillbirth, neonatal death, preterm delivery, caesarean section, low birthweight babies, babies with low Apgar scores, or babies who require admission to neonatal intensive care units. The results of different trials, in different places, were remarkably similar. On the evidence available so far, programmes which offer additional support to women who are at high risk during pregnancy are unlikely to improve the outcome of the current pregnancy.

An important question remains about the potential benefits for subsequent pregnancies, since the benefits, if any, of additional pregnancy support may be long term. Also, as with drugs, 'dosage' (length and quality of treatment) is an important consideration: the effective support intervention may need to be prolonged, comprehensive, and intensive. A few months of enhanced care may not be sufficient to overcome the effects of years of adversity.

Giving women more control during pregnancy may have an inherent value. Two small trials have evaluated allowing women to carry their own case notes. This simple reversal of usual practice had no apparent harmful effects and was associated with an increased likelihood of feeling in control during pregnancy. The results of these trials suggest that consideration should be given to a policy of allowing women to carry their own records during pregnancy, and also that other forms of care which offer women greater control during the childbearing period should be evaluated.

5 Caregivers

Both access to care and the extent to which care meets the social and psychological needs of women depend to a large extent on the nature and training of those who provide care during pregnancy and childbirth.

As technical advances became more complex, care has come to be increasingly controlled by, if not carried out by, specialist obstetricians. The benefits of this trend can be seriously challenged. Direct comparisons of care given by a qualified midwife with medical back-up with medical or shared care show that midwifery care was associated with a reduction in a range of adverse psychosocial outcomes in pregnancy, and with reductions in the use of acceleration of labour, regional analgesia/anaesthesia, operative vaginal delivery, and episiotomy. No differences have been demonstrated in the rates of labour induction, pharmacological analgesia, or caesarean section. Midwifery care also resulted in fewer babies weighing less than 2500 grams, needing resuscitation, or needing admission to special care units.

It is inherently unwise, and perhaps unsafe, for women with normal pregnancies to be cared for by obstetric specialists, even if the required personnel are available. Because of time constraints, specialists caring for women with both normal and abnormal pregnancies have to make an impossible choice: to neglect the normal pregnancies in order to concentrate their care on those with pathology, or to spend most of their time supervising biologically normal processes, in which case they would rapidly lose their specialist expertise.

Midwives and general practitioners, however, are primarily oriented to the care of women with normal pregnancies and are likely to have more detailed knowledge of the particular circumstances of individual women. The care that they can give to the majority of women whose pregnancies are not affected by any major illness or serious complication will often be more responsive to their needs than that given by specialist obstetricians.

Women have repeatedly stressed the importance of receiving care during pregnancy and childbirth from the same caregiver, or from a small group of caregivers with whom they can become familiar. Evidence from a controlled trial shows that women who had continuity of caregivers were less likely to use pharmacological analgesia or anaesthesia during labour and birth, to have labour augmented with oxytocin, to have a labour length of more than 6 hours, or to have a baby with a 5 minute Apgar score below 8. They were also more likely to feel well prepared for labour, perceive the labour staff as caring, feel in control during labour, and feel well prepared for child care.

6 Conclusions

When resources of money, time, and energy are limited, the possibilities of making choices that promote health are also limited. People may behave in a way that seems irrational to an outsider, but which is the best choice for them. Pregnant women may have other priorities besides care, such as finding the time and money to provide for children already in the household. A pregnant woman does not leave her work, community, and family responsibilities behind when she steps into the clinic or the doctor's office.

Persons providing maternity care share the collective responsibility for ensuring that effective care is not only known, but is also available, accessible, and affordable to all women who require it. Social and psychological support of pregnant women should be an integral part of all forms of care given during pregnancy and childbirth.

This chapter is derived from the chapters by Sheila Kitzinger (6), Madeleine Shearer (7), Margaret Reid and Jo Garcia (8), Raymond G. De Vries (9), Sarah Robinson (10), Michael Klein and Luke Zander (11), John Parboosingh, Marc J.N.C. Keirse, and Murray Enkin (12), Marc J.N.C. Keirse (13), Jo Garcia, Marie-Josephe Saurel-Cubizolles, and Beatrice Blondel (14), and Diana Elbourne, Anne Oakley, and Iain Chalmers (15) in EFFECTIVE CARE IN PREGNANCY AND CHILDBIRTH.

References to primary sources and more complete data for the statements made in this chapter can be found in the source chapters and/or in the following reviews from the *Cochrane collaboration pregnancy and childbirth database*:

Hodnett, E.D.
— Continuity of caregivers during pregnancy and childbirth. Review no. 07672.
— Support from caregivers for socially disadvantaged mothers. Review no. 07674.
— Support from caregivers during at-risk pregnancy. Review no. 04169.
— Women carrying their own case-notes during pregnancy. Review no. 03776.

Renfrew, M.J.
— Midwife vs. medical/shared care. Review no. 03295.

---

4 Antenatal classes

1 Introduction   2 Content of antenatal classes   3 Effects of antenatal classes   4 Conclusions

1 Introduction

'Natural childbirth' and 'psychoprophylaxis' began as alternatives to what was perceived as overmedicalized obstetrics, with its liberal use of pain-relieving drugs and operative delivery. Many different programmes appeared at about the same time, all with a single common aim: the use of psychological or physical non-pharmaceutical modalities for the prevention of pain in childbirth.

Modern antenatal classes have expanded their horizons beyond that simple objective. Most classes today have additional aims including good health habits, stress management, anxiety reduction, enhancement of family relationships, feelings of 'mastery', enhanced self-esteem and satisfaction, successful infant feeding, smooth postpartum adjustment, and advice on family planning. A major objective is to enhance the woman's sense of confidence as she approaches childbirth.

Because of their complex, often disparate, aims and ideologies, one cannot make general statements about the effects of antenatal classes as if they were a single entity. Research on the effectiveness of antenatal classes over the years reflects their changing emphasis. The early studies focused on the effects of class attendance on labour pain, use of medication, and other qualities of labour. Today the emphasis has shifted to study of the psychological effects, parenting behaviour, and the effectiveness of specific teaching, counselling, or labour coping techniques.

2 Content of antenatal classes

Most antenatal classes involve the use of group sessions to establish a community of experience, and include voluntary muscular relaxation, specific breathing patterns to serve as a focus of attention or distraction, and verbal suggestion for both pain reduction and for appropriate behaviour. They may also include specific massage techniques, suggestions for positions that the mother may wish to try in labour, and other physical comfort measures. They aim to provide accurate and reliable information about pregnancy and birth, and about the experiences that women may undergo or encounter.

The information content of modern antenatal classes may include the relation of pregnancy symptoms to underlying mechanisms, and suggest ways of alleviating these symptoms. The emotional shifts of pregnancy may be explored, and issues of sexuality and of relations with the partner and other children may be discussed as well.

Antenatal classes allow an opportunity to review the mechanisms of labour and birth in adequate detail, and to explain medical and obstetrical terminology as well as the use of tests and other interventions. Information need not come from the instructor alone. Discussion with other participants allows for the reassurance and sense of community that comes from sharing experience and information.

In addition to knowledge and information, most antenatal classes attempt to impart skills for coping with the stress of labour. These often include a variety of relaxation techniques, various forms of attention-focusing and distraction techniques, numerous comfort measures, various types of controlled breathing patterns, and the teaching of labour-support skills to the partners (husbands or others) of the pregnant woman (see Chapter 34).

Finally, antenatal classes can be a vehicle for attitude modification. On the one hand, they may lead towards increased self-reliance and questioning of professional routines and recommendations. On the other hand, they may lead towards increased acceptance of, and compliance with, prescribed medical regimens.

3 Effects of antenatal classes

Antenatal class attendance results in the use of significantly less pain-relieving medication during labour. No other important effects of antenatal classes have been clearly demonstrated. Non-randomized cohort studies have reported a variety of other beneficial effects of antenatal classes, but the self-selection of the study and control groups introduces such major biases that their results must be largely discounted.

There are few studies comparing the pain-relieving effects of different methods of childbirth preparation. The two major methods, Read's natural childbirth and Lamaze's psychoprophylaxis, have never been compared systematically. Because today's antenatal educators learn from a variety of sources, they are less likely to identify themselves with a particular method. Thus it is unlikely that direct comparisons of alternative methods will be carried out.

The use of less pain medication does not necessarily mean that women have less pain. Medication relates only partially to the pain experienced by the woman. Other factors, such as availability (many hospitals do not have 24 hour anaesthesia services), quality of labour support, wish of the mother, or hospital customs, may be major determinants of whether she will receive medication, what kind, and how much.

For these reasons, many investigators have attempted to evaluate the pain of labour on the basis of measures other than medication use. Two findings have been consistently reported: first, there is a wide variation in the pain experienced by women; second, the average level of pain felt during labour is high.

The benefits of antenatal education are difficult to document in a systematic manner. The adverse effects and potential hazards are even more elusive. The extent to which fear is created rather than alleviated by classes, and whether women succumb to peer or educator pressures to conform, or to refuse needed medication or intervention, is completely unknown. There has been little systematic evaluation of the extent to which negative feelings of anger, guilt, or inadequacy are engendered when the expectations of a woman or her partner, possibly raised by the antenatal classes, are not met. There has been equally little evaluation of the potential hazards of classes that teach women to comply with their caregivers' routines without adequate information.

As antenatal classes have become more popular, they have begun to reach more women from less privileged backgrounds. Whereas in years past, antenatal classes appealed primarily to middle class women or couples, they are now routinely offered in many clinics, health departments, and schools. The effects of antenatal classes depend not only on the characteristics of those who attend and the competence and skills of the teacher, but also, to a large extent, on the underlying objectives of the programme. Some classes are taught by independent childbirth educators or co-ordinated by large consumer groups. Others are offered by official health agencies, and still others by doctors for their own patients or by hospitals for the women who plan to deliver there. The curricula outlined for these classes may be similar, and there may be little difference in the information taught or the skills imparted. Nevertheless, there may be great differences in the attitudes that are encouraged. As a general rule community-sponsored childbirth education classes are structured to incorporate the interests of parents into the curriculum. Hospital-based classes may be directed at explaining existing policies to parents, rather than to question them, to offer alternatives, or to help parents decide their own birth plans.

It is possible that the actual existence of antenatal classes is more important than the details of what is taught — that 'the medium is the message'. The number of women attending antenatal classes is now substantial. The full impact of childbirth education cannot be assessed solely by its effect on the individual woman giving birth, for there may be indirect effects that engender significant changes in the ambience in which all women give birth. Once a critical mass of mothers becomes aware of the fact that options are available to them, major changes in obstetrical practice may ensue.

If information on risks, benefits, and alternatives to conventional care remains a major focus among a large proportion of antenatal classes, we may expect increasingly influential and well-informed consumer involvement in the future patterns of childbirth practices. However, if the ideology of classes shifts toward an acceptance of conventional obstetric practices, the group consciousness among expectant parents may fade, reducing their impact and their influence on the direction of maternity care.

4 Conclusions

The widespread popularity of antenatal classes testifies to the desire of expectant parents for childbirth education. As there are benefits in terms of amount of analgesic medication used and in some aspects of satisfaction with childbirth, and as significant adverse effects have not been demonstrated, such classes should continue to be available. The objectives of the classes must be made clear to the participants, and unrealistic expectations of what the classes can achieve must be avoided. A variety of different types of classes, whose aims are explicitly stated, may help women or couples choose the programme most likely to meet their needs.

This chapter is derived from the chapters by Penny Simkin and Murray Enkin (20) and by Penny Simkin (56) in EFFECTIVE CARE IN PREGNANCY AND CHILDBIRTH.

References to primary sources and more complete data for statements made in this chapter can be found in the source chapters and/or in the following review from the *Cochrane pregnancy and childbirth database:*

Howell, C.J.
— Biofeedback in prenatal class attenders. Review no. 05620.

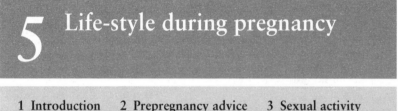

5 Life-style during pregnancy

1 Introduction   2 Prepregnancy advice   3 Sexual activity
4 Smoking   5 Alcohol   6 Work   7 Conclusions

1 Introduction

A pregnant woman is subject to a variety of prescriptions and proscriptions to modify her customary or desired life-style, in the guise of 'advice'. Unlike ordinary advice, however, there is frequently no option of refusal. Those believed to be authorities on reproduction, such as physicians, midwives, and childbirth educators, can give advice that appeals powerfully to the pregnant woman's desire for a perfect pregnancy and a perfect child. The effectiveness of this advice must be questioned and evaluated as rigorously as every other intervention carried out during pregnancy.

2 Prepregnancy advice

The appeal of prepregnancy advice is easy to understand. Whether a fetus is normally formed or malformed is usually determined by the time of the first antenatal visit. Antenatal care may permit detection of an abnormality, but preconceptional precautions may help to avoid it. Other major complications of pregnancy, such as preterm labour, which have proved difficult to influence during pregnancy may also be influenced by prepregnancy advice. It would seem plausible that care should start earlier, even before pregnancy begins.

A few measures have been found to be effective and useful. The most important among these is the use of folic acid supplementation to prevent neural tube defects (see Chapter 6). Prepregnancy assessment and advice can be vitally important for women with known problems, such as diabetes (see Chapter 20), or a family history of congenital abnormalities (see Chapter 9).

Except for these, and ensuring that the woman has been immunized against rubella and will not be taking any unnecessary drugs, what sensible advice can be given? Smokers need practical assistance for quitting rather than advice. Suggestions to adopt a prudent diet, although probably beneficial as general dietary guidelines, have not been shown to prevent malformations or low birthweight. Supplementation with trace minerals and vitamins other than folic acid cannot be justified in the present state of knowledge. Evidence on the effects of physical activity, work, exercise, and travel is still inconsistent.

The possible unwanted side-effects of prepregnancy advice include reduced self-confidence, reduced self-reliance, and increased anxiety for the woman. For advice-givers, the side-effects may be a reduced awareness of the social factors that underlie individual health behaviour associated with adverse pregnancy outcomes and a misguided though sincere belief that they have the answers.

It would be ill advised to make any judgements about routine prepregnancy advice except to say that, on the basis of present knowledge, its beneficial effects are likely to be extremely modest and it cannot automatically be regarded as harmless.

3 Sexual activity

Advice on the subject of sexual activity in pregnancy is often poorly given: inexact, inexplicit, euphemistic, misleading, allowing no opportunity for clarification and discussion of alternatives, and even being downright dangerous in terms of unintended side-effects. Moreover, there are few or no data to support the various forms of advice given. The many studies reported on the effects of sexual activity in pregnancy are methodologically unsound and contradictory. It is clear that on the basis of available evidence any prohibition of sexual activity is inappropriate. A wide range of changes in sexual feelings occur normally during pregnancy, including a marked increase or a marked decrease in desire.

4 Smoking

The evidence that cigarette smoking may have harmful effects on the fetus is strong. It is quite clear that maternal smoking reduces birthweight. The effect of smoking on other perinatal outcomes is more controversial.

Smoking cessation programmes have a definite place in antenatal care. They can be effective for a small minority of smokers in terms of reducing the amount smoked, in decreasing the proportion of women who continue to smoke, and in increasing mean birthweight and reducing the proportion of low birthweight babies. Behavioural strategies in particular significantly reduce the proportion of smokers who continue smoking through pregnancy, compared with standard antenatal care or with personal advice supplemented by written material. These strategies are much more effective than either feedback or advice. However, such programmes must be used with understanding, sensitivity, and compassion. Much anti-smoking 'advice' and propaganda ignores the problem of physical and psychological addiction, the meaning of smoking for the women involved, and the guilt and anxiety felt by those who continue to smoke in the face of exhortations to give it up. Much health promotion material for use in pregnancy is characterized by its particularly strident tone.

The campaign against smoking in pregnancy has not been free of unwanted side-effects: many smokers spend the whole of their pregnancy in a state of guilt and inadequacy. We do not know what the effects of such chronic stress and anxiety might be on the course of pregnancy and labour, or on the ultimate relationship with the child. We do know that more than half of the women who smoke worry about smoking during pregnancy and that 10 per cent of smokers actually smoke more heavily in pregnancy. The global nature of much anti-smoking exhortation means that any bad outcome of pregnancy (including malformations and mental retardation) may be retrospectively blamed on smoking, even when this could not have been the cause. Sometimes health professionals unwittingly reinforce the self-blame.

Recognition of the social and environmental context in which individuals take up or continue certain behaviour has led some people to condemn health education activities addressed to individuals as 'victim blaming'. Interventions that focus on self-help and behavioural strategies are less likely to be perceived in this way, and, in relation to quitting smoking, are soundly based. They are more effective than advice.

Obstetricians, family physicians or general practitioners, and midwives should support the population strategies towards a progressive reduction in cigarette smoking in the whole of society: to increase cigarette excise taxes, to ban all forms of tobacco advertising; to make public areas non-smoking, and to develop smoking policies for institutions and workplaces. The aim should be to make healthy choices easy choices.

5 Alcohol

The damaging effects of excessive alcohol consumption in pregnancy are well known. They include fetal growth retardation; mental retardation and a dysmorphic syndrome with variable features (at high levels of consumption), and altered neonatal behaviour. Developmental abnormalities are associated only with regular consumption of at least 28.5 ml alcohol (two standard drinks) per day, although one case has been reported following a single massive exposure to alcohol in the early weeks of pregnancy.

Campaigns to increase public awareness of the dangers of alcohol during pregnancy run the risk of arousing anxiety in some women already pregnant, partly because of the uncertainty about the safe lower limit for alcohol intake and also because of the possibility that the most dangerous time for dysmorphic effects may be the first trimester, sometimes even before women know that they may be pregnant. The very first weeks of pregnancy are often reported to be a period of high anxiety and depression that may increase drinking to relieve tension.

Policy development on alcohol and pregnancy requires, first of all, clarification as to the degree of risk around conception for low levels of regular alcohol consumption (fewer than two standard drinks a day) and for regular but infrequent 'binge' drinking. No formal trials of interventions to reduce high levels of consumption in pregnancy have been reported, and it may be that better detection of heavy drinkers should be the priority.

6 Work

Professional guidelines on work during pregnancy too often neglect any mention of housework and child care as work, whether with regard to exposure to toxic chemicals (pesticides, household spray cleaners?) or lifting heavy weights (a toddler plus a folding push chair? a handicapped seven-year-old?). For example, although women who have previously given birth to infants weighing less than 2 kg are strongly advised not to work, no one has as yet suggested that such women be provided with free child care and daily household help throughout pregnancy. Equally, discussions of whether pregnant women should work usually pay scant attention to the implications for family health and welfare of the concomitant reduction in family income. In fact, the benefits of paid employment are rarely mentioned.

The main cause of confusion results from regarding paid employment as a single category, lumping together women working with much less physical effort or stress than they would have at home with women whose work involves standing all day, carrying heavy loads, or exposure to extremes of temperature or humidity.

General advice on employment in pregnancy is clearly inappropriate. Where working conditions involve occupational fatigue, women's requests for a change of work during pregnancy should be supported by those providing antenatal care. Apart from this situation, it is extremely difficult to assess the net benefits and risks.

7 Conclusions

With the implied promise that it will help her have a perfect birth, a perfect baby, and become a perfect mother, a pregnant woman is exhorted to lead a selfless healthy life, uncontaminated by sex, cigarettes, alcohol, employment, or anxiety. The evidence for most of these exhortations is slight. Where the evidence is stronger, the flaw has been in the way that research and prescription fail to take into account the real lives and responsibilities of women.

This chapter is derived from the chapter by Judith Lumley and Jill Astbury (16) in EFFECTIVE CARE IN PREGNANCY AND CHILDBIRTH

References to primary sources and more complete data for statements made in this chapter can be found in the source chapter and/or in the following reviews from the *Cochrane pregnancy and childbirth database*:

Lumley J.
— Periconceptional folate (4 mg/day) vs placebo in high-risk mothers. Review no. 06488.
— Periconceptional multivitamins (no folate) vs placebo in high-risk mothers. Review no. 06489.
— Periconceptional multivitamins (incl folate 0.8 mgm) vs placebo in high-risk mothers. Review no. 06490.
— Periconceptional folic acid vs placebo. Review no. 07680.
— Strategies for reducing smoking in pregnancy. Review no. 03312.
— Advice as a strategy for reducing smoking in pregnancy. Review no. 03394.
— Feedback as a strategy for reducing smoking in pregnancy. Review no. 03395.
— Counselling for reducing smoking in pregnancy. Review no. 03396.
— Behavioural strategies for reducing smoking in pregnancy. Review no. 03397.

6 Dietary modification in pregnancy

1 Introduction

The relation between the diet of the mother and the well-being of the fetus and infant continues to be a matter of uncertainty and controversy. Observational studies have generated uncertain and conflicting conclusions because they encompass many other aspects of the pregnant woman's life that vary with diet and nutrition. Whenever limits to nutritional intake are imposed by economic, educational, social, or other constraints, there are likely to be accompanying stresses, such as exposure to infection, the need for physical labour, inadequate housing, or family disruption.

For many dietary interventions, adequately controlled studies have not been large enough to allow firm conclusions to be drawn. Nevertheless, much important information has been obtained.

2 Pre- and periconceptional nutrition

The protective effect of folate supplementation against neural tube defects (spina bifida and anencephaly) was first suggested by cohort and case–control studies, and has now been confirmed by randomized controlled trials. For women who are at increased risk because they have previously carried an affected baby (either at birth or earlier termination), periconceptional supplementation with folic acid can reduce the risk of a recurrence by more than two-thirds. No similar protective effect has been demonstrated for periconceptional vitamins without folate (folic acid).

Women who have had a fetus with a neural tube defect should be counselled about the increased risk in subsequent pregnancies and offered a folic acid supplement (4 mg/day) if they intend to have another pregnancy. Supplementation should begin before conception and continue through the first three months of pregnancy. Protection is not 100 per cent; some neural tube defects still occur in supplemented women, but the decrease is substantial. The recommendation may not apply to women with epilepsy, to women taking the anticonvulsant valproic acid, or to women with vitamin B_{12} deficiency. Trials of lower doses of folic acid are required, as there may be unknown harmful effects of the higher dose. In addition, if a lower dose is found to be effective, it might be possible to achieve it by dietary change and avoid the need for supplementation.

Periconceptional supplementation with multivitamins including a much smaller amount of folic acid (0.8 mg) also results in a significant reduction in the prevalence of neural tube defects in women who have not previously had an affected baby. The prevalence of all congenital malformations is also significantly lower in women who take this supplement, although no differences have been demonstrated for other specific malformations, such as cleft lip and/or cleft palate. To date, no protective effect has been demonstrated from a lower dose of periconceptional folic acid supplementation (0.36 mg/day), with or without multivitamins, although larger trials will be necessary to determine whether or not this low a dose is adequate.

Supplementation with multivitamin preparations including folic acid (0.8 mg), beginning before pregnancy and continued throughout the first trimester, can now be recommended for the prevention of first occurrences of neural tube defects. Ways of ensuring that dietary intake of folic acid reaches this level within affordable, available, and palatable food sources should be sought within each country and for each ethnic minority. This is a priority for public health nutrition advice.

3 Diet and fetal growth

Two main conclusions can be drawn from studies on dietary modification in pregnancy. First, severe dietary restriction can cause marked decreases in birthweight. The low birthweight observed in these studies resulted from impairment of fetal weight gain, as no effect was found on gestational age or on the rate of preterm birth. During famine, mean birthweight can be depressed by as much as 550 grams, and iatrogenic dietary manipulation and restriction can have almost as marked an effect. Trials of dietary restriction in pregnant women with high weight for height or high weight gain have been too small to demonstrate or exclude an effect on fetal growth; they also failed to demonstrate any significant effect on other outcomes. Although the extent to which major suppression of fetal weight is associated with perinatal mortality and morbidity is unknown, there can be no justification for allowing pregnant women to go hungry, or for imposing dietary restriction or major manipulation of the dietary constituents upon them.

Second, attempts at nutritional supplementation, while well intentioned, have not always had the desired effect. Trials of high-protein nutritional supplementation provide no evidence of benefit on fetal growth; comparisons of supplements with similar energy content but different protein concentrations show lower mean birthweights and a higher incidence of small for gestational age births in the higher protein intervention groups. Balanced energy and protein supplementation results in only a small increase in average birthweight (about 30 grams) and a small decrease in the incidence of small for gestational age birth. Nutritional advice appears to be moderately effective in increasing pregnant women's energy and protein intakes, but the implications for fetal, infant, or maternal health cannot be judged from the available trials. Given the rather modest health benefits demonstrated with actual protein/energy supplementation, the provision of such advice is unlikely to be of major importance. The modest increase in birthweight associated with supplementation does not appear to be associated with long-term benefits for child growth or development.

4 Diet and pre-eclampsia

Attempts to prevent pre-eclampsia by modification of protein or calorie intake continue to influence antenatal care, despite the fact that the evidence and arguments on which they are based are unconvincing. Evidence available at present provides no justification for telling women to restrict their diet to reduce 'high' weight gain. Equally, there is no evidence to support the alternative view that eating sufficient amounts of a good diet will reliably protect against pre-eclampsia.

Controlled trials of prophylactic fish oil in pregnancy show a promising reduction in the incidence of proteinuric pre-eclampsia and of preterm delivery, but there are insufficient data to show a decrease in the incidence of hypertension, or any measure of perinatal mortality or morbidity.

We know too little about the effects of high or low salt consumption on the development of pre-eclampsia to be able to offer well-informed advice. The effects of calcium supplementation are discussed in Section 7 below.

5 Special diets to avoid antigens

Special diets to avoid antigens have been prescribed for women at high risk of giving birth to an atopic child (based on a history of atopic disease in the mother, the father, or a previous child). The evidence available from controlled trials so far shows that prescription of an antigen-avoidance diet to a high-risk woman during pregnancy is unlikely to reduce substantially her risk of giving birth to an atopic child. It is also possible that such a diet might have an adverse effect on maternal and/or fetal nutrition. Further trials, involving larger numbers of women and babies with longer follow-up, are needed.

6 Haematinic supplements

The normal haematological adaptations to pregnancy are frequently misinterpreted as evidence of iron deficiency that needs correcting. Iron supplements have been given with two objectives in view: to try to return the haematological values towards the normal non-pregnant state, a strange objective when millions of years of evolution have determined otherwise, and to improve the clinical outcome of the pregnancy and the future health of the mother. The first objective can certainly be accomplished; the key question is whether or not achieving the 'normalized' blood picture benefits the woman and her baby. Routine iron supplementation raises and maintains serum ferritin above 10 μg/litre and results in a substantially lower proportion of women with a haemoglobin level below 10 or 10.5 grams per cent (below 6–6.5 mmol/litre) in late pregnancy. Routine folate supplementation as a haematinic after the first few weeks of pregnancy substantially reduces the prevalence of low serum and red cell folate levels, and of megaloblastic haematopoiesis. As yet, neither iron nor folate supplementation after the first trimester have shown any detected effect on the following substantive measures of maternal or fetal outcome: proteinuric hypertension, antepartum haemorrhage, postpartum haemorrhage, maternal infection, preterm birth, low birthweight, stillbirth, or neonatal morbidity. Women do not feel any subjective benefit from having their haemoglobin concentration raised.

A possible advantage claimed for a high level of haemoglobin in pregnancy is that the woman would be in a stronger position to withstand haemorrhage. There is no evidence to support this claim. Indeed, as a low haemoglobin in healthy pregnant women generally implies a large circulating blood volume, it is at least possible that women with a *low* haemoglobin might better withstand a given loss of blood.

There are few data derived from communities in which nutritional anaemia from either iron or folate deficiency is prevalent. Trials are needed in these populations to establish the most appropriate strategies for combatting the deficiencies.

Whether routine iron supplementation causes any harm in well-nourished communities is still unclear, but it is clearly wasteful. The evidence suggests that, except for genuine anaemia, the best reproductive performance is associated with levels of haemoglobin that are traditionally regarded as pathologically low. There is cause for concern in the findings of two well-conducted trials that iron supplementation resulted in an increase in the prevalence of preterm birth and low birthweight. Perhaps there is an adverse effect on fetal growth due to the increased viscosity of maternal blood that follows the iron-induced macrocytosis and increased haemoglobin concentration, which may impede uteroplacental blood flow.

An individual's haemoglobin concentration depends much more on the complex relation between red-cell mass and plasma volume than on deficiencies of iron or folate. The advent of electronic blood counters has given an opportunity for more appropriate criteria to be applied to the diagnosis of anaemia. Mean cell volume may be the most useful; it is not closely related to haemoglobin concentration and declines quite rapidly in the presence of iron deficiency. A low haemoglobin without other evidence of iron deficiency requires no treatment.

If there is evidence of genuine iron deficiency, iron treatment is needed, and the usual approach is to give iron salts by mouth. There is no convincing evidence that the addition of copper, manganese, molybdenum, or ascorbic acid improves the efficiency with which the iron is used.

The cause of megaloblastic anaemia in pregnancy is almost always folate deficiency, and treatment with folic acid supplementation is rapidly effective.

7 Other vitamin or mineral supplementation

Vitamin D deficiency may occur in women whose diet is relatively low in the vitamin, such as vegetarians and those who either remain indoors or whose clothing leaves little exposed skin, particularly in relatively sunless climates. Controlled trials in vulnerable populations show a reduction in neonatal hypocalcaemia (hyper-irritability) with vitamin D supplementation. No significant effects on other substantive outcomes have been reported.

The little evidence available on vitamin B_6 supplements in pregnancy suggests that they may protect against dental decay in the mother when given in the form of lozenges, but no effect has been found on other clinical outcomes.

A number of well designed trials have assessed the effects of calcium supplementation on important measures of maternal morbidity and perinatal morbidity and mortality. Calcium supplementation during pregnancy appears to reduce substantially the risk of women developing hypertension and proteinuric pre-eclampsia. There is insufficient evidence to draw conclusions about other important outcomes, such as intrauterine growth retardation and perinatal death. This form of supplementation appears promising, but further large trials are necessary before any firm conclusions can be drawn as to its role in routine care.

The trials evaluating routine magnesium supplementation during pregnancy suggest a reduced incidence of preterm birth and low birthweight, but the data are not entirely free of bias and there is insufficient evidence at present to justify routine magnesium supplementation during pregnancy.

The available data from controlled trials provide no convincing case for routine zinc supplementation during pregnancy. There is some suggestion of a benefit from selective supplementation.

Iodine supplementation in a population with high levels of endemic cretinism results in an important reduction in the incidence of the condition, with no apparent adverse effects.

8 Conclusions

There is no evidence that dietary restriction of any sort confers any benefit to pregnant women or their offspring.

All women who might become pregnant should ensure an adequate intake of folic acid, at least around the period of conception, either through supplementation or diet. Women who have had a fetus with a neural tube defect should be counselled about the increased risk in subsequent pregnancies and offered a folic acid supplement (4 mg/day) if they intend to have another baby. Supplementation should begin before conception and continue through the first three months of pregnancy. (This recommendation does not apply to women with epilepsy, to women taking the anticonvulsant valproic acid, or to women with vitamin B_{12} deficiency.)

Nutritional supplementation with supplements of relatively low protein density may result in a modest increment in fetal growth, but no other substantive benefits have as yet been documented. High protein dietary supplements should be avoided.

Antigen-avoidance diets have not shown any convincing benefit in the prevention of atopy.

Routine haematinic supplementation with iron has not been shown to confer any benefit to either mother or baby, except to build up the woman's iron stores. Routine supplementation may be of benefit in populations in which iron deficiency in pregnancy is a common problem.

No recommendations can be made, on currently available evidence, about the role of fish oils during pregnancy and the place, if any, for routine supplementation with minerals such as zinc, calcium, or magnesium.

Vitamin D supplementation at the end of pregnancy should be considered in vulnerable groups, such as Asian women in northern Europe and possibly others in climates with long winters. Iodine supplementation should be provided in populations with high levels of endemic cretinism.

While there is an obvious need for further research into the best means of promoting optimal nutrition in pregnancy, hungry women cannot wait for the results of such studies. They must have access to both adequate amounts of food and informed care.

This chapter is derived from chapters by David Rush (17), Jane Green (18), and Kassam Mahomed and Frank Hytten (19) in EFFECTIVE CARE IN PREGNANCY AND CHILDBIRTH.

References to primary sources and more complete data for statements made in this chapter can be found in the source chapters and/or in the following reviews from the *Cochrane pregnancy and childbirth database*:

Duley, L.
— Prophylactic fish oil in pregnancy. Review no. 05941.
— Any PG precursors for pre-eclampsia prevention/treatment. Review no. 05942.
— Low vs high salt intake in pregnancy. Review no. 05939.
— Routine calcium supplementation in pregnancy. Review no. 05938.

Keirse, M. J. N. C.
— Routine magnesium supplementation in pregnancy. Review no. 04008.

Kramer, M.S.
— Energy/protein restriction in pregnant women with high weight-for-height or weight gain. Review no. 07139.
— Nutritional advice in pregnancy. Review no. 07138.
— High protein supplementation in pregnancy. Review no. 07142.
— Balanced protein/energy supplementation in pregnancy. Review no. 07141.
— Isocaloric balanced protein supplementation in pregnancy. Review no. 07140.
— Maternal antigen avoidance in pregnancy in women at high risk for atopic offspring. Review no. 07273.

Lumley, J.
— Periconceptional folate (m/day) vs placebo in high-risk mothers. Review no. 06488.
— Periconceptional multivitamins (no folate) vs placebo in high-risk mothers. Review no. 06489.

— Periconceptional multivitamins (including folate 0.8 mg) vs placebo. Review no. 06490.
— Periconceptional folic acid (0.36 mg) vs placebo. Review no. 07680.

Mahomed, K.
— Routine iron supplementation in pregnancy. Review no. 03157.
— Routine folate supplementation in pregnancy. Review no. 03158.
— Routine iron and folate supplementation in pregnancy. Review no. 03159.
— Vitamin D supplementation in pregnancy. Review no. 06610.
— Pyridoxine (Vitamin B6) supplements in pregnancy. Review no. 06507.
— Routine zinc supplementation in pregnancy. Review no. 06944.
— Maternal iodine supplements in areas of deficiency. Review no. 06508.

SCREENING

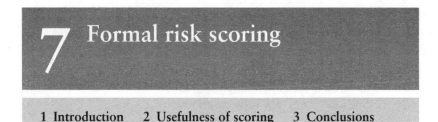
7 Formal risk scoring

1 Introduction

Identification of pregnancies that are at greater than average risk is a fundamental component of antenatal care. The primary purpose of a risk-scoring system is to permit classification of individual women into different categories, for which different actions can then be planned. The great variety of scoring systems that are used in pregnancy and childbirth aim to identify individuals at high risk and to discriminate between levels of risk. In theory, a process of rationally based risk scoring should be more accurate than the rather nebulous process of clinical impression that is part of daily clinical practice. As for many other tests, the validity and utility of risk scoring systems remain to be determined.

There are many practical difficulties involved in the use of scoring systems. Often the risk markers are poorly defined. For example, does 'bleeding' also include spotting or bleeding from a local lesion in the vagina or cervix? The need to dichotomize continuous variables that are as different from each other as blood pressure (how high?) or smoking (how much?) imposes a rigidity that can often be counter productive. With formalized risk scoring, a woman may be assigned to a high-risk group because of fixed definitions of the risk markers, whereas a capable clinician could have assessed the situation more sensitively by using clinical judgement.

The reproducibility of scoring systems tends to be low. Apart from the still small number of antenatal clinics where details of each pregnant woman are transferred interactively to a computer, scoring is done by a busy clinician with pen and paper. To be feasible, there should be not too many questions, values should be restricted to integers, and addition is preferable to multiplication in deriving the score.

Some scoring systems require women to be scored only once. At the opposite extreme, some require reassessment at each antenatal visit. Reassessment allows the inclusion of complications appearing in the current pregnancy and revision of the score upwards or downwards depending on the course of the pregnancy. Little is known about the benefits of re-evaluation.

Scoring is more effective in second or later pregnancies than in women pregnant for the first time. The poor predictive value of the scoring systems for nulliparae is, at least in part, inherent in the choice of risk markers, many of which relate to characteristics of the past obstetric history.

An ideal scoring system should allow assignment to a risk group in time for appropriate action to be taken. Scoring systems will perform better in assessing the risk if they are implemented late in pregnancy or if they allow for readjustment during pregnancy. This leads to the paradox that the most precise risk prediction is made at a time when there is little or no further need for it, whereas the potentially more useful early identification of risk is relatively imprecise.

Both the positive and the negative predictive values of all scoring systems are poor. Depending on the cut-off point and the test chosen, only between 10 and 30 per cent of the women who are allocated to the high-risk groups actually experience the adverse outcome for which the scoring system declares them to be at risk. Evaluation of risk scores for preterm birth and low birthweight revealed that between 20 and 50 per cent of mothers who deliver preterm or low-birthweight infants have low risk scores.

2 Usefulness of scoring

It may be useful to the clinician to know which pregnancies under his or her care are most likely to result in an adverse outcome. To the individual woman, being labelled as 'high risk' will be beneficial only if something can be done either to decrease the risk or to reduce its consequences.

Although often referred to as risk factors, most of the elements that are incorporated in the scores are merely risk markers, indicating that there is a statistical association with the outcome that is being scored for. These risk markers do not cause the outcome. The most important of them, such as parity, prepregnancy weight, height, and past reproductive performance, cannot be altered by any intervention. For the individual woman who is labelled as being at increased risk, both the threat of adverse outcome and the inability to change its markers may create anxiety.

The most powerful way to test the effectiveness of formal risk-scoring systems is to mount randomized controlled trials in which formal risk scoring is a component of the antenatal care of some women, while a control group receive the usual antenatal care without formal risk scoring. No such trials have been published.

A number of observational studies have claimed a reduction in preterm births following introduction of systematic scoring. They attributed this improvement to better selection of those women who require treatment and to better 'systematization of interventions'. In many settings there was an increase in the frequency with which interventions of dubious value were performed. Although some authors expressed the belief that their prevention policy had played a part in the overall reduction of preterm birth rates, most of the improved results occurred in the women scored to be at low risk.

3 Conclusions

Formal risk-scoring systems are a mixed blessing for the individual woman and her baby. They may help to provide a minimum level of care and attention in settings where these are inadequate. In other settings, however, formal risk scoring results in a variety of unwarranted interventions. The introduction of risk scoring into clinical practice carries the danger of replacing a potential risk of adverse outcome with the certain risk of dubious treatments and interventions.

The potential benefits of risk scoring have been widely publicized, but the potential harm is rarely mentioned in the current literature. Such harm can result from unwarranted intrusion in women's private lives, from superfluous interventions and treatments, from creating unnecessary stress and anxiety, and from allocating scarce resources to areas where they are not needed.

This chapter is derived from the chapter by Sophie Alexander and Marc J.N.C. Keirse (22) in EFFECTIVE CARE IN PREGNANCY AND CHILDBIRTH.

References to primary sources and more complete data for statements made in this chapter can be found in the source chapter.

8 Imaging ultrasound in pregnancy

1 Introduction   2 Selective use of ultrasound   3 Routine
early ultrasonography   4 Routine late ultrasonography
5 Placental grading   6 Women's reactions to ultrasound in
pregnancy   7 Potential hazards of obstetric ultrasound
8 Conclusions

1 Introduction

Ultrasound imaging has been used increasingly in obstetrical care since
its introduction in medicine. Improvement in resolution and quality of
ultrasound imaging equipment has been rapid; progress from the first
detection of the gross abnormality of anencephaly in 1972 to the
current sophisticated diagnoses of subtle fetal anomalies has taken only
a few years.

Whether ultrasound imaging should be used routinely for prenatal
screening or only used selectively for specific indications has not, as yet,
been firmly established.

2 Selective use of ultrasound

There is a clear difference between selective and routine use of ultrasound. The time taken, the detail inspected, and perhaps the seniority
of the ultrasonographer will vary with the reason for the examination.
Identification of fetal presentation, for example, takes seconds and can
be carried out by a minimally trained technician. However, thorough
investigation of the cause of raised alpha-fetoprotein levels may require
considerable time and expertise. Routine examinations must be accomplished quickly for practical reasons and, therefore, some fetal malformations are less likely to be detected than when there are specific
reasons to anticipate their presence. The selective examination should
be tailored to answer a specific question posed by the person who requests the examination.

There can be no doubt about the value of ultrasound in many specific
clinical situations. It has the ability to establish rapidly and accurately
whether a fetus is alive or dead, and to predict whether a pregnancy is

likely to continue after threatened miscarriage. Gestational age can be accurately estimated from early measurements of fetal size in the first or early second, trimesters. In the investigation of possible fetal malformation, ultrasound can often visualize the malformation and can facilitate other diagnostic techniques, such as amniocentesis and chorion villus sampling. It can assess fetal size and growth in the second half of pregnancy with reasonable accuracy. It can localize the placenta in cases of suspected placenta praevia. Other situations in which the selective use of ultrasound can provide invaluable help include confirmation of suspected multiple pregnancy, assessment of amniotic fluid volume in suspected polyhydramnios or oligohydramnios, confirmation of fetal position, and assistance for other procedures such as cervical cerclage or external cephalic version.

The great value of selective ultrasound to answer specific questions does not provide information as to whether or not routine ultrasound screening of all women during pregnancy would be worthwhile. The greatest controversy surrounding obstetrical ultrasound has been whether its use should be extended from specific indications to the routine screening of all pregnant women, either early (usually 16–19 weeks, but sometimes earlier) or later (usually 32–36 weeks) in pregnancy.

3 Routine early ultrasonography

The benefits expected of routine ultrasonography in early pregnancy are better gestational age assessment, earlier detection of multiple pregnancies, and detection of clinically unsuspected fetal malformation at a time when termination of pregnancy is possible. Data from the controlled studies show that these expectations have been largely fulfilled.

When compared with selective ultrasonography in early pregnancy, routine ultrasound examination results in a reduced rate of induction of labour for apparent post-term pregnancy, presumably as a result of better gestational dating. There are fewer undiagnosed twins at 26 weeks gestation among the screened group and also, it seems, fewer low birthweight babies, although the reason for this is unclear. None of these effects has so far been shown to improve fetal outcome.

If the screening examination is performed early in pregnancy, some clinically unsuspected non-viable pregnancies (for example blighted ova and hydatidiform moles) may be detected. However, a satisfactory inspection of fetal anatomy to detect malformation cannot be performed before 18 weeks, and if inspection of the heart is to be included, examination closer to 22 weeks may be necessary. Only one of the controlled trials included the specific aim of detecting malformed fetuses. In this trial, the screened group had a lower perinatal mortality (but no increase in the proportion of live births) because of early detection and selective termination of pregnancies in which the baby had a malformation. Varying standards of ultrasound expertise make this a particularly difficult issue about which to make general statements (see also Chapters 9, 14, 16, and 17).

4 Routine late ultrasonography

The main purpose of routine scanning in late pregnancy is to identify unsuspected growth-retarded fetuses who may benefit from elective delivery. The randomized trials of routine anthropometry in late pregnancy suggest an increased incidence of antepartum hospital admissions and of induction of labour, with no improvement in perinatal outcome. There were no detectable effects on the incidence of low Apgar score, admission to the special care nursery, or perinatal mortality. These trials provide no support for routine ultrasonography in late pregnancy for fetal measurement (see also Chapter 12).

5 Placental grading

One randomized trial of reporting placental 'texture' grading was conducted in a maternity unit in which routine ultrasound examinations were performed at 30–32 and 34–36 weeks' gestation. Reports of placental appearances were made available to clinicians caring for women assigned to the experimental group but not for those in the control group. Knowledge of the result of placental grading was associated with increased use of other fetal assessment techniques and with a tendency to increased use of elective delivery for fetal compromise. There was a better pregnancy outcome among women whose physicians knew their results, with less frequent meconium staining of amniotic fluid, fewer babies with low Apgar scores at 5 minutes, and fewer deaths of normally formed babies during the perinatal period. Several unexplained intrauterine deaths in the group of women whose physicians did not know their results were associated with (unreported) placental grades which were thought to be predictive of fetal compromise.

The results of this study strongly suggest that knowledge of placental appearances can result in clinical action to improve pregnancy outcome. Although firm recommendations cannot be made on the basis of a single study, it might be worthwhile for ultrasonographers to report the placental grade at any late pregnancy ultrasound examination. At the very least the findings should not be ignored, and warrant confirmation (or refutation) by further controlled research.

6 Women's reactions to ultrasound in pregnancy

An ultrasound examination has the potential to be a fascinating and happy experience for prospective parents, but real or mistaken diagnoses of fetal abnormality on ultrasound can lead to psychological devastation. Women's reactions to ultrasonography during pregnancy have not received the systematic attention from researchers that they deserve.

A majority of the women interviewed in the available studies valued ultrasonography in early pregnancy because it confirmed the reality of the baby for them, and because the examination often led to a reduction in anxiety and an increase in confidence. A study conducted in the United States found that American women felt that nearly half the value of ultrasonography in an uncomplicated pregnancy was outside the realm of medical decisions, such as knowing the sex of the child or having an early picture to show their children.

Women's views on the desirability of routine ultrasonography during pregnancy, in addition to being influenced by what is actually available, are also influenced by differing perceptions of the potential benefits and concerns about the possible adverse effects of ultrasound. The only generalization that can be made on the basis of the available research is that women's views vary about the indications for ultrasonography. The obvious implication for practice is that it is important for ultrasonographers to take this variation into account.

Another important message is that the experience of having a scan, even if the findings are normal, can be unpleasant because of uncommunicativeness on the part of the ultrasonographer. Some ultrasonographers, technicians in particular, may be put under professional constraints not to communicate freely with the women they are examining. Whether as a result of these constraints or for other reasons, uncommunicativeness can eliminate the potential psychological benefit of the examinations. The risk of this adverse effect is likely to be increased when the human resources available for ultrasonography are stretched, as they often are when ultrasonography is routinely performed on every pregnant woman.

7 Potential hazards of obstetric ultrasound

Any consideration of the use of diagnostic ultrasound in obstetrical practice must weigh potential benefits against potential risks. There has been surprisingly little well-organized research to evaluate possible adverse effects of ultrasound exposure on human fetuses. Based on the available follow-up of randomized trials, there is no evidence of a greater risk of impaired school performance at age eight to nine or of dyslexia following routine imaging ultrasonography during the second and third trimesters of pregnancy, but there are suggestions of an increased incidence of left-handedness.

Two apparently well-designed and well-conducted case–control studies have sought a relationship between ultrasound exposure and childhood malignancy. Both were reassuring with one possible exception: neither study showed any difference in exposure between the cases and controls up to the age of five, but in one of the studies children dying of leukaemia or cancer over the age of five were more likely than controls to have been exposed to ultrasound as fetuses. This difference was not seen in the other study, which was statistically more powerful.

Evidence from cohort studies is very limited. In one such study, babies exposed to ultrasound as fetuses were more likely than controls to have abnormal grasp and tonic neck reflexes but were comparable in respect of 122 other parameters for which associations were sought. In another study, many outcomes were examined and no association with ultrasound exposure was found for the majority of them.

These two non-randomized cohort studies and the randomized trials have been used to investigate whether or not the reduction in birthweight after ultrasound exposure suggested by some animal studies can be found in humans. The data are not entirely conclusive in that an increase in the incidence of low weight for gestation cannot be excluded at present.

The randomized controlled trials conducted to date have been far too small to have a reasonable chance of identifying an effect of ultrasound exposure on any rare adverse outcome.

8 Conclusions

The value of selective ultrasound for specific indications in pregnancy has been clearly established. The place, if any, for routine ultrasound has not been determined as yet. In view of the fact that its safety has not been convincingly established, such routine use should be considered experimental for the present, and should not be implemented outside the context of randomized controlled trials.

Many obstetric units already practice routine ultrasonography in early pregnancy. For those considering its introduction, the benefit of the demonstrated advantages would need to be set against the theoretical possibility that the use of ultrasound during pregnancy could be hazardous and against the need for additional resources. At present, there is no sound evidence that ultrasound examination during pregnancy is harmful.

The available randomized trials do not support the use of routine ultrasonography in late pregnancy for fetal measurement. The only imaging ultrasound technique of late pregnancy that appears to improve outcome is placental grading, and this finding requires confirmation.

During ultrasound examinations at any time in pregnancy, mothers should see the monitor screen, have their baby's image pointed out, and receive as much information as they desire.

This chapter is derived from the chapter by Jim Neilson and Adrian Grant (27) in EFFECTIVE CARE IN PREGNANCY AND CHILDBIRTH.

References to primary sources and more complete data for statements made in this chapter can be found in the source chapter and/or in the following reviews from the *Cochrane pregnancy and childbirth database*:

Neilson, J.P.
— Routine ultrasonography in early pregnancy. Review no. 03872.
— High vs low feedback to mother at fetal ultrasonography. Review no. 04001.
— Fetal exposure to ultrasound (effects on school performance). Review no. 06888.
— Routine fetal anthropometry in late pregnancy. Review no. 03873.
— Routine ultrasound placentography in late pregnancy. Review no. 03874.

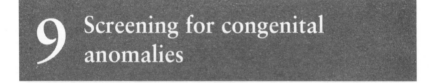

9 Screening for congenital anomalies

1 Introduction   2 Genetic counselling   3 Methods of screening and diagnosis   3.1 *Ultrasound*   3.2 *Cytogenetic techniques*   3.2.1 *Amniocentesis*   3.2.2 *Chorion villus sampling*   3.3 *Serum alpha-fetoprotein*   4 Conclusions

1 Introduction

Prenatal screening for congenital abnormalities and genetic disorders has become increasingly important since the development of amniocentesis in 1969. The planning of a genetic screening programme involves defining the populations in which screening procedures are justifiable. This requires careful consideration of a number of factors, including the prevalence of the condition in the population to be tested, the severity of the disorder, how successfully available tests separate those with the condition from those without it (sensitivity and specificity), and the costs.

Costs are not wholly financial. It is equally important to weigh the human costs. Although screening programmes may bring reassurance to some women who are tested, they may generate anxiety for others by merely raising the question of abnormality. The consequences of erroneous diagnoses, both positive and negative, warrant particularly careful consideration.

2 Genetic counselling

The prevention and treatment of genetic disease is still a new branch of medicine, but genetic counselling is becoming an increasingly important component of health care. The list of disorders that are amenable to prenatal diagnosis continues to grow, particularly with advances in molecular genetics.

Counselling prior to prenatal testing is important. The central issue is, of course, one of risk. What is the risk of producing an abnormal child? What is the risk of the investigating procedure? It is clear that one individual may interpret risk figures very differently from another. Moreover, parental decisions may depend not only on the actual level of risk, but on whether or not they could imagine handling the consequences of having an abnormal child. There is still a regrettable dearth of good-quality psychological and sociological research into the field of prenatal diagnosis.

The decision to screen and the action taken as a result of a screening test should be the decision of the individuals themselves after they have been made thoroughly aware of the situation. Therefore every screening programme must provide adequate time for thorough counselling when it is required. The common practice is for women undergoing amniocentesis to be scheduled for counselling on the same day that the procedure is carried out. Often it would be preferable for counselling to be given earlier, to allow time for couples to think carefully without feeling pressured into reaching a decision. However, this latter policy may pose other problems, as many women may find it difficult to make an additional hospital visit, particularly when this involves substantial amounts of travelling.

Personal or religious beliefs will influence whether screening is undertaken at all. There is seldom any point in carrying out prenatal diagnosis for chromosome studies when the couple would refuse a termination under any circumstances. Nevertheless, no one should be made to feel that once they have undergone screening they are bound to follow a rigid course of action. The couple should feel free to exercise whatever options they choose.

Genetic screening tests often yield information that is relevant to other family members. Usually there is no barrier to a free exchange of information within the family, but occasionally individuals will wish to keep the results of tests to themselves. This puts the counsellor in a difficult position. He or she must preserve confidentiality, but at the same time may be concerned about relatives at risk of genetic disease who should be traced and tested. The woman's preference must be respected.

3 Methods of screening and diagnosis

3.1 *Ultrasound*

Ultrasound can be employed in three ways to assist the identification of fetal malformations: to visualize the malformation; to facilitate other diagnostic techniques, such as amniocentesis and chorion villus sampling, and to allow fetal measurement (thereby maximizing the performance of other tests that require accurate knowledge of gestational age).

A large number of abnormalities can now be detected by modern ultrasound imaging. Some defects, such as anencephaly, are easily identified; some, such as spina bifida, are moderately difficult to detect; some, such as certain cardiac anomalies, may be very difficult to identify. The presence of one defect may suggest the presence of others and/or a chromosomal abnormality. Detection rates may vary with the quality of the equipment and expertise of the ultrasonographer. Whether the examination has been performed because of high-risk features (such as a previously malformed baby or raised alpha-fetoprotein levels) or as part of a screening examination (when less time and possibly less expertise would be available) may also influence the detection rate. Ultrasound can be of particular help in demonstrating to parents who have previously had a malformed baby that their fetus in the current pregnancy does not have the same defect.

Termination of pregnancy will be acceptable to many couples when the fetus has a lethal abnormality such as anencephaly, or a defect likely to result in major handicap such as spina bifida with hydrocephalus. Difficulties can arise when defects have less predictable sequelae, and also because diagnostic errors may occur. Prior consultation with surgical colleagues should help to reduce unnecessary elective early delivery (and the morbidity resulting from iatrogenic immaturity) of babies with conditions that will not benefit from early surgery.

It is vital that ultrasound does not lead to a diagnosis of abnormality in a baby that is normally formed. Such 'false positive' diagnoses are particularly tragic if they lead to the termination of a normal pregnancy.

The finding of a malformation does not mandate termination of the pregnancy. Diagnosis of a malformation may help some parents prepare for the birth of an impaired child. However, some parents may suffer a prolonged and devastating upset through such information. Skilled counselling is needed to help them plan ahead for the care of their child. Imparting information about important defects requires personal sensitivity and the availability of people who can be supportive.

The use of ultrasound during amniocentesis can minimize the risk of placental and fetal contact during the insertion of the needle into the amniotic cavity. In addition, the presence of multiple pregnancy may be identified, fetal life confirmed, and gestational age estimated. Both the pregnant woman and the operator may be reassured by seeing that the fetus appears to be unaffected by the procedure.

3.2 Cytogenetic techniques

Studies of newborn infants show a world-wide frequency of chromosome disorders of about 6 per 1000 births. The total population load of chromosome abnormalities is far greater still, as the majority of affected embryos are spontaneously miscarried early in pregnancy. Over half of all clinically recognizable spontaneous miscarriages are chromosomally abnormal.

Over the past fifteen years there has been a steady rise in demand for prenatal diagnosis for chromosome disorders. The majority of referrals for prenatal chromosome diagnosis have been of older women (aged 35 and over) who have an increased risk of carrying a fetus with Down's syndrome (trisomy 21) and most other chromosomal abnormalities. Up to the age of about 29 there is little effect of maternal age on the birth frequency of Down's syndrome (the incidence ranges from approximately 0.5 to 1.0 per 1000 live births). Between the ages of 30 and 34 the frequency begins to rise; by age 35 it is 2 to 3 per 1000 live births, and at the age of 40 it is 8 or 9 per 1000. Until recently, most laboratories used the age of 35 as the cut-off point for offering prenatal diagnosis, a decision governed partly by available resources. Policies are changing with the introduction of new tests that offer improved detection rates. Risk assessment based on measurements of one or more of serum alpha-fetoprotein, human chorionic gonadotrophin, and oestriol, together with maternal age, can further improve the selection of women for cytogenetic study. Invasive techniques in current use for the prenatal diagnosis of chromosome disorders, for example for women at high risk of trisomy, are amniocentesis, chorion villus sampling, and (occasionally) fetal blood sampling (cordocentesis).

3.2.1 *Amniocentesis* The overall safety of early second trimester amniocentesis has been well established from several large studies. The chief hazard is the additional risk of miscarriage associated with the procedure, which is approximately 0.5–1 per cent. In addition, there is a risk of about 0.5 per cent of a baby with very low birthweight and of neonatal respiratory problems.

Amniotic fluid cell chromosome studies have two major drawbacks. First, it usually takes two to three weeks from the time the sample is taken until the result is available. Many women find this long wait in itself distressing. Second, amniocentesis is not usually carried out before weeks 14–16 of pregnancy. Thus, if termination is indicated it has to be carried out at a relatively late stage in pregnancy. Culture failure occurs in about 2 per cent of samples. In these cases a repeat sample becomes necessary, and the pregnancy may be distressingly far advanced by the time that the result is available.

A controlled trial of genetic amniocentesis in over 4000 women at low risk of an abnormality showed that the procedure was associated with a high incidence of fetomaternal bleeding, an almost threefold increase in the miscarriage rate, and, perhaps most noteworthy, a significant increase in the incidence of babies with very low birthweight and of respiratory distress syndrome.

3.2.2 *Chorion villus sampling* Chorion villus sampling constitutes an attractive alternative to amniocentesis or fetal blood sampling since it can be performed in the first trimester. It involves the use of a cannula (needle), or biopsy forceps under ultrasound guidance, to take a small biopsy of villi from the chorion frondosum. In general, the transabdominal route results in a lower failure rate, less bleeding, and fewer miscarriages than the transcervical route. Some small trials have attempted to evaluate differences between different types of cannulas, or between cannulas and biopsy forceps for chorion villus sampling; no clear differences have emerged. No benefit has been demonstrated from betamimetic administration prior to chorion villus sampling.

The advantages of first-trimester diagnosis are obvious, but for some women these are likely to be offset by obstetrical and cytogenetic problems. Direct comparisons of first-trimester chorion villus sampling with second-trimester amniocentesis show that complications are uncommon with both procedures, but that chorion villus sampling is associated with significantly more need for a repeat test, more bleeding following the test, and more false-positive diagnoses. Stillbirths and neonatal deaths were more common among the women allocated to chorion villus sampling, and fewer of these women were able to achieve a term delivery or have a normal birthweight baby.

The possibility that chorion villus sampling may rarely cause face or limb abnormalities, particularly when performed very early in pregnancy, is not proved but there is sufficient concern for some experts to recommend that chorion villus sampling should not be used before 10 weeks, thereby reducing some of its advantages.

Apart from technical difficulties, there are other important aspects of prenatal diagnosis by chorion villus sampling that require consideration. A number of chromosomally unbalanced conceptuses are miscarried spontaneously in early pregnancy. Thus a significant proportion of the chromosomally abnormal fetuses that are detected through chorion villus sampling may have been destined for spontaneous miscarriage before amniocentesis would have been carried out.

3.3 Serum alpha-fetoprotein

Serum alpha-fetoprotein determination has made screening of the general population feasible where there is a high incidence of neural tube defects. When the result from a serum test shows an elevated level of alpha-fetoprotein the woman should undergo either a detailed ultrasound examination or amniocentesis to allow the more sensitive amniotic fluid alpha-fetoprotein assay to be carried out. When appropriate ultrasound expertise is available, amniocentesis is unnecessary.

4 Conclusions

Genetic screening and diagnosis now has a well-established place in modern obstetrical care. It should be offered as an option to those women or couples who are deemed to be at significant risk. The potential benefits and potential adverse effects should be made known to them, so that they can make a properly informed choice.

The indications for genetic screening require further clarification including, in particular, surveys of women's views of the desirability of the screening and of the psychological effects of both positive and negative results.

As new techniques such as chorion villus sampling are introduced, their safety in comparison with established techniques must be evaluated by large-scale randomized trials. This also applies when an established test starts to be used at a different gestational age, for example first-trimester amniocentesis (a technique of current interest).

The benefits of earlier exclusion or diagnosis of some fetal disorders afforded by first-trimester chorion villus sampling compared with late amniocentesis must be set against the greater risks of the former. Women considering prenatal diagnosis must be fully informed about the risks and benefits of the alternative procedures.

This chapter is derived from the chapter by Michael Daker and Martin Bobrow (23) in EFFECTIVE CARE IN PREGNANCY AND CHILDBIRTH.

References to primary sources and more complete data for statements made in this chapter can be found in the source chapter and/or in the following reviews from the *Cochrane pregnancy and childbirth database*:

Grant, A.M.

— Genetic amniocentesis at 16 weeks gestation. Review no. 04002.
— Ultrasound guidance during 2nd trimester amniocentesis. Review no. 06588.
— IV betamimetic on the success of amniocentesis. Review no. 06611.
— Chorion villus sampling compared with amniocentesis. Review no. 06007.
— Transabdominal vs transcervical CVS. Review no. 06005.
— Betamimetic administration prior to CVS. Review no. 06587.
— Malleable steel cannula vs Portex cannula for TC CVS. Review no. 06583.
— Aluminium cannula vs Portex cannula for transcervical CVS. Review no. 06584.
— Malleable steel cannula vs aluminium cannula for TC CVS. Review no. 06585.
— Silver cannula vs Portex cannula for transcervical CVS. Review no. 06582.
— Aspiration vs forceps for transcervical CVS/ Review no. 06006.

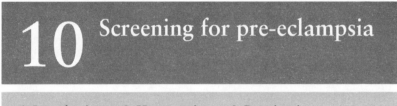
10 Screening for pre-eclampsia

1 Introduction

Hypertensive disorders in pregnancy comprise at least two distinct entities. One, pregnancy-induced hypertension, is a disorder mainly, but not exclusively, of nulliparae which appears during the course of pregnancy and is reversed by delivery. Pregnancy-induced hypertension associated with proteinuria is a more severe form and is commonly referred to as pre-eclampsia. The other condition is pre-existing hypertension unrelated to, but coinciding with, pregnancy. It may be detected for the first time in pregnancy, but does not regress after delivery. The two conditions may occur together in the same woman. They are an important cause of maternal, fetal, and neonatal morbidity and mortality.

Since the cause of pregnancy-induced hypertensive disorders is not understood, definition and diagnosis are based on the signs that are considered to be most characteristic: hypertension, proteinuria, and oedema. These are signs, not the disease itself; they constitute secondary features of an underlying circulatory disorder. As signs they are non-specific; they can be induced by pregnancy itself as well as by a variety of conditions unrelated to, but coinciding with, pregnancy. Hypertension and proteinuria are usually asymptomatic and therefore must be detected by screening.

2 Hypertension

The diagnosis of hypertension is made only when blood pressure passes a predefined threshold. Many pregnancies develop normally despite the raised blood pressure, indicating that adequate uteroplacental and maternal organ flows are maintained. A certain degree of hypertension may well be beneficial to maintain perfusion pressures in the face of an elevated vascular resistance.

Many factors, such as age, parity, and race, cause marked variability in blood pressure between individuals, while others, such as time of day, level of activity, emotions, and posture, may result in variations within the same pregnant woman. Measurements by doctors or midwives in antenatal clinics are usually higher than those obtained at home ('white coat hypertension').

Although measurement of blood pressure is the mainstay of screening, diagnosis, and decision-making in pregnant women, the inherent technical and sampling errors, whether by auscultation or automatic devices, constitute an important limitation of the accuracy and precision of this measurement. Errors in blood-pressure measurement cannot be abolished by spending money on automated equipment. The mercury sphygmomanometer and stethoscope are still compatible with good antenatal care, and will probably remain so in the foreseeable future. Continuous ambulatory monitoring of blood pressure may improve diagnosis and subsequent care, but awaits formal evaluation.

Hypertension in pregnancy can be defined as either a diastolic pressure above a predetermined cut-off point or a rise from a woman's pre-existing blood pressure level. Pregnant women with a diastolic blood pressure between 90 and 100 mmHg in the second half of pregnancy experience an increased incidence of proteinuria and perinatal death. For that reason, a diastolic blood-pressure level somewhere between 90 and 100 mmHg may be considered to be a threshold between women at low risk and women with an increased risk of pregnancy complications.

A diagnosis of hypertension thus defined is not a diagnosis of a disease, but a marker of an increase in risk and an indication for careful monitoring of mother and fetus. It is clinically important to realize that, in view of the physiological blood-pressure changes in pregnancy, a diastolic blood pressure of 90 mmHg in mid-pregnancy is more abnormal than such a pressure would be if it occurred for the first time at term.

Pregnancy-induced hypertensive disorders rarely occur before 20 weeks' gestation. Hypertension and/or proteinuria diagnosed before 20 weeks will usually be due to pre-existing chronic hypertension or renal disease. Hypertension may also be diagnosed for the first time during labour; such hypertension will often be transitory, due to effort and/or anxiety.

The differential diagnosis between pregnancy-induced hypertension and pregnancy associated with chronic hypertension can be difficult. In comparison with pregnancy-induced hypertension or pre-eclampsia, which usually occurs in young nulliparous women, women with chronic hypertension tend to be older and parous. Many women with chronic or renal hypertension show an even greater physiological fall in blood pressure during the first half of pregnancy than do normotensive women, with an exaggerated rise in the third trimester. The earlier in pregnancy that hypertension is noted, the more likely it is to be chronic hypertension.

Chronic hypertension is a major predisposing factor for pre-eclampsia ('superimposed pre-eclampsia'). The maternal and fetal risks of chronic hypertension in pregnancy seem to be mainly attributable to the development of superimposed pre-eclampsia; the majority of chronically hypertensive women who do not develop pre-eclampsia have a normal perinatal outcome.

3 Proteinuria

Renal protein excretion increases in normal pregnancy, and proteinuria is not considered abnormal until it exceeds 300 mg per 24 hours. An increase in protein output will usually, but not always, lead to higher concentrations in random urine samples. The volume and concentration of urine will affect protein concentration and may give rise to erroneously high or low results of tests on random urine specimens.

Proteinuria may be a temporary phenomenon due to pregnancy-induced renal lesions, or it may be an expression of pre-existing renal disease coinciding with pregnancy. In the first case it should disappear at some time after delivery; in the latter it will remain present.

Proteinuria is a late sign in pregnancy-induced hypertension and is associated with an increased risk of poor fetal outcome. The overall correlation between the degree of elevation of blood pressure and the occurrence of proteinuria is weak. However, the magnitude of protein loss correlates positively with the severity of renal lesions. Thus urine testing is a vital part of the screening process for hypertensive disorders in pregnancy.

In practice, screening for proteinuria is usually done with reagent strips or 'dipstick' tests, which will start detecting protein (albumin) concentrations of approximately 50 mg /litre. Protein concentrations depend on urine volume and specific gravity. Dipsticks may give up to 25 per cent false-positive results with a trace reaction, and 6 per cent false-positive results with a one + reaction on testing of random specimens from women with normal 24 hour total protein excretion.

The definitive test for proteinuria in pregnancy is determination of total protein excretion in a 24 hour urine collection using a reliable quantitative method (for example, Esbach's method). This is too complicated to be used for screening, but may be considered whenever significant proteinuria is detected on screening of a random urine specimen. Alternative approaches to the measurement of 24 hour protein excretion, such as determination of the protein/creatinine index in random urine specimens, await validation.

4 Oedema

Moderate oedema occurs in 50–80 per cent of healthy normotensive pregnant women. This physiological oedema of pregnancy is often confined to the lower limbs, but it may also occur in other sites, such as the fingers or face, or as generalized oedema. Most pregnant women note that the rings on their fingers become tight in the course of the third trimester. Physiological oedema usually develops gradually, and is associated with a smooth rate of weight gain. The finding that pregnant women with generalized oedema without hypertension or proteinuria have larger babies than women without obvious oedema strongly suggests that oedema is part of the normal maternal adaptation to pregnancy.

Oedema affects approximately 85 per cent of women with pre-eclampsia. It may appear rather suddenly, and may be associated with a rapid rate of weight gain. It cannot be differentiated clinically from oedema in normal pregnancy. Pregnant women without oedema, with-early-onset oedema, or with late-onset oedema all have a similar incidence of hypertension.

The combination of hypertension and oedema, or hypertension and increased maternal weight gain, is associated with a lower fetal death rate than hypertension alone. Pre-eclampsia without oedema ('dry pre-eclampsia') has long been recognized as a dangerous variant of the condition, with a higher maternal and fetal mortality than pre-eclampsia with oedema.

As oedema in pregnancy is common and does not define a group at risk, it should not be used as a defining sign of hypertensive disorders in pregnancy.

5 Biochemical and biophysical tests

A number of tests have been devised to demonstrate the presence or absence of an abnormal vascular responsiveness before the clinical onset of pregnancy-induced hypertension. These include the cold-pressor test, the flicker fusion test, the isometric exercise test, the roll-over test, and infusions of catecholamines or vasopressin. Evaluation of these tests has shown them to be worthless, and they are of historical interest only. The angiotensin sensitivity test is useful in a research setting, but is too complicated and time consuming to be of clinical use.

There is insufficient evidence to warrant the use of uric acid levels as a screening test to predict the later development of pregnancy-induced hypertension. In women with established pre-eclampsia, however, serum uric acid levels appear to reflect fetal prognosis. For that reason uric acid levels can be used in the care of women with pregnancy-induced hypertension to monitor the course of the disease.

6 Conclusions

Hypertension and pre-eclampsia are usually asymptomatic; screening and diagnosis depend mainly on careful determination of blood pressure and proteinuria.

Pregnancy-induced hypertension may occur at any time in the second half of pregnancy. It rarely occurs before 28 weeks of pregnancy, but when it does, it frequently leads to pre-eclampsia with its associated high rate of perinatal morbidity and mortality. However, pregnancy-induced hypertension occurs much more frequently late in the third trimester, when antenatal visits are usually increased in frequency but the maternal and fetal risks of late-onset disease are much smaller. Therefore, although routine antenatal screening for hypertensive disorders before 28 weeks' gestation may have a low productivity in terms of the number of positive diagnoses per visit, it has a high potential in terms of prevention of maternal and fetal morbidity and mortality. For that reason the number of antenatal visits of nulliparae in the second trimester should not be reduced without further evidence of safety.

Simple blood pressure measurement remains an integral part of antenatal care; it should be performed in a standardized fashion by a skilled midwife or nurse, or by the attending physician.

The appearance of proteinuria in a previously non-proteinuric woman with pregnancy-induced or pre-existing hypertension is associated with a marked increase in maternal and fetal risk. For that reason, screening for proteinuria using a dipstick remains a valuable tool.

None of the other tests advocated for screening, prediction, and early diagnosis of pregnancy-induced hypertensive disease have been shown to be clinically useful. Determination of haematocrit values, serum uric acid concentrations, platelet counts, and, to a lesser extent, plasma antithrombin III concentration and factor-VIII-related antigen may be of value in monitoring the progress of established hypertensive disorders in pregnancy.

This chapter is derived from the chapter by Henk C.S. Wallenburg (24) in EFFECTIVE CARE IN PREGNANCY AND CHILDBIRTH.

References to primary sources and more complete data for statements made in this chapter can be found in the source chapter.

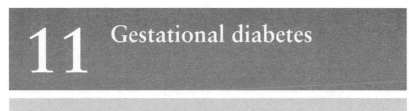

11 Gestational diabetes

1 Introduction

The concept of gestational diabetes evolved from the earlier concept of prediabetes, which held that much of the pathology associated with overt diabetes develops before the appearance of insulin dependence. The glucose tolerance test became the mainstay of this diagnosis, as it was believed to uncover a defect in glucose homeostasis that could only be demonstrated after a glucose challenge. It was not until 1973 that an attempt was made to link prediabetes (an abnormal glucose tolerance test in the absence of overt disease) to perinatal outcome. Although this link is remarkably tenuous, it gave rise to the concept of gestational diabetes as a disease entity to be searched for and treated. Caregivers and women became anxious lest 'gestational diabetes' develop, and various forms of glucose challenge screening were introduced and carried out to identify the condition.

2 Risks of 'gestational diabetes'

From the evidence available, the small increase in perinatal mortality associated with abnormal glucose tolerance appears to be predicted as much by the indication for glucose tolerance testing (such as obesity, large fetus, previous stillbirth, or malformation) as by the test result. Thus glucose intolerance is simply a marker for other underlying conditions that adversely influence perinatal outcome.

Even as only a marker for increased perinatal mortality, the glucose tolerance test could still be a useful indicator of risk. The question remains as to whether or not identification and treatment of women with gestational diabetes can prevent some of the associated perinatal deaths.

The 'adverse outcome' most frequently associated with gestational diabetes is 'fetal macrosomia' (a larger than average baby). The adverse outcomes of caesarean section, shoulder dystocia, and trauma derive from this primary outcome. Up to 30 per cent of mothers with an abnormal glucose tolerance test have a baby with a birthweight of more than 4000 grams. However, clinical judgement based on assessment of prepregnant weight, weight gain, and a pregnancy past 42 weeks, without any reference to glucose tolerance, is more predictive of fetal macrosomia than is the glucose tolerance test. Wide application of glucose tolerance testing to pregnant women would thus be of limited value in identifying women at increased risk of fetal macrosomia.

3 Therapy in 'gestational diabetes'

There is no convincing evidence that treatment of women with an abnormal glucose tolerance test will reduce perinatal mortality or morbidity. Trials of dietary regulation for 'gestational diabetes' do not demonstrate a significant effect on any outcome, including macrosomia. Trials comparing the use of insulin plus diet with diet alone show a decrease in macrosomia, but no significant effect on other outcomes such as use of caesarean section, the incidence of shoulder dystocia, or perinatal mortality. There is also no evidence that such treatment reduces the incidence of neonatal jaundice or hypoglycaemia. One trial actually assessed the use of elective caesarean section for 'gestational diabetes'. The result was a statistically significant increase in maternal morbidity, with no benefit shown for the baby. In one trial, no significant differences were found in maternal or neonatal outcome by use of elective early induction of labour.

4 Effects of glucose tolerance testing

The diagnosis of 'gestational diabetes', as currently defined, is based on an abnormal glucose tolerance test. At least 50–70 per cent of the time this test is not reproducible, and the increased risk of perinatal mortality and morbidity said to be associated with this 'condition' has been considerably overemphasized. As no clear improvement in perinatal mortality has been demonstrated with insulin treatment for gestational diabetes, screening of the pregnant population with glucose tolerance testing is unlikely to make a significant impact on perinatal mortality or morbidity.

An abnormal glucose tolerance test is associated with a two- or threefold increase in the incidence of macrosomia, but the majority of macrosomic infants will be born to mothers with a normal glucose tolerance test.

In addition, there is a great potential for doing more harm than good by performing a glucose tolerance test. A positive test labels the woman as having a form of diabetes. Her pregnancy is likely to be considered as 'high-risk', invoking an extensive and expensive programme of tests and interventions of unproven benefit. A negative glucose tolerance test also has a potential for harm by falsely reassuring the physician and the woman that the risk, engendered by the indication for the test, has been removed.

As no benefit has yet been established for glucose screening during pregnancy, the method used for this screening is irrelevant. However, it should be noted that a glucose polymer has been shown to be more acceptable than glucose to women undergoing such screening, and is associated with less nausea and headache. If any value of screening for minor degrees of glucose intolerance during pregnancy should ever be demonstrated, the possible advantages of using a glucose polymer rather than glucose for this screening should be reviewed.

5 Conclusions

All forms of glucose tolerance testing should be reviewed. Women in whom overt diabetes is suspected should be followed with fasting or blood glucose estimations two hours after meals throughout pregnancy.

The available data provide no evidence to support the wide recommendation that all pregnant women should be screened for 'gestational diabetes', let alone that they should be treated with insulin. Until the risk of minor elevations of glucose during pregnancy have been established in appropriately conducted trials, therapy based on this diagnosis must be critically reviewed. The use of injectable therapy on the basis of the available data is highly contentious, and in many other fields of medical practice such aggressive therapy without proven benefit would be considered unethical.

This chapter is derived from the chapter by David J.S. Hunter and Marc J.N.C. Keirse (25) in EFFECTIVE CARE IN PREGNANCY AND CHILDBIRTH.

References to primary sources and more complete data for statements made in this chapter can be found in the source chapter and/or in the following reviews from the *Cochrane pregnancy and childbirth database*:

Walkinshaw, S.A.
— Dietary regulation for 'gestational diabetes'. Review no. 06649.
— Diet plus insulin vs diet alone for 'gestational diabetes'. Review no. 06650.
— Glucose polymer vs glucose for screening for/diagnosing 'gestational diabetes'. Review no. 06652.
— Elective caesarean for 'gestational diabetes'. Review no. 06648.

12 Assessment of fetal growth, size and well-being

1 Introduction

A wide range of tests of fetal well-being have been introduced during the last thirty years, and have enjoyed waves of popularity.

Both biochemical tests (which monitor the endocrine function of the placenta or the fetoplacental unit) and biophysical methods of monitoring (which provide diverse information about fetal growth and physiological function) have the theoretical ability to detect changes in fetal well-being that may occur over hours, days, or weeks. No known method of assessment can predict sudden events such as cord prolapse or placental abruption which may also cause fetal damage or death.

Two general assumptions underlie the contention that antenatal monitoring is clinically useful: first, that these methods can detect or predict fetal compromise; second, that with appropriate interpretation and action, they can reduce the frequency or severity of adverse perinatal events or prevent needless interventions.

Tests have been used both in specific clinical situations with a high estimated fetal risk and as screening tests with the primary purpose of preventing those otherwise unpredictable fetal problems that occur from time to time. 'High-risk' circumstances include diabetes or pre-eclampsia, multiple and post-term pregnancy, and, most of all, when the fetus is thought to grow poorly. Because of the frequency with which such babies are monitored and because of the controversy over what constitutes 'growth retardation', it is worthwhile re-examining the definitions and underlying pathophysiological concepts.

2 Size and growth

Fetal size and fetal growth are often confused in clinical practice. It is common to see 'birthweight for gestation age' standards described as 'fetal growth charts', and a weight for gestation below some arbitrary centile referred to as 'intrauterine growth retardation'. There are two main reasons why this confusion may lead to false conclusions. First, *birth*weights at a given length of gestation are not a good reflection of *fetal* weights at the same gestation. Second, some authors have made inferences about fetal growth by comparisons with cross-sectionally derived mean or median birthweights at consecutive weeks of gestation. This approach is grossly misleading. Infants below a particular centile are 'light for gestational age' without necessarily being 'growth retarded'; this distinction between size and growth is critical. Growth cannot be estimated without two or more measurements of size; the argument that size at some time in late pregnancy and a presumed zero size at conception can represent those two measurements is clinically useless. What the clinician would like to know is whether fetal growth has deviated from its normal progression.

The term 'intrauterine growth retardation' should be restricted to those fetuses with definite evidence that growth has faltered. Such infants may not necessarily be 'light for gestational age'; a fetus whose weight falls from the 90th to the 30th centile in a short period of time is almost certainly in greater peril than a fetus who has maintained a position on the 5th centile.

Genuine fetal growth retardation is attributed to an inadequate supply of nutrition to the fetus by a malfunctioning placenta (or, more accurately, the blood supply to the placenta). The fetus reponds to this hostile environment by invoking adjustments that maximize the chances of survival. These include redistribution of blood flow (more to brain and heart, less to liver and kidneys) and limiting unnecessary movements. These phenomena provide the basis of some tests of fetal assessment.

3 Abdominal examination

The oldest clinical method of estimating fetal size — abdominal palpation — is so inaccurate as to be little better than a blind guess. Twenty per cent of such assessments just before birth are not within 450 grams of the actual birthweight, and the errors are worse at the extremes of the range where the information is most needed.

A more quantitative approach is to measure the increase in size of the maternal abdomen, which must, to some extent, reflect uterine growth. The two most widely practised techniques are the measurement of fundal height (the distance between the upper border of the pubic symphysis and the uterine fundus) and the measurement of abdominal girth at the level of the umbilicus.

There have been several studies of fundal height as an indicator of fetal size, but little investigation of the potential of this measurement for assessing growth. This is understandable, given the considerable inter- and intra-observer variation in measuring fundal height. Nevertheless, several studies have shown quite good sensitivity and specificity of fundal height for predicting low birthweight for gestation. The ability to predict low birthweight is not the same as the ability to detect growth retardation, but fundal height may be useful as a screening test for further investigation. Abdominal girth measurement has not been adequately evaluated.

4 Fetal movement counting

Reduction or cessation of fetal movements may precede fetal death by a day or more. In theory, recognition of this reduction, followed by appropriate action to confirm fetal jeopardy and expedite delivery, could prevent fetal death. This is the basis for using counts of fetal movements as a test of fetal well-being. The most commonly used method is to ask the mother to make a daily record on a chart of the time at which she has noticed ten kicks.

Not all late fetal deaths are even theoretically preventable in this way. Some are not preceded by a reduction in fetal movements. For others, there may be insufficient time between the reduction and fetal death to allow clinical action. Still others may be preceded by conditions that are recognizable in their own right, such as pre-eclampsia or intrauterine growth retardation. In the latter circumstances, fetal movement counting might still be theoretically useful as a supplement to other hospital-based tests of fetal well-being.

However, the cause of most antepartum late fetal deaths is unknown. These deaths are unpredictable, and therefore the extent to which they can be prevented by current forms of antenatal care is limited. Screening by fetal movement counting, because it can be performed each day, has theoretical advantages over other tests of fetal well-being, all of which are either difficult or impossible to perform daily for practical reasons.

Two randomized controlled trials have addressed the question of whether clinical actions taken on the basis of fetal movement counting improve fetal outcome, with the largest involving over 68 000 women. These trials collectively provide no evidence that routine formal fetal movement counting reduces the incidence of intrauterine fetal death in late pregnancy. Routine counting results in more frequent reports of diminished fetal activity, greater use of other techniques of fetal assessment, more frequent antepartum admission to hospital, and an increased use of elective delivery for decreased movement and hence an increase in the use of resources without compensating benefit. The practice does not appear to result in either a significant increase or decrease in feelings of anxiety among mothers compared with selective use of formal counting in specific high risk situations.

There is a marginal possibility that an occasional woman might benefit from fetal movement counting in terms of prevention of late fetal death. However, the wider social, psychological, and economic implications affect all women, and they must be taken into account.

5 Biophysical tests

5.1 *Ultrasound measurements*

Ultrasound imaging has the potential to assist assessment of fetal growth and well-being by measuring fetal growth over a period of time, by assessing fetal size at a single point in time to confirm or refute the clinical impression that a fetus is small for gestational age, by assessing amniotic fluid volume (and therefore indirectly fetal urine production), by investigating the appearance of the placenta, and by examining the movement and behaviour of the fetus.

The use of fetal measurements to assess whether or not the fetus is growing satisfactorily may be considered in the general framework of diagnostic tests, and appraised from the standpoints of their test properties: sensitivity, specificity, and predictive values. It is not clear, however, precisely what one is trying to predict. There is no absolute postnatal criterion of growth retardation that can be used to assess the validity of the 'test'. In the absence of an appropriate outcome measure, authors frequently use some measure of 'relatively low' birthweight, such as being below the 10th centile for gestational age. Such observational studies indicate that abdominal measurements are better than head measurements, which is consistent with the 'brain–sparing' effect seen in intrauterine growth retardation.

There has been very little research aimed at producing true growth curves based on repeated measurements over time of the same fetus. Such information would provide a far sounder basis for assessing fetal growth, and is arguably the only valid approach for detecting growth retardation. This approach is particularly useful when gestational age is unknown or uncertain. Ultrasonography, it would be argued, rather than 'birthweight for gestational age' should be considered as the 'gold standard' for measuring fetal growth. Despite this, controlled trials show that routine ultrasound measurement of fetal size in late pregnancy results in an increased rate of antenatal hospital admission, and possibly of induction of labour, with no evidence of any substantive benefit to the baby. No adequately controlled trial data are available from specifically high risk pregnancies.

'Light-for-dates' fetuses form a heterogeneous group within which individual risk varies greatly. Attempts have been made to analyse patterns of growth in the hope of uncovering the underlying pathogenesis and of providing an estimate of the risk to the individual fetus. 'Light-for-dates' fetuses can be divided into two groups, with the first showing arrest of previously normal growth and the second showing early departure from normal limits of growth that continues until delivery. The first of these patterns has been attributed to 'utero-placental insufficiency' and the latter to low growth potential. This latter group includes infants that are inherently abnormal (particularly chromosomally abnormal), some that have suffered a major insult (e.g. from rubella) during the critical period of organogenesis, and some that are small because of their genetic endowment. Comparative measurements of the fetal head and abdomen have been suggested as a means of further differentiating these two groups. While there is a higher incidence of intrapartum fetal distress and operative delivery in the asymmetrical growth group, the perinatal mortality and incidence of low Apgar scores are similar and high in both groups.

The significance of ultrasonically detected patterns of fetal growth requires further study, and seeking correlations with the results of Doppler studies would be of interest. Thus far, the evidence indicates that abdominal measurements are superior to head measurements in predicting 'light-for-dates' babies, but little is known about serial measurements and particularly about their relationship to neonatal outcome.

A single trial, reported in 1987, examined the value of ultrasound placental appearances. Reporting placental 'texture' to clinicians providing antenatal care resulted in less meconium stained amniotic fluid in labour, fewer babies with low 5 minute Apgar scores, and, most importantly, fewer deaths of normally formed babies than occurred for women whose clinicians were not provided with the report of placental grading. It appears that knowledge of placental grading can lead to appropriate clinical action that can improve pregnancy outcome. This study deserves to be repeated. In the meantime, it seems advisable to report the placental grade following ultrasound examinations for specific clinical indications during the third trimester.

No information on the use of amniotic fluid volume estimation is available from randomized trials.

5.2 Doppler ultrasound

Doppler ultrasound has been used for a number of years to identify and record fetal heart pulsation and, in adults, to assess blood flow in compromised vessels. The use of the technique to demonstrate blood velocity waveforms in the fetal umbilical artery was first described in 1977. Alterations in fetal umbilical blood flow may occur as an early event in conditions of fetal compromise. For this reason, Doppler studies could provide important information on the pathophysiology of compromised pregnancies (particularly of fetal growth retardation) and be a useful technique for evaluating fetal well-being in high-risk pregnancies.

Information from trials supports this hypothesis. Combined evidence from several trials in high risk pregnancies (complicated mainly by fetal growth retardation or maternal high blood pressure) shows that there are fewer stillbirths and neonatal deaths among normally formed babies when the results of Doppler velocimetry are made available to the clinicians. A number of the deaths in the control groups in these studies seem to have resulted from uteroplacental insufficiency, and these might well have been avoided by Doppler study and the clinical action prompted by it. The use of Doppler ultrasound in high-risk pregnancies so far appears to have little or no effect on the rate of admission to hospital during pregnancy or of elective delivery. Nor is there any effect on the incidence of fetal distress and caesarean section during labour, or on the condition of the newborn baby other than a greater likelihood of being born alive.

In contrast with its effectiveness in decreasing perinatal mortality among women at identified high risk, Doppler ultrasound appears to have little, if any, effect on pregnancy outcome when used as a screening test in unselected pregnancies. This should not be surprising; when used in a low-risk population, the predictive power of any test is low, and the benefits from adequately responding to the few true-positive tests can be more than counterbalanced by the harm done in response to the inevitable high proportion of false-positive tests.

5.3 Contraction stress testing

Continuous recording of fetal heart rate and uterine activity was first developed for use in labour in an attempt to identify the fetus at risk of death or morbidity due to intrapartum asphyxia. Because many fetal deaths occur prior to the onset of labour, the stimulation of contractions with oxytocin for short periods of time was proposed to allow observation of the fetal heart rate under labour-like conditions in pregnancies at risk.

This technique, which subsequently became known as the 'oxytocin challenge test' or 'contraction stress test', has no demonstrated benefits and suffers from a number of disadvantages. It is time consuming, requires an intravenous infusion, and has the potential to harm the fetus. Its use is contraindicated in some pregnancies at risk, for example when there is antepartum bleeding, placenta praevia, a history of preterm labour, or preterm rupture of membranes.

The nipple-stimulation stress test is similar in purpose to, and has been directly compared with, the oxytocin challenge test. The pregnant woman is encouraged to stimulate her nipples with her fingers, palms, a warm moist face cloth, or a heating pad either directly or through her clothing. Contractions can be stimulated effectively, but the mechanism, previously assumed to be oxytocin release, remains unknown. The nipple-stimulation stress test has some of the same disadvantages as the oxytocin challenge test with the additional problem that, while stimulation is easily discontinued, there is a time lag of more than 3 minutes between stimulation and peak uterine response. Excessive uterine activity occurs in over half the women who undergo this test, and this will result in fetal bradycardia in 7–14 per cent. Cases of severe uterine tetany associated with fetal heart rate abnormalities have also been reported. Because of these concerns and because it offers no clear advantages over other techniques, the nipple-stimulation stress test should be relegated to the history books.

5.4 Non-stress cardiotocography

The evaluation of fetal heart rate patterns without the added stress of induced contractions was first proposed in 1969. This 'non-stress test' has been widely incorporated into antenatal care for both screening and diagnosis.

There is no universally accepted technique for performing non-stress antepartum cardiotocography. Various durations and frequencies of monitoring are used, and these can have a powerful influence on the predictive properties of the test. Additional manoeuvres, such as abdominal stimulation, sound stimulation, glucose infusions, postprandial repeat tests, and follow-up oxytocin challenge tests for suspected abnormal records have been suggested or used. None of these have been shown to improve the predictive value of the test.

Many factors may interfere with the interpretation of the non-stress test. Like the contraction stress test, the non-stress test requires sophisticated equipment which may occasionally malfunction. Fetal and maternal movements may produce artefacts since ultrasound monitoring detects movement rather than sound. During fetal rest periods, which not uncommonly last for more than thirty minutes, normal physiological reduction in heart rate variability may be confused with pathological change. Medications taken by the mother are often transferred to the fetus, and, particularly in the case of drugs with a sedative effect on the central nervous system, may in themselves produce heart rate patterns that can be interpreted as abnormal. Likewise, the gestational age of the fetus has a strong influence on the frequency of false positive non-reactive tests, with 'abnormal' patterns being more frequently described in the preterm fetus. When all these factors are taken into consideration, as many as 10–15 per cent of all records may be unsatisfactory for interpretation.

A variety of methods for interpreting non-stress cardiotocograms have been described, all of which include evaluation of some or all of the following characteristics: baseline fetal heart rate; various interpretations of fetal heart rate variability; accelerations of fetal heart rate associated with spontaneous and/or stimulated movements of the fetus; decelerations associated with spontaneous uterine contractions. The more complicated systems assign scores to some or all of these parameters, sometimes subsequently grouping the scores. The most commonly used method is to divide traces into reactive (normal) and non-reactive (abnormal), based on the presence or absence of adequate baseline heart rate variability and heart rate accelerations with fetal movement. It is now well documented that even when a fixed method is used, interpretation of a trace may vary when an individual observer reads the same trace at different times or when the same trace is read by different observers.

In addition to the difficulties with test interpretation, there is a danger inherent in any form of screening when the likelihood of the fetus being in difficulty is small. A higher proportion of positive tests will be false positive when the probability of an adverse outcome is small. Intervention based on the results of a 'positive' non-stress test in a low-risk group of women will often do more harm than good.

This risk is real rather than only theoretical. In each of the four trials of non-stress testing that have been reported, perinatal deaths from causes other than malformations were *more* common in the groups in which clinicians had access to the test results. Collectively, the increase in perinatal deaths among the women tested was importantly (more than threefold) and statistically significantly higher. There was no demonstrable effect on caesarean section rates, incidence of low Apgar scores, abnormal neonatal neurological signs, or admission to special care nurseries. These analyses provide no support at all for the use of antepartum non-stress cardiotocography, as employed in these studies, as a supplementary test of fetal well-being in 'high risk' pregnancies. One can only speculate as to why cardiotocography continues to be used in such an extensive way, and why the results from the only four randomized trials that have been published are so widely disregarded by many obstetricians.

Antepartum cardiotocography is essentially an assessment of immediate fetal condition. Unless evidence emerges to the contrary, its clinical use would seem best restricted to situations in which acute fetal hypoxaemia may be present (e.g. sudden reduction of fetal movement or antepartum haemorrhage).

5.5 Fetal biophysical profile

The 'biophysical profile' was derived from a study of serial ultrasound examinations and antenatal cardiotocography (non-stress test) in high-risk pregnancies. Combining five biophysical 'variables' considered to be of prognostic significance (fetal movement, tone, reactivity, breathing, and amniotic fluid volume) into a score reduced the frequency of false-positive and false-negative results compared with the non-stress test alone. An additional advantage of the biophysical profile over the non-stress test is that it permits assessment of the possibility of major congenital anomalies. This may be important, as detection of a serious anomaly may on occasion help to avoid a caesarean section where the baby is clearly abnormal.

Only two controlled trials of biophysical profile testing have been performed. Both were conducted in women referred to units specializing in fetal biophysical assessment. They compared care based on biophysical score results with that based on non-stress test results following a management protocol. In both studies the biophysical profile score was a better predictor of low 5 minute Apgar scores than the non-stress test. The biophysical profile was both more sensitive and more specific in predicting overall abnormal outcome than the non-stress test.

Despite the better predictive value of the biophysical score than the non-stress test, its use did not result in any improvements in outcome for the baby. Outcomes measured included perinatal death, fetal distress in labour, low Apgar score, and low birthweight for gestational age. Compared with cardiotocography alone, biophysical profile testing showed no obvious effect (either beneficial or deleterious) on these outcome measures. The available evidence provides no support at all for the use of biophysical profile as a test of fetal well-being in high risk pregnancies. However, the number of women included in these studies is so small that any estimates of effect are extremely imprecise.

6 Biochemical tests

6.1 Oestriol

The enthusiasm for oestrogen assays that prevailed in the 1960s and 1970s was based on the observation that perinatal mortality rates were twice as high in women with low oestriol excretion than in the general population. However, the usefulness of the tests was marred by the facts that they were not sensitive enough to detect the majority of pregnancies destined to have an adverse outcome, and that a great many women with normal pregnancies falsely appeared to be at risk.

Among the many studies reported, there has been only one randomized controlled trial. In this trial, knowledge of oestriol levels had no detectable effect on either perinatal mortality or the rate of elective delivery. Similar conclusions were reached from a comparison of pregnancy outcomes within the same institution in two consecutive periods with and without the use of oestriol assays. Thus there is no evidence to suggest any benefit from oestriol assays.

6.2 Human placental lactogen

Like oestriol, human placental lactogen measurements have been used in the hope of predicting and possibly averting a variety of poor outcomes. There has been only one randomized intervention study of human placental lactogen measurement. In this trial, human placental lactogen assays were performed on all women at each visit, but in half of them (the control group) the results were not reported to the clinician. In the experimental group abnormal values were reported promptly to a perinatology fellow. If the results of a repeat measurement were low and the fetus was considered to be mature, action was taken to expedite delivery.

The results of this trial suggest, on first inspection, that revealing the results of human placental lactogen measurements to a clinician armed with a predetermined intervention programme statistically significantly reduced fetal and perinatal mortality. Although the data from this trial are often cited to indicate that human placental lactogen measurements are beneficial for the surveillance of high-risk pregnancy, they relate only to the 8 per cent (4 per cent in each group) of pregnancies that had abnormal human placental lactogen values. Data on the large majority of pregnancies (92 per cent) that did not belong to that category were not reported and are no longer available. One cannot exclude the possibility that the apparent benefit in the small minority with abnormal placental lactogen values was not offset by negative effects in the majority of pregnancies with normal tests.

6.3 Other biochemical tests

Other biochemical markers of fetal growth and well-being have not proved popular. Unexplained high levels of maternal serum alpha-fetoprotein (at the screening period in mid pregnancy) help to predict later intrauterine growth retardation but without sufficient sensitivity to justify the test in the absence of a policy of detection of fetal neural tube defects (see Chapter 9).

Alpha-fetoprotein is of fetal origin, but several proteins of placental origin have also been assessed as biochemical tests in obstetric practice. These never achieved the popularity of oestriol and human placental lactogen, even in the heyday of biochemical assessment, nor has there been much enthusiasm for the use of placental protein measurements more recently. It may be that there is some scope for further research in the field of placental proteins, but there are no indications as yet that any of these measurements are of clinical use.

Various other hormones, enzymes, and other substances have generated interest, although this interest has often been short lived. These include oxytocinase, alkaline phosphatase (thermostable and leucocyte), and progestogens. These tests have waned in popularity. The available studies do not present data in such a way that either their predictive properties or their usefulness can be determined.

7 Conclusions

Tape measurement of symphysis–fundus height is simple, inexpensive, and widely used during antenatal care. Fundal height could be used as a screening device for referral of women to an obstetrician for further assessment, but many small fetuses will be missed and many perfectly well-grown fetuses will be considered worryingly small. Nevertheless, in our present state of knowledge it would be unwise to abandon the practice.

Ultrasound techniques have the capacity to detect abnormalities of fetal growth but have not been effectively exploited. There is a need for prospective studies to examine the differential growth of fetal parts, in large populations, to see whether patterns of growth that are associated with fetal compromise and later infant morbidity can be defined. Such studies should be carried out with scheduled repeated measurements. It is particularly crucial that better measures of outcome should be defined. Intervention based on assessments of size or growth should be evaluated in randomized trials before they are accepted into general obstetrical practice. For the present, the evidence provides no support for routine ultrasonography for fetal measurement in late pregnancy.

Biochemical tests of fetal well-being are expensive and have a low predictive value for adverse outcome. While they have added immensely to knowledge of placental and fetal physiology, none of them have been shown to be clinically useful. Their use should be restricted to research, and they should not be employed in clinical practice.

Monitoring of fetal movements by the mother is a simple and inexpensive test of fetal well-being that can be performed daily. However, there is no evidence that a policy of routine fetal movement counting will have beneficial results. If used at all, it should be used in individual circumstances, for example where a woman perceives diminished fetal movements. The results should prompt other diagnostic tests rather than more definitive obstetric intervention. The possibility of congenital malformation should be considered before delivery is expedited.

Biophysical tests are employed today with the same enthusiasm that characterized biochemical testing in the past. These tests have greatly increased our understanding of fetal behaviour and development, but with the exception of Doppler studies of umbilical artery waveform in high-risk pregnancies, and possibly placental grading, their use has not been demonstrated to confer benefits in the care of an individual woman and her baby. For this reason and despite their widespread clinical use, most biophysical tests of fetal well-being should be considered of experimental value only rather than as validated clinical tools. They should be acknowledged as such and, at the very least, further extension of their clinical use should be curtailed until or unless they can be demonstrated to be of benefit in improving the outcome for mother or baby.

The role of non-stress cardiotocography as either a screening or a diagnostic test seems questionable because of its relatively high cost and poor predictive properties. The biophysical profile may have greater potential as a diagnostic test for women in whom there is a high risk of fetal problems, but the usefulness of this approach is still not established.

Doppler ultrasound has been evaluated more rigorously and extensively than any other test of fetal health or fetoplacental function. The encouraging results justify the use of Doppler ultrasound during high-risk pregnancy to guide clinical care, but there is no evidence that any benefit is derived from routine Doppler screening in unselected pregnancies.

This chapter is derived from the chapters by Douglas Altman and Frank Hytten (26), Jim Neilson and Adrian Grant (27), Diana Elbourne (28), Sophie Alexander, Rosalind Stanwell-Smith, Pierre Buekens, and Marc J.N.C. Keirse (29), and Patrick Mohide and Marc J.N.C. Keirse (30) in EFFECTIVE CARE IN PREGNANCY AND CHILDBIRTH.

References to primary sources and more complete data for statements made in this chapter can be found in the source chapters and/or in the following reviews from the *Cochrane pregnancy and childbirth database*:

Neilson, J.P.
— Routine serial symphysis-fundal height measurement. Review no. 06889.
— Routine fetal anthropometry in late pregnancy. Review no. 03873.
— Hormonal placental function tests. Review no. 03883.
— Routine formal fetal movement (FM) counting. Review no. 04364.
— Cardiotocography for antepartum fetal assessment. Review no. 03881.
— Maternal glucose ingestion with tests of fetal well-being. Review no. 07193.
— Maternal intravenous glucose with tests of fetal well-being. Review no. 07194.
— Fetal vibroacoustic stimulation before labour. Review no. 07192.

— Fetal manipulation with antepartum cardiotocography. Review no. 07195.
— Doppler ultrasound in high risk pregnancies. Review no. 03889.
— Doppler ultrasound screening of unselected pregnancies. Review no. 07357.
— Doppler ultrasound (all trials). Review no. 07337.
— Doppler ultrasound of umbilical artery plus biophysical scoring. Review no. 06004.
— Doppler ultrasound vs CTG in suspected IUGR. Review no. 07358.
— Routine ultrasound placentography in late pregnancy. Review no. 03874.

Neilson, J.P. and Alfirevic, Z.
— Biophysical profile for antepartum fetal assessment. Review no. 07432.
— Routine third trimester cardiotocography plus ultrasound. Review no. 07196.
— Ultrasound fetal assessment in post-term pregnancy. Review no. 07433.

PREGNANCY PROBLEMS

1 Introduction

Although an uncomplicated pregnancy is generally considered to be a state of health rather than disease, it is frequently accompanied by symptoms that at other times or in other circumstances might be thought to be signs of illness. Nausea and vomiting, tiredness, heartburn, constipation, haemorrhoids, leg cramps, varicosities, and vaginitis can be the cause of significant discomfort and unpleasantness. They can result in women having to change their normal patterns of behaviour significantly. Their alleviation is an important aspect of antenatal care.

2 Nausea and vomiting

Nausea and vomiting are the most frequent, the most characteristic, and perhaps the most troublesome symptoms of early pregnancy. The causes of nausea in pregnancy are still unknown, and the variety of treatments that have been recommended reflect the multitude of theories about the underlying causes. As might be expected for a self-limiting condition, uncontrolled studies of these 'treatments' have yielded rather spectacular, if spurious, results. In contrast with the results obtained in uncontrolled trials, those from controlled trials have been less impressive.

An unconventional approach using acupressure at the Neiguan (P6) point in the wrist has been evaluated in two placebo-controlled randomized trials; the trials were small, but showed a clear effect in reducing the prevalence of persistent nausea. This non-pharmaceutical approach deserves further evaluation.

Several trials have demonstrated a variety of antihistamines to be better than placebo. Simple antihistamines, phenothiazines, and piperazines, although they sometimes provoke troublesome side-effects such as drowsiness and blurring of vision, are generally considered to be safe during pregnancy. However, there have been no major epidemiological studies to look for their adverse effects on the fetus. One small trial suggested that vitamin B_6 (pyridoxine) might be effective, and ginger has also been shown to be more effective than placebo. A trial of dramamine showed it to be less effective alone than when combined with benzylamine. ACTH was not shown to be more effective than placebo for excessive vomiting (hyperemesis).

By far the most widely used drug used to treat nausea in pregnancy was, until recently, Debendox (marketed as Bendectin in the United States and Canada, and as Lenotan in some other countries). Indeed, Debendox was the most widely used prescription drug of any kind taken in pregnancy. Controlled trials showed it to be highly effective. In June 1983, Debendox was removed from the market as a direct result of litigation brought against the manufacturers, claiming that the drug had caused congenital malformations when used in pregnancy. The litigation was brought despite overwhelming evidence *against* Debendox's being a teratogen. The removal of Debendox has almost certainly led to an increased use of other medications for treating nausea and vomiting, about which, for any single product, much less human research has been conducted.

At the time of its recall, Debendox had been used by over 30 million women world-wide. In many countries between a quarter and a third of all pregnant women used Debendox. When Debendox was used to treat nausea it was usually in the first trimester during embryological development. Because of concern about the teratogenic risks of other drugs, Debendox was often the only prescription drug used in pregnancy. If we assume an overall congenital malformation incidence rate of 3.5 per cent at birth, then, by chance alone, exposure to Debendox will have occurred in over one million babies born with a congenital malformation. In the inevitable search for what may have produced the child's malformations, it is not surprising that many mothers implicated Debendox, and, perhaps prompted by over-eager lawyers, some chose to sue.

The three small placebo-controlled trials that have been published provide strong evidence that Debendox provided considerable relief for nausea and vomiting in pregnancy. Despite the wide variety of populations studied and methods employed, there is widespread agreement among the nineteen epidemiological studies of Debendox that the drug is *not* associated with an increased risk of congenital malformations. Its abrupt withdrawal from the market has provided a unique further test of its safety. Despite the reduction in use of Debendox, from about one-third of pregnancies to none, there has been no correlated reduction in the reported incidence of any group of malformations.

With the withdrawal of the most widely studied anti-nausea drug, Debendox, from the market, nausea in pregnancy is being treated by antihistamines that appear to be efficacious, as shown by the early trials; however, their safety has not been as extensively studied.

3 Tiredness

Many women report feelings of extreme tiredness, particularly in the early months of pregnancy. Despite the adverse effect that this has on women's lives, there is a dearth of research on this condition. Interesting questions relate to factors which might help, such as taking time off work and the relationship between tiredness and nausea. In the absence of controlled research, it seems sensible to reassure women that tiredness is a normal symptom, and to advise those who are excessively tired to rest whenever they can, and to seek help with home and work commitments whenever this is feasible.

4 Heartburn

Heartburn affects about two-thirds of all women at some stage of pregnancy. It is another so-called 'minor' disorder of pregnancy, but it causes more discomfort and distress than do many more serious conditions. It is commonly associated with eating, stooping, or lying down. The most clear-cut precipitating factor is posture.

Self-medication with proprietary antacids is the most commonly employed treatment, and this often provides adequate relief. There is little evidence of differences in efficacy between the various preparations available. No trials of H_2 suppressants such as cimetidine or omeprozole have been reported in pregnant women, and the safety of these drugs in pregnancy has not been established.

Other pharmacological approaches to the treatment of heartburn appeared to show beneficial effects in single randomized trials. One trial showed prostigmine to be more effective than placebo; another showed dilute hydrochloric acid, which may counteract the effects of regurgitated bile on the stomach and oesophagus, to be more effective than placebo. Some women who had not obtained relief from their symptoms with alkali obtained relief from acid, and vice versa. Methodological weaknesses in these two trials preclude over-reliance on their findings.

In situations where simple and sensible measures, such as avoiding fatty or spicy foods and minimizing bending over or lying flat after eating, have failed, the evidence suggests that antacids should be prescribed initially. Individuals should choose their preferred product among different preparations, since therapeutic compliance will be increased when the product is found to be palatable as well as effective. If symptoms fail to respond to antacids, prostigmine or dilute hydrochloric acid may be tried, although suitable preparations may be hard to find. Lemon juice or acidic lemonade may perhaps be as effective, and more palatable. There does not appear to be any hazard from the occasional use of any of the agents discussed.

5 Constipation

Constipation is a troublesome problem for many women during pregnancy, particularly during the last trimester. Women who are habitually constipated usually become more so during pregnancy. The frequency of constipation among pregnant women will reflect their dietary habits, fluid intake, and pattern of physical exercise. In observational studies, management of constipation using physiological approaches (alterations in diet, fluid intake, or exercise) has been found to bring relief in at least one-third of the women. More recently, interest has focused on using modest supplemental intake of dietary fibre to reduce constipation. In a randomized trial comparing dietary bran supplement with no supplementation, women in the fibre-supplemented groups increased their number of bowel movements and experienced less constipation than the untreated women.

Nevertheless, many women may require laxatives if physiological approaches afford no relief. Laxatives are usually classified by their mode of action. Hydrophilic bulking agents (polysaccharide and/or cellulose derivatives) and detergent stool softeners (the dioctylsulphosuccinates) are relatively free of adverse effects because they are inert and not absorbed. Some laxatives such as the diphenylmethanes (for example bisacodyl and phenolphthalein), the anthraquinones (aloe, cascara, and senna), and castor oil operate by their irritant action on the intestine. The most common maternal side-effects include cramping or griping, increased mucus secretion, and excessive catharsis with resultant fluid loss. Chronic use can result in loss of normal bowel function and laxative dependence. These irritant laxatives are all absorbed systemically to some extent. Most of them probably cross the placenta, but there is little information about possible effects on the fetus.

Saline cathartics (magnesium, sodium, and potassium salts) and lubricants (such as mineral oils) should not be used during pregnancy, the former because of the danger of inducing electrolyte disturbances and the latter because they interfere with absorption of fat-soluble vitamins.

Both bulking agents and stool softeners are safe for long-term use in pregnancy. If these preparations fail to relieve symptoms, irritant laxatives such as standardized senna or bisocodyl should be used on a short-term basis. Saline cathartics and lubricant oils should not be used at all.

6 Haemorrhoids

Effective prevention or treatment of constipation will, of course, help to reduce the severity of haemorrhoids, another common and painful symptom of pregnancy. In the absence of any sound research about the best means of preventing or treating this condition, advice similar to that given to non-pregnant sufferers may be appropriate, such as rest, elevation of the legs, and avoiding constipation.

7 Vaginitis

7.1 *Candidiasis*

Vaginal candidiasis (thrush or yeast infection) is a frequent problem during pregnancy, and causes an intensely irritating itchy vaginal discharge. The infection is found between two and ten times more frequently in pregnant women than in non-pregnant women and is more difficult to eradicate during pregnancy. The condition usually clears spontaneously soon after delivery.

The clinical diagnosis of vaginal candidiasis is neither specific nor sensitive. For some women, it may be difficult to distinguish between symptoms of candida and the increase and changes in normal vaginal secretion in pregnancy. Typical symptoms include an irritating vaginal discharge and pruritus (itch). Examination reveals reddened mucosa of the labia minora, introitus, and lower third of the vagina, with white patches and a thin discharge containing white flakes.

Congenital candida infections can affect infants of very low birthweight and manifest with pneumonia and skin infections. However, these infections are rare considering the high frequency of vaginal candidal carriage in pregnant women. In view of the low incidence of adverse effects on babies born to asymptomatic mothers, screening of pregnant women for candida is not indicated.

A variety of different local antifungal agents as well as different dosages and frequencies of administration have been studied. Imidazoles are more effective than nystatin. There is no evidence that a fourteen day course of antifungal therapy is more effective in curing candidiasis than a seven day course, but the available evidence does not support courses shorter than seven days for symptomatic candidiasis during pregnancy. Recently, single-dose vaginal treatments have been used; there are no comparative trials in pregnancy.

The initial treatment of symptomatic vaginal candidiasis should consist of clotrimazole (an imidazole) because of its proven superior efficacy over nystatin. Repeat courses may be required, given the tendency of the infection to recur. No treatment is indicated in asymptomatic infection.

Oral drugs, such as fluconazole, are now also being prescribed. In theory, they may be more effective than vaginal treatments, as they are effective against bowel as well as vaginal organisms. However, this form of treatment has not been tested during pregnancy and it cannot be assumed to be safe.

7.2 Trichomoniasis

The protozoan Trichomonas vaginalis is frequently isolated from vaginal secretions during pregnancy. Infection with trichomonas in pregnancy may cause severe symptomatic vaginitis in some women. Whether or not it can have adverse effects on the course of pregnancy or on the newborn is uncertain. Unsubstantiated comments have suggested that failure to treat severe trichomonas vaginitis leads to preterm birth. One randomized controlled trial reported no statistically significant difference in birthweight or gestational age at birth whether asymptomatic Trichomonas vaginalis infection was treated with a single dose of metronidazole or left untreated. Colonization of the neonate following vaginal delivery appears to be rare.

As Trichomonas vaginalis is often associated with other sexually transmitted organisms of significance in pregnancy, these should be specifically sought when it is identified.

Up to 50 per cent of women carrying the organism are asymptomatic. Vaginal discharge is the most common complaint, but the classically described green frothy discharge is found in only a small proportion of women. Microscopic examination of a wet preparation of vaginal secretions is simple to perform and highly specific, but its sensitivity compared with culture may be as low as 50 per cent. To maximize the sensitivity of a wet preparation, a drop of vaginal discharge diluted in saline should be examined under the microscope immediately. Cooling the secretions results in loss of the characteristic jerky movements of the organism.

Culture is the best method to diagnose trichomonas, but the methodology is time consuming and not generally available. Papanicolaou smears have been used to diagnose the disease, but the sensitivity of this method compared with culture is only 40 per cent.

Metronidazole is a highly effective treatment of infection with Trichomonas vaginalis. Both a single 2 gram dose and an extended seven-day regimen of 250 mg three times a day are curative in more than 90 per cent of women. Because compliance is high with single-dose therapy, this regimen is preferred clinically. The sexual partner should be treated concomitantly with the woman.

Metronidazole readily crosses the placenta. Although it has been found to be carcinogenic in rodents and mutagenic for certain bacteria, there is no evidence of teratogenicity in humans after its administration to pregnant women. However, most obstetricians refrain from its use during the first trimester of pregnancy.

Clotrimazole, an antifungal agent, has been shown to be effective *in vitro* against *Trichomonas vaginalis*. It has promise as a local palliative treatment during pregnancy.

8 Leg cramps

Leg cramps (painful spasms of the calf muscles) are experienced to some extent by almost half of all pregnant women, particularly in the later months of pregnancy. The symptom tends to occur at night, and may recur repeatedly for weeks or months, causing considerable distress. The cause and mechanism of these cramps are still not clear. Sometimes based on rather astonishing analogies, unsupported hypotheses, and uncontrolled studies, a number of drugs have been widely prescribed for treatment and prophylaxis. Quinine, Benadryl, vitamin D, and dietary calcium have been widely acclaimed to be of benefit on the basis of uncontrolled studies.

Sodium chloride tablets were demonstrated to be more effective than placebo, no treatment, or calcium lactate in one controlled trial, but these results have not been confirmed. Placebo-controlled trials of calcium failed to show any improvement in symptoms among women who took 1 gram of calcium twice daily for three weeks.

Calcium salts are still widely prescribed for the syndrome of nocturnal calf cramps in pregnancy, despite the lack of evidence from controlled trials that they have any benefit beyond that of a placebo. Unfortunately, no further trials of the efficacy of increased sodium intake appear to have been carried out. It is probable that the observed benefits may be restricted to women who are sodium deficient. Massage and stretching the affected muscles are stated to afford relief during an attack, and these innocuous measures are surely worth trying.

9 Varicosities and oedema

The standard treatment for symptoms of varicosities during pregnancy is support stockings. These seem to provide relief although they have not been tested in randomized trials. One small but well controlled trial provided evidence that rutosides provide better relief than placebo for symptoms of varicosities, including cramps, tiredness, and oedema (swelling). Further trials are required to confirm these findings and to provide reassurance that the use of rutosides in pregnancy has no harmful effects on the fetus. Trials should also be designed to compare drugs with physical remedies such as elastic support hosiery.

Other methods of providing symptomatic relief of oedema, such as intermittent external pneumatic compression and immersion in a bath for almost an hour, can result in short term reduction in leg volume or increased urine output, but they are of little practical help. The main importance of oedema is the discomfort experienced by women, and there is no evidence that these measures provide symptomatic relief.

10 Other symptoms

Small trials have attempted to assess the effectiveness of a cream to prevent stretch marks (striae gravidarum), of sunscreen cream for melasma, of antihistamines compared with aspirin for itching in late pregnancy, and of special pillows for backache. None had adequate methodology or sample size to clearly establish whether or not the treatments provided any benefit or relief. The single trial of the Ozzlo pillow, however, suggested that it was somewhat better than a standard pillow in alleviating backache in late pregnancy. This treatment may be worth considering since backache can be very troublesome in late pregnancy and no other remedy is known to be effective.

11 Conclusions

Many of the symptoms commonly experienced during pregnancy can be relieved by simple physiological approaches. Where medication is deemed to be necessary, antihistamines appear to be the drugs of choice for nausea and vomiting, although no single product has been satisfactorily tested for efficacy in enough trials and few studies are available to inform us about possible teratogenic risks. Acupressure has shown promise and should be further investigated. No information is available to help in the alleviation of tiredness. Where possible, women should be encouraged to rest, and to seek help with home and work responsibilities.

When symptoms of heartburn fail to respond to postural and diet measures, antacids should be taken. Dilute acidic drinks may be worth trying if these fail to bring relief. For constipation in pregnancy, modification of the diet, including increasing dietary fibre and fluid intake, should be considered before resorting to laxatives. Bulking agents, if necessary combined with stool softeners, should be used if dietary measures do not provide sufficient relief. Irritant laxatives (such as standardized senna) should be reserved for short-term use in refractory cases.

Symptoms of candidal vaginitis respond well to short courses of clotrimazole, which will often give relief from the symptoms of trichomonas vaginitis as well. No treatment is indicated in asymptomatic infection. Metronidazole, which is most effective for trichomonas, readily crosses the placenta and probably should be withheld during the first trimester.

No pharmaceutical treatment for leg cramps has yet been firmly based on scientific evidence. At present there is little evidence to recommend any treatment for oedema which is either effective or acceptable. Despite the lack of objective evidence, support stockings remain the standard treatment for troublesome oedema.

Prevention and treatment of the so-called 'minor', but often extremely unpleasant, symptoms of pregnancy have received little systematic study in clinical trials. Because of their wide prevalence and the significant discomfort that they cause, such systematic study is urgently required.

This chapter is derived from the chapters by Michael Bracken, Murray Enkin, Hubert Campbell, and Iain Chalmers (32), and Elaine Wang and Fiona Smaill (34) in EFFECTIVE CARE IN PREGNANCY AND CHILDBIRTH.

References to primary sources and more complete data for statements made in this chapter can be found in the source chapters and/or in the following reviews from the *Cochrane pregnancy and childbirth database*:

Jewell, M.D.
— Debendox (Bendectin) for nausea in pregnancy. Review no. 03351.
— Benzylamine and dramamine for nausea in pregnancy. Review no. 03350.
— Phenothiazines for nausea in pregnancy. Review no. 03388.
— ACTH for nausea in pregnancy. Review no. 03352.
— Piperazines for nausea in pregnancy. Review no. 04428.
— Vitamin B_6 (pyridoxine) for nausea in pregnancy. Review no. 07703.
— Ginger treatment for hyperemesis gravidarum. Review no. 06812.
— P6 acupressure to treat nausea. Review no. 06520.
— Antacid therapy for heartburn in pregnancy. Review no. 06855.
— Dilute hydrochloric acid for heartburn in pregnancy. Review no. 06856.
— Compound antacid preparations for heartburn in pregnancy. Review no. 06858.
— Prostigmine injection for heartburn in pregnancy. Review no. 06857.
— Treatments for constipation in pregnancy. Review no. 06894.

Young, G.L.
— Clotrimazole for vaginitis in pregnancy. Review no. 06859.
— Imidazoles vs nystatin for vaginal candidiasis. Review no. 06810.
— 7-day vs 14-day treatment for vaginal candidiasis. Review no. 06811.

— 4-day vs 7-day treatment for vaginal candidiasis. Review no. 03151
— Calcium for leg cramps in pregnancy. Review no. 04385.
— Sodium chloride for leg cramps in pregnancy. Review no. 06891.
— Rutosides for varicosis. Review no. 06518.
— Treatments for leg oedema in pregnancy. Review no. 06893.
— A cream to prevent striae gravidarum. Review no. 06519.
— Sunscreen cream for melasma in pregnancy. Review no. 06890.
— Special vs standard pillow for backache in late pregnancy. Review no. 06895.
— Antihistamines vs aspirin for itching in late pregnancy. Review no. 06892.

14 Prevention of miscarriage

1 Introduction   2 Confirmation of fetal life   3 Hormone administration for the prevention of miscarriage
3.1 *Diethylstilboestrol*   3.2 *Progestogens*   3.3 *Human chorionic gonadotrophin*   4 Other medications   5 Bed-rest and hospitalization   6 Conclusions

1 Introduction

Miscarriage is the unintended loss of a pregnancy before the period of viability. Caregivers and family may sometimes consider this not to be as tragic a loss as fetal demise later in pregnancy, but it often results in a similar degree of mental suffering and anguish for the woman and her partner. It is not surprising that a number of interventions have been proposed and used in efforts to prevent miscarriage, particularly for women perceived to be at greater than average risk. Chief among these has been the use of various hormones and the prescription of bed-rest either at home or in hospital. The choice of care should depend on evidence of effectiveness and on the preferences of the woman concerned.

2 Confirmation of fetal life

Ultrasound has the ability to establish rapidly and accurately whether a fetus is alive or dead, and to predict whether a pregnancy is likely to continue when there is a threatened miscarriage. This ability has rationalized the care of women with threatened miscarriage in early pregnancy. The gestational sac can be visualized by six weeks menstrual age, and the fetus by seven weeks. As soon as the fetus can be demonstrated with ultrasound, it can be measured and its viability confirmed by detection of heart movement. Fetal life is confirmed by observation of heart pulsation, and fetal death by its absence. Except in very early pregnancy, there should be no doubt about the diagnosis. Blighted ova, which constitute the largest group of early pregnancy failures, are diagnosed by the inability to detect a fetus on careful examination. When the sac is small, the pregnancy without an embryo has to be differentiated from the normal, very early pregnancy by a repeat ultrasound examination. Missed abortion can be diagnosed by absence of heart movement. The small group of pregnancies in which the embryo is alive but destined to miscarry cannot be predicted with certainty by ultrasound. These constitute less than 15 per cent of the total number of miscarriages. Although a reduction in amniotic fluid volume, diminished fetal activity, or the presence of large intrauterine haematomas may suggest a poor prognosis, there are no specific ultrasound features.

3 Hormone administration for the prevention of miscarriage

Over the last fifty years, hormones have been given to pregnant women in attempts to prevent miscarriage, as well as fetal death, preterm delivery, and other adverse outcomes of pregnancy. Studies in the mid-1930s suggested a link between abnormal hormone levels and complications of pregnancy. The conclusion that inadequate hormone secretion meant that additional hormones had to be administered is a classic example of the danger of applying pathophysiological reasoning to clinical practice without appropriate evaluation.

3.1 Diethylstilboestrol

The synthetic hormone diethylstilboestrol was administered to pregnant women on a wide scale for over thirty years. On the basis of animal studies and uncontrolled observations in humans, it was thought that it would be effective in preventing a variety of adverse outcomes. This hypothesis appeared to be supported by studies using observational data, but studies with contemporary controls did not substantiate the postulated beneficial effects. By the mid-1950s these controlled trials, individually and collectively, indicated that diethylstilboestrol did not reduce the risk of miscarriage, pre-eclampsia, low birthweight, preterm birth, stillbirth, or neonatal death, nor increase the likelihood of a woman's pregnancy resulting in a surviving infant.

Thus strong evidence was available by the mid-1950s to challenge the claims made on behalf of diethylstilboestrol. The drug should have been abandoned at that time, or prescribed only in the context of further trials. Despite this lack of evidence of benefit, obstetricians continued to use diethylstilboestrol until the 1970s, when several cases of vaginal adenocarcinoma (a very rare form of cancer in women less than 50 years old) were reported in young women whose mothers had received diethylstilboestrol while pregnant with them.

The randomized cohorts of women and their children generated by three of the trials have been studied between twenty and thirty-five years after receiving diethylstilboestrol or placebo. The data suggest that there may be an increased risk of breast cancer in the women, and they clearly demonstrate an increased incidence of psychiatric illness and urogenital abnormalities in the children of both sexes. These abnormalities include benign tumours, vaginal adenosis, vaginal/cervical ridges, abnormal cervical smears, amenorrhoea/oligomenorrhoea, and infertility among the daughters, and low sperm density and other urogenital abnormalities among the sons. The long term adverse effects of the use of diethylstilboestrol in pregnancy that later became evident could have been minimized if more attention had been paid to the results of the controlled trials which showed the drug to be ineffective.

3.2 Progestogens

There are reasonable theoretical grounds for suspecting that a low serum progesterone may be a cause of pregnancy loss in the first 20 weeks of gestation. Although this hypothesis has been tested in several randomized trials, there is no evidence from the available data to suggest that progestogens reduce the risk of miscarriage, stillbirth, or neonatal death in women either with bleeding (threatened miscarriage) or with a history of recurrent miscarriage. However, the trials have not been large enough to exclude an important effect in either direction (increase or decrease in the risk of miscarriage), and in many of the studies ultrasound examinations were not made. These could have shown that the fetus was already dead, so that there may have been few pregnancies in which the progestogens could have been effective.

Although the progestogen follow-up studies have been largely anecdotal and uncontrolled, there have been suggestions from some studies that fetal exposure to the drugs may increase the risk of oesophageal atresia, cardiac, neurological, neural tube and other major malformations, and masculinization, or 'tomboyishness' in girls. However, other studies have failed to detect these adverse effects, and so the safety of progestogens, like their postulated benefits, remains an open question.

3.3 *Human chorionic gonadotrophin*

Data from the three small controlled trials that have examined the effects of human chorionic gonadotrophin on the risk of miscarriage in women with a past history of repeated early pregnancy loss suggest that this treatment may be effective in preventing recurrent miscarriage. The results of these trials must be interpreted with great caution because of methodological weaknesses in the studies, as well as the small number of women who participated. They must be replicated on a larger scale before any recommendation can be made about the use of human chorionic gonadotrophin in practice.

4 Other medications

The effect of indomethacin treatment for threatened miscarriage has been tested in one small trial, and no effect on the rate of miscarriage was demonstrated. Small trials evaluating tocolytic and antispasmodic medications also failed to demonstrate any beneficial effect of these agents on the rate of miscarriage.

Several uncontrolled studies have suggested a beneficial effect of immunotherapy on the risk of miscarriage in women with an unexplained history of recurrent miscarriage (no identifiable cause, same partner). Controlled trials of immunotherapy (the maternal injection of paternal lymphocytes) suggest a beneficial effect on the risk of miscarriage, but these results must be interpreted with great caution as they are not mutually consistent.

5 Bed-rest and hospitalization

Bed-rest is a common prescription for women whose pregnancies are complicated by a number of conditions, including a history of recurrent miscarriage or early bleeding in the present pregnancy. Women with these problems may be either advised to rest in bed at home or admitted to hospital, to facilitate bed-rest and to permit closer investigation and surveillance of their pregnancy.

The extent to which women are advised to rest in bed at home or in hospital varies considerably, but such advice is very common in some places. The intervention is not innocuous. Both confinement to bed at home and hospitalization during pregnancy may result in great financial and social costs for pregnant women and their families, particularly those with children. Antenatal hospitalization is often a disruptive and stressful experience involving separation of women from their families at a time of great anxiety. In addition, adoption of this policy has brought substantial costs to the health services.

The only reported attempt to undertake any form of controlled evaluation of the effectiveness of bed-rest in the management of threatened miscarriage was made over thirty-five years ago. The results of this study gave no support to the view that a policy of advising bed-rest reduces the risk of miscarriage after bleeding occurs in early pregnancy.

Bed-rest is sometimes advised for many days if spotting or bleeding is persistent, and this may cause considerable family disruption. Yet in a substantial proportion of these pregnancies the fetus is already dead. The presence of a non-viable pregnancy can now be demonstrated by ultrasound and no form of care can preserve the pregnancy.

Even with a viable embryo, there is no valid basis for advising bed-rest. Therefore the preferences of individual women should be the deciding factor in whether or not they should rest in bed. Some women may feel they wish to rest. Women should be encouraged to do whatever feels best for them.

6 Conclusions

In the present state of knowledge, hormone administration of any type in pregnancy should be used only within controlled clinical trials until the ratio of benefits to hazards has been more clearly established. Future trials should involve only those pregnancies in which ultrasonography has confirmed that the fetus is alive. A similar caveat must be made with regard to indomethacin, tocolytics, and antispasmodics.

The use of immunotherapy (maternal injection of paternal lymphocytes) for women with an unexplained history of recurrent miscarriage appears to show some promise, but should not be used outside the context of further randomized controlled trials. A greater understanding of the theoretical basis for immunotherapy may help guide further research in this field.

Hospitalization during pregnancy is costly and disruptive for many families. Discussion with individual women will make it clear that a prescription for rest, either at home or in hospital, would sometimes be welcome. As there is no strong evidence that this is likely to have harmful effects, women's views should be taken into account. By the same token, women with bleeding in early pregnancy should not be coerced into resting in bed at home or being hospitalized against their better judgement.

This chapter is derived from the chapters by Jim Neilson and Adrian Grant (27), Peter A. Goldstein, Henry Sachs, and Thomas C. Chalmers (38), Caroline Crowther and Iain Chalmers (39), Adrian Grant (40), and G. Justus Hofmeyr (41) in EFFECTIVE CARE IN PREGNANCY AND CHILDBIRTH.

References to primary sources and more complete data for statements made in this chapter can be found in the source chapters and/or in the following reviews from the *Cochrane pregnancy and childbirth database*:

Chalmers, I.
— Diethylstilboestrol (DES) in pregnancy. Review no. 02891.

Neilson, J.P.
— Routine ultrasonography in early pregnancy. Review no. 03872.

Prendiville, W.J.
— Progestogens for threatened miscarriage. Review no. 05534.
— Progestogens to prevent miscarriage and preterm birth. Review no. 04398.
— Progestogens in pregnancy. Review no. 03287.
— 17alpha-hydroxy-progesterone caproate in pregnancy. Review no. 04399.
— HCG for recurrent miscarriage. Review no. 02890.
— Tocolytics/antispasmodics for threatened miscarriage. Review no. 03284.
— Indomethacin for threatened miscarriage. Review no. 03283.
— Immunotherapy for recurrent miscarriage. Review no. 06814.

15 Hypertension in pregnancy

1 Introduction

Two aetiologically distinct entities account for most hypertensive disorders in pregnancy. One is a disorder induced by pregnancy which, in this chapter, is referred to as 'pre-eclampsia' if it is accompanied by proteinuria and as eclampsia if it leads to convulsions or coma. The second is chronic hypertension that happens to coincide with pregnancy; sometimes this is associated with a known underlying condition such as renal disease. In addition, a combination of the two pathological conditions may occur, and this is referred to as 'superimposed pre-eclampsia'.

The reported incidences of maternal and fetal complications of hypertensive disorders in pregnancy vary widely. These conditions may have devastating consequences for mother and baby. Therefore, a number of medical and surgical regimens have been tried for the prevention and treatment of pre-eclampsia and eclampsia. One author reports that women with eclampsia have been 'blistered, bled, purged, packed, lavaged, irrigated, punctured, starved, sedated, anaesthetized, paralyzed, tranquillized, rendered hypotensive, drowned, been given diuretics, had mastectomies, been dehydrated, forcibly delivered, and neglected'.

Few of these many 'remedies' have been evaluated in controlled trials. This chapter assesses the available evidence about current approaches to either the prevention or treatment of pregnancy-induced hypertension, pre-eclampsia, and eclampsia.

2 Prophylaxis

2.1 *Diuretics*

There is no good evidence that excessive water retention, or even frank oedema, defines a group of women at particular risk of developing pre-eclampsia. Nevertheless, many attempts have been made, and indeed still are being made by some obstetricians and midwives, to prevent retention of salt and water in pregnancy by prescribing diuretics or a rigidly sodium-free diet in the belief that this will prevent pre-eclampsia.

The effects of the prophylactic use of diuretics in pregnant women with normal blood pressure (with or without oedema or excessive weight gain), and of their therapeutic use in moderate hypertension have been studied in twelve randomized trials. The results reflect the well-known ability of diuretics to reduce blood pressure, rather than any improvement in substantive outcomes. When the data on proteinuric pre-eclampsia or the risk of perinatal death are considered, there is no clear evidence of benefit. This may be because the treatment was ineffective. It may also be because the numbers studied were too small to detect some modest, but worthwhile, benefit. None of the serious side-effects of diuretic treatment in pregnancy that have been described in occasional case reports, were observed in these trials, which included nearly 7000 women and their babies. This suggests that the putative maternal and fetal risks of diuretic administration may have been overstated, perhaps as a result of selective case reporting.

2.2 *Antithrombotic and antiplatelet agents*

Changes in the blood-clotting system are well documented in established pregnancy-induced hypertension. The extent of the clotting disorders appears to be related to the severity of the disease, and early activation of the clotting system may contribute to the pathology of pre-eclampsia. For these reasons, the use of anticoagulant or antiplatelet agents has been considered for the prevention of pre-eclampsia and intrauterine growth retardation.

Aspirin is an antiplatelet agent that has been shown to prevent thrombotic occlusion of arteriovenous shunts and coronary artery bypass grafts, and to reduce death and reinfarction in unstable angina and after myocardial infarction and cerebral ischaemic attacks.

Results of the randomized trials of antiplatelet prophylaxis using either aspirin or aspirin combined with dipyridamole reported to date have been encouraging, but until larger studies have been completed it will remain unclear whether or not antiplatelet administration should be adopted in routine clinical practice.

Heparin has been used in uncontrolled studies involving single cases or small series of women. It requires subcutaneous (or intravenous) administration, making it an inconvenient form of treatment, and its use in severe cases may be associated with dangerous side-effects. Warfarin has also been used prophylactically in an attempt to prevent recurrent pre-eclampsia in multiparous women. Anecdotal reports of its use do not provide any evidence of maternal or fetal benefit, and there is some suggestion that it may have serious side-effects.

2.3 Dietary measures

As noted in Chapter 6, there is no evidence that modification of protein or calorie intake can protect against pregnancy-induced hypertension.

Controlled trials of prophylactic fish oil in pregnancy show a promising reduction in the incidence of hypertension, proteinuric pre-eclampsia, and preterm birth, but there are insufficient data to show a decrease in any measure of perinatal mortality or morbidity.

Calcium supplementation during pregnancy appears to reduce the risk of women developing hypertension. There is insufficient evidence to draw conclusions about more important outcomes, such as caesarean section, intrauterine growth retardation, or perinatal death. This form of supplementation requires evaluation in further large trials before we can draw any firm conclusions as to its role in antenatal care.

There is too little good information on the effects of high or low salt consumption on the development of pregnancy-induced hypertension for us to be able to offer well-informed advice. Thus far, there appears to be no justification for advocating any changes in salt intake during pregnancy.

3 Mild or moderate pregnancy-induced hypertension and pre-eclampsia

Chronic hypertension and mild or moderate pregnancy-induced hypertension carry little risk to the mother or the fetus unless severe hypertension, pre-eclampsia, or eclampsia ensue. For this reason, the aim of treatment of mild to moderate hypertensive disease in pregnancy has been to defer or prevent the development of severe hypertensive disease. Bed-rest and a variety of medications have been used.

3.1 Bed-rest

Women with pre-eclampsia are often either advised to rest in bed at home or may be admitted to hospital to facilitate bed-rest and to permit closer investigation and surveillance of their pregnancies. The extent to which women are advised to rest in bed at home or in hospital varies considerably, but such advice is very common in some places. The intervention is not innocuous. Both confinement to bed at home and hospitalization during pregnancy may result in financial and social costs for pregnant women and their families. Antenatal hospitalization is often a disruptive and stressful experience, and adoption of this policy has brought substantial costs to the health services.

The effectiveness of bed-rest for non-proteinuric hypertension has been evaluated in two controlled trials. Unfortunately, no clear picture of the value of the policy emerges. One of the trials suggests that hospitalization may have a beneficial effect on the evolution of pre-eclampsia, while the other tends to suggest the opposite. There are also contrasting patterns in the frequency of preterm birth between the two trials.

When proteinuria develops in addition to hypertension in pregnancy, the risks for both mother and fetus are substantially increased. Admission to hospital is then considered necessary for evaluation and increased surveillance to detect any deterioration in maternal or fetal condition as soon as possible. Whether the hospitalization should be linked with strict bed-rest is less clear. The two small trials that have addressed this question have not provided clear answers. To date, there is no good evidence to support a policy of strict bed-rest in hospital for a woman with pre-eclampsia.

A related question refers to the role of elective delivery in the care of women with pre-eclampsia. Is an aggressive policy, with early resort to elective delivery, more effective than a more conservative approach? The trends in neonatal outcome observed in the one small trial that addressed this question suggest that an aggressive policy may have adverse effects, but the trial was too small to provide reliable estimations.

3.2 Antihypertensive agents

Antihypertensive medication for either mild to moderate pregnancy-induced hypertension or for chronic hypertension in pregnancy prevents an increase in blood pressure, but its effects on other important outcomes are less clear. It may reduce the risk of perinatal death, but the evidence for this remains weak. There is no evidence that antihypertensive treatment with any of the drugs available prevents proteinuria, and there is insufficient evidence for reliable conclusions about the effects of antihypertensive treatment on other important end-points, such as caesarean section, preterm birth, or neonatal morbidity. Few children exposed to antihypertensive drugs *in utero* have been followed up beyond the perinatal period.

Methyldopa is the most widely used drug in women with mild to moderate pregnancy-induced hypertension, but beta-blockers, labetalol, and calcium channel blockers are rapidly being introduced into clinical practice. Although there are anecdotal reports in the literature on their use in women with mild to moderate hypertensive disease in pregnancy, there are no controlled trials of sufficient size to allow reliable assessment of their effects on serious outcome or on deterioration of the disease.

The use of methyldopa in women with moderate hypertension substantially reduces the risk of developing severe hypertension. There is a suggestion that its use may also reduce the risk of perinatal death, but evidence on this remains weak. In the same way as for diuretics, there is no evidence of an effect on the incidence of proteinuria, intrauterine growth retardation, preterm birth, or caesarean section. Clonidine is similar to methyldopa in most respects, except for a more rapid onset of action (about 30 minutes compared with 4 hours for methyldopa).

Administration of beta-blockers will reduce the incidence of severe hypertension, but it may increase the risk of intrauterine growth retardation. Beta-blockers are effective antihypertensive agents that reduce cardiac output, and it is possible that this effect would be unfavourable in pregnancy since adequate perfusion of the maternal and uteroplacental circulation depends on maintenance of the elevated cardiac output of pregnancy. Direct comparisons of beta-blockers with methyldopa do not indicate significant differential effects, but the trials have been too small to provide reliable information on their relative safety and efficacy.

Randomized trials of calcium-channel blockers in pregnancy-induced hypertension, compared either with placebo or with beta-blockers, have been too small to give useful estimates of their effects.

The small numbers of women studied in adequately controlled trials of antihypertensive medication for both the prevention and treatment of moderate hypertensive disease preclude definitive conclusions about their effects, even when the results of all trials are considered together. It seems likely that antihypertensive treatment prevents the development of severe hypertension in pregnancy, and for that reason it may reduce the number of hospital admissions and emergency deliveries. There is no clear evidence that antihypertensive treatment with any of the drugs available may defer or prevent the occurrence of proteinuric pre-eclampsia or associated problems such as fetal growth retardation and perinatal death. Nor is there good evidence about the safety of such treatments, in particular with respect to child development. At present there seems to be no reason to prefer any one particular antihypertensive agent over another.

3.3 *Other treatments*

Small trials have attempted to assess the efficacy of magnesium and of prostaglandin precursors in the treatment of pre-eclampsia. The trials are too small to provide reliable information about either the effectiveness or the safety of these approaches.

4 Severe pre-eclampsia and eclampsia

Although treatment of hypertension does not strike at the basic disorder, it may still benefit the mother and fetus. One of the important objectives in severe hypertension in pregnancy is to reduce blood pressure in order to avoid hypertensive encephalopathy and cerebral haemorrhage. For this reason, the aim in treating severely hypertensive pregnant women is to keep the blood pressure below dangerous levels (around 170/110 mmHg) and to maintain adequate circulating blood volume.

4.1 *Antihypertensive agents*

Hydralazine is the antihypertensive drug used most commonly in women with severe pregnancy-induced hypertension and pre-eclampsia, followed by beta-blockers and, more rarely, diazoxide. Methyldopa is also used in the treatment of severe hypertensive disease in pregnancy, although it has the disadvantage of a relatively slow onset of action (about 4 hours). Direct comparisons have been made of nifedipine versus hydralazine, labetalol versus hydralazine, labetalol versus diazoxide, and prostacyclin versus dihydralazine. To date, there is no evidence to justify a strong preference for any one of the various drugs that are available for treating severe hypertension in pregnancy.

Therefore, in clinical practice the choice should probably depend on the familiarity of an individual clinician with a particular drug. In general, maternal side-effects are not different from those in the non-pregnant state and are listed in pharmacology texts. All drugs used to treat hypertension in pregnancy cross the placenta, and so may affect the fetus directly, by means of their action within the fetal circulation, or indirectly, by their effect on uteroplacental perfusion.

From what is known about direct and indirect adverse effects on the fetus and neonate, hydralazine appears to be relatively safe. Labetalol may cause severe and long-lasting fetal and neonatal bradycardia, particularly after high doses. These effects may be clinically important in the presence of fetal and neonatal hypoxia. Diazoxide can provoke a precipitous fall in maternal arterial pressure, leading to a serious reduction in uteroplacental or even cerebral perfusion. In addition, hyperglycaemia has been reported in the newborn after maternal treatment with diazoxide. In view of the potential dangers this drug should not be used.

As yet, there is little clinical information on the fetal or neonatal effects of maternal use of calcium antagonists. Various problems have been observed in the neonatal period following their use in uncontrolled studies, but these effects could not be definitely attributed to maternal drug treatment.

4.2 *Plasma volume expansion*

Women with severe pre-eclampsia before delivery often have a reduced circulating plasma volume. This has led to a recommendation that plasma volume should be expanded with non-crystalloid solutions, such as dextran and salt-poor albumin, in attempts to improve the maternal systemic and uteroplacental circulation. Some recent uncontrolled studies suggest that rapid replenishment of intravascular volume may result in decreased arterial blood pressure in pregnant women with moderate third-trimester hypertension or pre-eclampsia. Although blood pressure was not restored to normal by volume expansion, these uncontrolled studies suggest that such treatment may be an effective adjunct to the administration of antihypertensive drugs and, by reducing the drug doses needed, might minimize the risks of maternal and neonatal side-effects.

It should be borne in mind, however, that intravascular volume expansion carries a serious risk of volume overload, which may lead to pulmonary and perhaps cerebral oedema in pre-eclamptic women in whom colloid osmotic pressure is usually low. Plasma volume expansion may be particularly dangerous after birth, when venous volume tends to rise. It should not be applied without careful monitoring.

There is still insufficient evidence from controlled trials to provide reliable guidelines for the use of plasma volume expansion for hypertension in pregnancy. Further studies are clearly needed to define its place, with or without additional antihypertensive treatment, in the care for women with severe pregnancy-induced hypertension.

4.3 *Anticonvulsant agents*

Anticonvulsant drugs are widely used in the management of eclampsia, as well as in severe hypertensive disease and pre-eclampsia, in an attempt to prevent the occurrence of eclamptic seizures. For eclampsia, the major question is which drug is best. For pre-eclampsia, there is uncertainty about if and when drug treatment is required.

In the United States barbiturates are still used in moderately severe pre-eclampsia, while parenteral magnesium sulphate is the treatment most usually used in fulminating pre-eclampsia and eclampsia. In contrast, magnesium sulphate is rarely used in Europe and Australia, where the most frequent choice is diazepam, followed by phenytoin, chlormethiazole, and barbiturates.

The current popularity of magnesium sulphate in the United States reflects experience with a regimen of intravenous and intramuscular administration of magnesium sulphate, intravenous hydralazine, and delivery within 48 hours. Only a small amount of magnesium appears to cross the blood–brain barrier after intravenous administration of magnesium sulphate; it has little effect on blood pressure and virtually no sedative effect. Magnesium can attain life-threatening levels as a result of excessive dosage or diminished excretion (for example due to renal failure). Since an overdose can cause death by cardiorespiratory arrest, women receiving magnesium sulphate must be closely monitored at all times and calcium should be readily available as an antidote. Magnesium readily crosses the placenta, and high magnesium concentrations in cord blood have been shown to be associated with depression of the baby.

The benzodiazepines chlordiazepoxide (Librium) and diazepam (Valium) were introduced in the treatment of eclampsia and severe pre-eclampsia in the late 1960s as an alternative to the 'lytic cocktails' popular at that time. Following intravenous administration to the mother, diazepam readily crosses the placenta to the fetus. Maternal administration of diazepam may lead to loss of beat-to-beat variability in the fetal heart rate, and thus interfere with the clinical interpretation of the cardiotocogram. Diazepam is very slowly metabolized and cleared in the neonate. Given in high doses to the mother before birth, diazepam causes delayed onset of respiration, apnoea, hypotonia, impaired response to cold, and poor sucking in the newborn infant. These well-documented effects may last for several days.

Chlormethiazole has been widely used as a sedative and anticonvulsant in the treatment of severe pre-eclampsia and eclampsia. It acts rapidly following intravenous administration, but its effectiveness has only been described in uncontrolled studies. Like the other anticonvulsants, chlormethiazole also crosses the placenta but, in contrast with diazepam, it is rapidly excreted by the fetus and the newborn.

The ideal anticonvulsant for use in eclampsia would be the one which is easiest to administer, is rapidly effective in arresting and preventing convulsions, has a wide safety limit for the mother, and is non-toxic and non-depressant to the baby.

Small controlled trials have compared the effects of magnesium sulphate with diazepam and with phenytoin, but these have been far too small to provide reliable estimates of their relative effects. The search for the ideal anticonvulsant to be used in eclampsia continues.

5 Conclusions

Although preliminary results with antiplatelet agents appear promising, there is as yet no evidence of any effective prophylactic measure against pre-eclampsia.

Discussion with individual women will make it clear that a prescription of rest, either at home or in hospital, would sometimes be welcome. However, hospitalization and bed-rest is an expensive and intrusive intervention, with as yet no clear evidence of value. Thus women with non-albuminuric hypertension should not be coerced into either resting in bed at home or being hospitalized against their better judgement.

Antihypertensive medication for mild to moderate hypertension in pregnancy prevents further increases in blood pressure. This in turn may reduce the number of hospital admissions, inductions, and emergency deliveries, although these effects have not been studied systematically. Trials to date have been too small to provide reliable information about other important outcomes. At present, there is insufficient evidence to assess at what level of hypertension the benefits of antihypertensive therapy outweigh the disadvantages.

For treatment of severe hypertensive disease, hydralazine appears to be effective in lowering blood pressure. Labetalol may produce a more controlled lowering of blood pressure than hydralazine or diazoxide, but this possibility requires further study.

The use of anticonvulsants is widespread, not only in the treatment of eclampsia but also in the management of severe pre-eclampsia. The choice of drug has been determined by historical factors rather than by evidence from controlled comparisons.

The potential benefits of antiplatelet agents, such as aspirin, in both the prevention and treatment of pre-eclampsia have received priority in further investigation, and better information about their role should be forthcoming soon. In view of the promising effects of beta-blockers and methyldopa in preventing the development of severe hypertension (and in view of their widespread use by many doctors), there is an urgent need for them to be evaluated in further large-scale trials.

The results of the small trials of plasma volume expansion conducted so far are promising. Further study might help to determine whether this would be a practicable and effective way of reducing the need for (or dose of) antihypertensive therapy in the treatment of severe pregnancy-induced hypertension.

Current treatment of hypertensive disorders in pregnancy appears to be largely based on clinical experience fed by anecdotal reports, rather than on reliable evidence from properly controlled trials of sufficient size. Given the large number of women who develop hypertensive disease in pregnancy, multicentre collaborative studies involving *much* larger numbers of women than have so far been studied should be mounted. These could assess the effects of the various treatments that are used more reliably than has been done so far.

This chapter is derived from the chapter by Rory Collins and Henk C.S. Wallenburg (33) in EFFECTIVE CARE IN PREGNANCY AND CHILDBIRTH.

References to primary sources and more complete data for the statements made in this chapter can be found in the source chapter, and/or in the following reviews from the *Cochrane pregnancy and childbirth database*:

Collins, R. and Duley, L.
— Diuretics for prevention of pre-eclampsia. Review no. 04393.
— Diuretics for prevention/treatment of pre-eclampsia. Review no. 03308.
— Diuretics in the treatment of pre-eclampsia. Review no. 04394.
— Any antihypertensive therapy for pregnancy hypertension. Review no. 04426.
— Methyldopa-based therapy for treatment of pre-eclampsia. Review no. 03997.
— Beta-blockers vs methyldopa in the treatment of pre-eclampsia. Review no. 03999.
— Beta-blockers in the treatment of pre-eclampsia. Review no. 03998.
— Labetalol vs hydralazine in severe pregnancy-induced hypertension. Review no. 04392.
— IV labetalol vs iv diazoxide in severe pre-eclampsia. Review no. 04400.

Collins, R.
— Antiplatelet agents for IUGR and pre-eclampsia. Review no. 04000.

Duley, L.
— Prophylactic fish oil in pregnancy. Review no. 05941.
— Any PG precursors for pre-eclampsia prevention/treatment. Review no. 05942.
— Low vs high salt intake in pregnancy. Review no. 05939.
— Routine calcium supplementation in pregnancy. Review no. 05938.
— Ca channel blockers in pregnancy hypertension. Review no. 07075.
— Ca channel blockers vs beta blockers in pregnancy hypertension. Review no. 07074.
— Ca channel blockers for postpartum hypertension. Review no. 07073.
— Magnesium for treatment of pregnancy hypertension. Review no. 07079.
— PG precursors in the treatment of pre-eclampsia. Review no. 05940.
— Clonidine vs methyldopa in the treatment of pre-eclampsia. Review no. 05732.
— Any antihypertensive therapy in chronic hypertension. Review no. 05733.

— Hospitalization for non-proteinuric pregnancy hypertension. Review no. 03371.
— Strict bed rest for proteinuric hypertension in pregnancy. Review no. 03373.
— Nifedipine vs hydralazine in severe pregnancy-induced hypertension. Review no. 05731.
— Prostacyclin vs dihydralazine in severe hypertension. Review no. 07662.
— Plasma volume expansion in pregnancy-induced hypertension. Review no. 05734.
— Phenytoin sodium vs magnesium sulphate in eclampsia. Review no. 05682.
— Phenytoin vs magnesium sulphate in severe pre-eclampsia. Review no. 05943.
— Magnesium sulphate vs diazepam in eclampsia. Review no. 05683.
— Aggressive vs expectant management of pre-eclampsia. Review no. 07080.

16 Fetal compromise

1 Introduction   2 Hospitalization for bed-rest for suspected impaired fetal growth   3 Abdominal decompression
3.1 *Prophylaxis in normal pregnancy*   3.2 *Treatment of the compromised fetus*   4 Betamimetics   5 Miscellaneous treatments   5.1 *Maternal oxygen therapy*   5.2 *Hormone therapy*   5.3 *Calcium channel blockers*   5.4 *Other measures*
6 Conclusions

1 Introduction

In general, the care of women in whom fetal compromise is suspected or diagnosed involves a choice between pre-emptive delivery and conservative care in order to avoid potentially serious consequences for the baby. A number of conservative measures have been considered in controlled trials.

2 Hospitalization for bed-rest for suspected impaired fetal growth

Despite the fact that bed-rest is widely used, only one trial has been reported to assess its effects in women considered, on the basis of ultrasound examination, to have a small for gestational age fetus. This trial failed to show any effect of hospitalization on any clinical outcome.

There is at present no good evidence that bed-rest in hospital promotes fetal growth, but the numbers studied to date are too small to exclude that possibility with any degree of certainty.

3 Abdominal decompression

Suboptimal maternal blood flow to the placenta is hypothesized to be an important cause of both impaired fetal growth and pregnancy-induced hypertension, but there are no well validated methods for increasing uteroplacental circulation. The concept that repeated brief decompression of the abdominal region may increase the flow of blood to the placenta has intrigued a number of enthusiasts, but has not found acceptance in mainstream obstetrical practice.

Abdominal decompression is achieved with the use of a garment or large plastic bell on the abdomen, from which air can be evacuated to form a partial vacuum under the woman's control. It was originally used during the first stage of labour to relieve the resistance of the abdominal wall musculature to the forward movement of the contracting uterus. Physiological data suggest that the technique may reduce intrauterine pressure, increase maternal placental blood flow, increase fetal movements, and increase fetal heart rate accelerations and variability. The observation of unanticipated effects such as relief of pain, shortening of the duration of labour, and improved fetal oxygenation prompted further investigation of the effects of abdominal decompression on the outcome of normal and abnormal pregnancies.

3.1 Prophylaxis in normal pregnancy

Abdominal decompression came into clinical use as a prophylactic measure in the early 1960s on the basis of the results of several poorly controlled studies. These appeared to show that it improved fetal well-being and subsequent intellectual development. Two prospective studies followed in which attempts were made to compare the outcome in women subjected to abdominal decompression with comparable control groups. None of the available data suggest that prophylactic abdominal decompression affects duration of pregnancies or infant birthweight. They give no support to suggestions that prophylactic abdominal decompression has a beneficial effect on the condition of the infant at birth.

One carefully controlled trial studied the effects of abdominal decompression on child development. Although the developmental scores were slightly higher in the study group at 28 days and at three years of age, the differences were neither clinically nor statistically significant.

These studies provide convincing evidence that antenatal abdominal decompression used in uncomplicated pregnancies does not improve any of the outcomes measured. Thus there is no support for the use of antenatal abdominal decompression as a prophylactic procedure.

3.2 Treatment of the compromised fetus

In contrast with the lack of evidence that prophylactic abdominal decompression in normal pregnancies has any beneficial effects, there is some evidence to suggest that the technique may have a place in the management of pregnancies in which the fetus is likely to be compromised.

Abdominal decompression appears to slow the progression of pre-eclampsia. In addition, a trial showed abdominal decompression to be associated with a statistically significantly larger weekly growth in the fetal biparietal diameter. These differences between experimental and control groups were reflected in somewhat fewer inductions of labour for 'placental insufficiency' and less fetal distress during labour in the women who had received decompression.

Observer bias and possibly reporting bias may account for some or all of the putative effects of abdominal decompression noted above. The assessment of birthweight is less subject to large observer biases. Abdominal decompression was associated with a substantial reduction in the incidence of low birthweight and an increase in mean birthweight (2800 grams versus 2296 grams) in two of the three trials that have been conducted. The available data suggest a reduction in the incidence of both depressed Apgar scores and perinatal mortality.

The studies are thus suggestive of a beneficial effect of abdominal decompression on fetuses with impaired growth, but their methodological shortcomings limit the confidence that can be placed in these conclusions. It may be important to assess whether or not abdominal decompression affects uteroplacental blood flow. The advent of Doppler flow measurements offers an opportunity to assess this more rigorously than has been done in the past.

4 Betamimetics

Small trials mounted to assess the effects of betamimetics for women with suspected impaired fetal growth have not provided sufficient data to determine clinical effects. There were no significant differences between groups for any of the outcomes measured, but because of the small numbers studied, clinically important differences may have been missed. While there is a possibility that betamimetic therapy may promote fetal growth either by improving uteroplacental circulation or by increasing blood sugar and plasma insulin levels, there is insufficient evidence to support their use except in the context of randomized clinical trials.

5 Miscellaneous treatments

5.1 *Maternal oxygen therapy*

One trial of maternal oxygen therapy for suspected impaired fetal growth has shown a reduction in perinatal mortality in the oxygenation group. Methodological shortcomings and a preponderance of extremely preterm babies in the control group preclude reliance on this finding.

Available biochemical data suggest that oxygen therapy for suspected impaired fetal growth may have both beneficial and harmful effects. In view of the suggestion from some studies that oxygen therapy may decrease uterine blood flow, the effectiveness of oxygen therapy would need to be demonstrated in well-controlled clinical trials before further recommendations can be made.

5.2 *Hormone therapy*

No clinically relevant outcomes are available from the trials of various types of hormone therapy for suspected impairment of fetal growth. Thus we have no evidence to support the use of hormones for this purpose.

5.3 *Calcium channel blockers*

The only controlled trial to assess the clinical effect of calcium channel blockers on suspected impaired fetal growth showed the mean birthweight to be higher in babies of the women in the actively treated group than in those of the placebo group. There was also a non-significant trend towards lower perinatal mortality.

There is inadequate evidence from this trial to support the routine use of calcium-channel blockers in pregnancies with an increased risk of impaired fetal growth. The results of this small trial are sufficiently encouraging to justify further trials with sufficient numbers to assess effects on substantive outcomes.

5.4 *Other measures*

Plasma volume expansion with hydroxyethyl starch, transcutaneous electrostimulation, and intravenous calf blood extract have been tried as conservative treatments for suspected placental insufficiency or impaired uterine growth. The available studies did not consider clinically relevant outcomes, and there is no evidence to justify the clinical use of any of these measures.

6 Conclusions

The effectiveness of bed rest in hospital for suspected impaired fetal growth is unproven, but has not been adequately evaluated. Because of the enormous financial and personal costs of prolonged hospitalization, this form of care should be used only in the context of well-designed controlled trials to test its effectiveness.

There is some evidence that abdominal decompression may be of value in certain abnormal states of pregnancy, but the studies reported to date are not sufficiently strong to support its use outside the context of methodologically sounder controlled trials. Nevertheless, there are so few options for managing the compromised fetus other than elective delivery that it is important to subject abdominal decompression to further evaluation. Doppler blood flow measurement may offer an opportunity to assess its effect on uteroplacental and fetal blood flow.

There is a reasonable theoretical basis for investigating the usefulness of betamimetic therapy for suspected impaired fetal growth, but as yet no evidence for clinical use of betamimetics for this purpose.

Preliminary evidence suggests an effect of calcium channel blockers on fetal growth, but further research is needed to evaluate its effectiveness with respect to substantive pregnancy outcomes.

This chapter is derived from the chapters by Caroline Crowther and Iain Chalmers (39), Adrian Grant (40), and G. Justus Hofmeyr (41) in EFFECTIVE CARE IN PREGNANCY AND CHILDBIRTH.

References to primary sources and more complete data for the statements made in this chapter can be found in the source chapters and/or in the following reviews from the *Cochrane pregnancy and childbirth database*:

Hofmeyr, G.J.
— Hospitalization for bed-rest for suspected impaired fetal growth. Review no. 05288.
— Prophylactic abdominal decompression in pregnancy. Review no. 03421.
— Abdominal decompression for suspected fetal compromise/ pre-eclampsia. Review no. 03300.
— Betamimetics for suspected impaired fetal growth/pre-eclampsia. Review no. 05289.
— Maternal oxygen therapy in suspected impaired fetal growth. Review no. 06436.
— Hormone therapy in suspected impaired fetal growth. Review no. 06437.

— Calcium channel blockers in potential impaired fetal growth. Review no. 06645.
— Plasma volume expansion for suspected impaired fetal growth. Review no. 07640.
— Transcutaneous electrostimulation for suspected placental insufficiency (diagnosed by Doppler studies). Review no. 07641.
— Calf blood extract for suspected impaired fetal growth. Review no. 05290.

17 Multiple pregnancy

1 Introduction   2 Advice and support   3 Hospitalization and bed-rest   4 Medication   5 Cervical cerclage   6 Home uterine activity monitoring   7 Delivery   8 Conclusions

1 Introduction

Multiple pregnancy poses particular problems for women and for caregivers because of the increased risks of preterm labour, poor fetal growth, perinatal death, and, in the long term, cerebral palsy. In addition, women with multiple pregnancy are likely to experience the common unpleasant symptoms of pregnancy, such as heartburn, haemorrhoids, and tiredness, to a greater degree than women with a singleton pregnancy. They may also be more anxious about the pregnancy and their ability to cope with more than one new baby.

2 Advice and support

Women with a multiple pregnancy will need advice and support from caregivers to help them deal with the common unpleasant symptoms of pregnancy, such as haemorrhoids, heartburn, and backache (see Chapter 13). They may be particularly anxious about the pregnancy, the delivery, and their ability to cope with the practical and financial demands of more than one new baby. Helping women to find support, such as special antenatal classes for women with a multiple pregnancy or contacting a support group, may help.

3 Hospitalization and bed-rest

Prolonged bed-rest in multiple pregnancy, with the aim of increasing the duration of gestation, improving fetal growth, and decreasing perinatal mortality, has been advocated for over half a century. There is still no consensus as to when (or if) bed-rest should be started or discontinued, nor whether rest should be in hospital or at home. In several countries it is common practice to admit women with a multiple pregnancy to hospital for rest for varying periods between 29 and 36 weeks of gestation. The practice is not innocuous. The general considerations about the use of bed-rest (described in the chapter about miscarriage) apply equally strongly to its use in multiple pregnancy.

Hospitalization and bed-rest in multiple pregnancy was introduced into clinical practice without adequate evaluation and the policy has still not been fully evaluated. A few trials have been conducted recently, and further controlled evaluations are necessary to clarify the effects of this intervention. More information is available from twin than from higher multiple pregnancies.

There is some suggestion from these trials that routine hospitalization of women with twin pregnancies may result in a decreased risk of maternal hypertension, but the impact on more relevant outcomes has been negligible. Little effect was shown on the incidence of low birthweight. In other respects, the data suggest that routine hospitalization may have adverse effects. The risk of (spontaneous) preterm birth and babies with very low birthweight appears to be *increased* by routine hospitalization. No differences have been detected in the incidence of depressed Apgar score, admission to special care nurseries, or perinatal mortality.

Some obstetricians have suggested that hospitalization for bed-rest in twin pregnancies should be applied only for women deemed to be at higher than average risk of preterm birth. Although this more conservative advice is possibly justified, there is remarkably little good evidence to support it. Only one such selective policy has been evaluated in a randomized trial. Comparison between the hospitalized and control groups of women with early cervical dilatation tend to suggest beneficial effects of hospitalization with reduced incidence of very low birthweight and preterm delivery. The differences could easily reflect the play of chance, however, and were not reflected in reduced perinatal mortality. They do not provide a basis for widespread adoption of the policy.

Only one published trial exists of bed-rest in triplet pregnancies. This suggests reduction, by routine hospitalization, of a number of adverse outcomes including preterm delivery, perinatal death, and low birthweight. The trial was small, the findings were compatible with chance, and further research is required.

4 Medication

Trials have been conducted with a number of oral betamimetic agents, including isoxuprine, ritodrine, salbutamol, and terbutaline in various doses, for the prevention of preterm labour in women with multiple pregnancy. Despite the diversity of agents and the varying doses used, the results are consistent. No effect of prophylactic betamimetic administration has been detected on preterm birth, low birthweight, or perinatal mortality. Although prophylactic betamimetic agents have not succeeded in postponing delivery and/or improving fetal growth, the four trials which provide information on the incidence of respiratory distress syndrome suggest that its frequency may be reduced significantly. No such effect has been found with prophylactic betamimetics in singleton pregnancies, and it might be a chance finding.

In the light of the theoretical dangers of chronic fetal exposure to betamimetic agents, prophylactic administration of these drugs should be confined to their use in well-controlled trials.

5 Cervical cerclage

In normal pregnancy the uterine cervix is thought to assume a sphincter-like function to retain the contents of the uterus. A congenital or traumatically acquired weakness of the cervix, or the unusual physiological circumstance of multiple pregnancy, are factors that may render the cervix incapable of performing this function as efficiently as usual. Belief in such 'incompetence' of the cervix is the basis for performing the operation of cervical cerclage.

The data available from controlled trials of cervical cerclage in twin pregnancy are too few to be clinically useful. They are compatible with both a large beneficial effect and a large adverse effect of the operation. Cervical cerclage does affect other aspects of clinical care and carries some specific risks. It should not be adopted specifically for twin pregnancy outside the context of further controlled trials.

6 Home uterine activity monitoring

Trials of home uterine activity monitoring in multiple pregnancy have been small, and not enough detail is available to evaluate the potential sources of bias. There are suggestions that babies born to mothers using home uterine activity monitoring for twin pregnancy may be less likely to weigh less than 1500 grams, or to be admitted to a special care nursery. Because of the high potential for bias, these data must be viewed with caution. Home uterine activity monitoring should not be adopted outside the context of adequately controlled trials.

7 Delivery

Virtually no data from controlled trials are available to help determine the best mode or type of delivery for women with multiple pregnancy. A single trial has assessed the effect of caesarean section for delivery when the second twin was in a non-vertex presentation. As would be expected, maternal morbidity was increased with caesarean section. No offsetting advantages in terms of decreased fetal or neonatal morbidity or mortality were found.

8 Conclusions

Additional support may be needed to help women with the emotional, practical, and financial demands of pregnancy and planning for more than one baby.

There is currently no sound evidence to support the widespread practice of routine hospitalization for bed-rest for women with a twin pregnancy. Whether or not such a policy would be justified in women at higher risk of preterm labour, such as with triplet pregnancy or with early cervical dilatation, remains to be established.

The use of oral betamimetics, home uterine monitoring, and cervical cerclage for women with multiple pregnancy cannot be justified outside the context of adequately controlled trials.

The indications for caesarean delivery with multiple pregnancy have not been established.

This chapter is derived from the chapters by Caroline Crowther and Iain Chalmers (39), Adrian Grant (40) and G. Justus Hofmeyr (41) in EFFECTIVE CARE IN PREGNANCY AND CHILDBIRTH.

References to primary sources and more complete data for statements made in this chapter can be found in the source chapters and/or in the following reviews from the *Cochrane pregnancy and childbirth database*:

Crowther, C.A.
— Hospitalization for bed-rest in multiple pregnancy. Review no. 03375.
— Hospitalization for bed-rest in twin pregnancy. Review no. 03376.
— Hospitalization for bed-rest in triplet pregnancy: Review no. 03377.
— Hospitalization for cervical dilatation in twin pregnancy. Review no. 03378.
— Effect of caesarean delivery of the second twin. Review no. 06467.

Grant, A.M.
— Cervical cerclage in twin pregnancy. Review no. 03280.

Keirse, M.J.N.C.
— Prophylactic oral betamimetics in twin pregnancies. Review no. 03462.
— Home uterine activity monitoring in twin pregnancies. Review no. 06661.

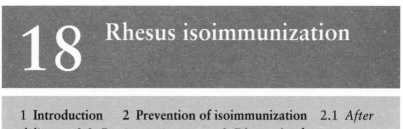

18 Rhesus isoimmunization

1 Introduction   2 Prevention of isoimmunization   2.1 *After delivery*   2.2 *During pregnancy*   3 Diagnosis of isoimmunization   4 Treatment of isoimmunization   5 Conclusions

1 Introduction

Isoimmunization against rhesus antigens, which may result in haemolytic disease of the fetus and newborn, was once a major cause of perinatal mortality, neonatal morbidity, and long-term disability and mental handicap. The condition is rarely seen today.

The discovery, introduction, and utilization of anti-D gammaglobulin has been one of the major obstetrical achievements of the last few decades. It should not be forgotten, however, that a large part of the reduction in the frequency of rhesus isoimmunization is the result of an increased proportion of one- and two-child families, in which the condition is unlikely to occur.

2 Prevention of isoimmunization

The effectiveness of anti-D immunoglobulin in the prevention of rhesus isoimmunization has been demonstrated in a number of trials conducted in different countries. The question is not whether women at risk of rhesus immunization should receive anti-D immunoglobulin, but which women should receive such prophylaxis, at what times, and in what doses.

There is a risk of isoimmunization in any situation in which rhesus-positive red blood cells enter the circulation of a rhesus-negative woman. The degree of this risk will vary with the amount of rhesus antigen to which she is exposed.

2.1 After delivery

The most common time for rhesus-positive fetal cells to enter the mother's circulation is at delivery of a rhesus-positive baby. Without the administration of anti-D immunoglobulin, rhesus-negative women who give birth to a rhesus-positive baby have a 7.5 per cent risk of developing rhesus antibodies within six months of delivery and a much larger risk (17.5 per cent) of showing evidence of sensitization in a subsequent pregnancy. With postpartum administration of anti-D immunoglobulin, this risk is reduced to 0.2 per cent.

ABO incompatibility, commonly believed to confer significant protection against the development of rhesus antibody formation, does not confer enough protection for it to be relevant in guiding clinical practice.

It is reasonably well established that 20 μg (100 IU) of anti-D immunoglobulin will neutralize the antigenicity of 1 ml of rhesus-positive red cells or 2 ml of whole blood. Hence, the usually administered dose of 300 μg (1500 IU), applied in some countries, is sufficient to protect against a fetomaternal bleed of 30 ml. Smaller doses might be given in the interests of economy, but in this case the amount of fetomaternal transfusion must be assessed. The relative cost effectiveness of giving a smaller dose of anti-D immunoglobulin along with assessment of the amount of fetomaternal transfusion, compared with routine administration of a higher dose, will depend on local circumstances and the relative costs of anti-D immunoglobulin and laboratory tests. Given the now ready availability of anti-D immunoglobulin, the cost-effectiveness of a smaller dose of anti-D immunoglobulin combined with screening for the degree of fetomaternal haemorrhage, compared with routine use of a larger dose of anti-D immunoglobulin, would seem questionable.

Fetomaternal haemorrhages of more than 30 ml can occur after even uncomplicated births. This is reported in about 0.5 per cent of deliveries, and these require larger doses of anti-D immunoglobulin to prevent immunization. After traumatic deliveries, caesarean section, and manual removal of the placenta, the risk of a large fetomaternal haemorrhage is increased. For this reason, counts of fetal cells in maternal blood to determine whether a larger dose of anti-D immunoglobulin should be given have become routine after delivery in some centres.

The best time to administer the anti-D immunoglobulin would be as soon as possible after birth, but immediate administration is not practical in view of the time required to determine the blood group of the baby. From the trials that have been conducted it would appear that an interval of up to 72 hours is compatible with effective prophylaxis.

2.2 During pregnancy

A small proportion of women develop rhesus antibodies during their first pregnancy; most such immunizations take place after 28 weeks of gestation. The antenatal administration of 100 µg (500 IU) of anti-D immunoglobulin at 28 and 34 weeks to rhesus (D) negative mothers has been shown to reduce the number of women with a positive Kleihauer test at both 32–35 weeks and at delivery, as well as the incidence of isoimmunization. The combination of antenatal administration of anti-D immunoglobulin to all unsensitized rhesus-negative women, and a further dose administered after delivery to all such women who give birth to a rhesus-positive child, will reduce the remaining incidence of rhesus isoimmunization from 0.2 to 0.06 per cent. The costs of such a programme may be high, and further studies will be required to determine the cost effectiveness.

Administration of anti-D immunoglobulin is required after procedures known to carry a risk of fetomaternal transfusion. Fetomaternal haemorrhage has been documented as a consequence of chorion villus sampling, amniocentesis, cordocentesis, placental biopsy, and external version of a breech presentation.

Placental trauma associated with either spontaneous or induced abortion can also cause fetomaternal bleeding. The incidence of fetomaternal bleeding in spontaneous abortion has been estimated at about 6–7 per cent during the first trimester of pregnancy. In the second trimester, incidences of 20 per cent and more have been reported.

Abdominal trauma, placenta praevia, placental abruption, or any form of uterine bleeding may occasionally cause fetomaternal transfusion. This may sometimes be suspected on the basis of the findings at cardiotocography prompted by these conditions. Fetomaternal transfusion may also occur without any obvious cause. Unexplained fetal or intrapartum death, or the birth of a pale distressed baby should raise the possibility that a fetomaternal transfusion has occurred.

Up to 13 weeks of pregnancy, a dose of 50–75 µg of anti-D gammaglobulin after termination of pregnancy or miscarriage is sufficient to ensure adequate protection. In the second trimester the standard postpartum dose is recommended.

3 Diagnosis of isoimmunization

All women should have routine assessment of their rhesus status in early pregnancy. Women who are rhesus (D) negative should be further screened for the presence of antibodies. The other rhesus antigens are far less immunogenic, but occasionally can cause serious clinical problems. Anti-Kell, anti-Kidd, anti-Duffy, and some of the more rare antigens can also on occasion cause haemolytic disease in the fetus and newborn. For this reason, many centres screen all pregnant women for other blood group antibodies in addition to the rhesus status.

The presence of antibodies indicates that the fetus may become affected if it carries the antigen to which the antibodies were formed; it does not indicate whether the fetus carries the antigen. For this reason it is helpful to determine whether the woman's partner is homozygous or heterozygous for the antigen. If the father is homozygous, the fetus will always carry the antigen; if the father is heterozygous, there is a 50 per cent risk that it will carry the antigen. Zygosity of the rhesus (D) antigen cannot be determined with certainty, but it can be estimated with a probability ranging between 80 and 96 per cent.

The level of an antibody titre does not always predict the presence or severity of disease, although in a first affected pregnancy there is a reasonable correlation between the antibody titre and the severity of disease. The major prognostic determinant of the severity of disease is the past obstetric history. The severity of the disease in previous pregnancies, in combination with serial antibody titres, will give a reasonably accurate assessment of the severity of the haemolytic disease in the current pregnancy about 95 per cent of the time. However, this approach is not sufficiently precise to determine the optimal time for intervention. For this, ultrasound, amniocentesis with spectrophotometric analysis of haemoglobin degradation products (bilirubin) in the amniotic fluid, or fetal haematocrit are required. Tests must be repeated at regular intervals, depending upon the severity and progression of the findings. The timing of intervention will depend primarily on this.

4 Treatment of isoimmunization

Pre-emptive delivery before the fetus is too severely affected to be effectively treated by postnatal therapy remains the mainstay of treatment for established isoimmunization. When the pregnancy cannot be carried to a stage of fetal maturity that would allow delivery of an infant who can be successfully treated by the currently available methods of intensive neonatal care, intrauterine transfusion is the treatment of choice.

By infusing rhesus (D) negative packed cells into the fetal circulation, fetal anaemia can be corrected and delivery may be postponed to a more advanced stage in pregnancy. The procedure can be started as early as 20 weeks of gestation and can be repeated as necessary.

Other treatments that have been used include plasmapheresis from early pregnancy onward, immunosuppression with promethazine, and desensitization by oral administration of rhesus (D) red blood cells. The value of these methods has not been evaluated in controlled trials.

5 Conclusions

Haemolytic disease of the fetus and newborn, while by no means the frequent condition that it once was, remains a problem that requires constant vigilance and attention to prophylactic care. Although effective prophylaxis is available, it must be used properly.

Postpartum prophylaxis with anti-D immunoglobulin is of overriding importance. It should be given to all rhesus (D) negative women who give birth to a rhesus (D) positive baby, or a baby whose rhesus (D) status cannot be determined, irrespective of their ABO status.

Anti-D immunoglobulin should also be administered to all rhesus (D) negative women during pregnancy when there is an increased risk of fetomaternal bleeding. Routine use at 28 weeks of pregnancy for all rhesus-negative women is of value as well, but the costs of such a programme are high.

Situations in which rhesus immunization does occur have become sufficiently rare, and its treatment is sufficiently complex to warrant regionalization of care for these women and babies. It is hoped that this may also facilitate adequate evaluation of the methods used for diagnosis and treatments, none of which have been subjected to controlled trials thus far.

This chapter is derived from the chapter by Jack Bennebroek Gravenhorst (35) in EFFECTIVE CARE IN PREGNANCY AND CHILDBIRTH.

References to primary sources and more complete data for the statements made in this chapter can be found in the source chapter and/or in the following reviews from the *Cochrane pregnancy and childbirth database*:

Crowther, C.A. and Keirse, M.J.N.C.
— Anti-Rh D prophylaxis postpartum (overall, irrespective of ABO status). Review no. 03314.
— Anti-Rh D prophylaxis postpartum < 72 hours in ABO compatible cases. Review no. 03315.
— < 200 μg anti-Rh-D postpartum vs higher doses. Review no. 03316.
— < 301 μg anti-Rh-D postpartum vs higher doses. Review no. 03317.
— Anti-Rh-D administration in pregnancy. Review no. 04027.

19 Infection in pregnancy

1 Introduction

In the past, infection was a major cause of both maternal and perinatal death, but deaths from bacterial infection are now rare in most parts of the developed world. Nevertheless, maternal infection and colonization with pathogenic organisms continue to cause problems for both mothers and babies. Various forms of infection are more prevalent among socio-economically disadvantaged pregnant women, and this may in part explain the greater frequency of adverse outcomes of pregnancy among these women.

In addition to bacteria, other organisms such as fungi, viruses, and protozoa may lead to serious disease during pregnancy and the perinatal period. A number of organisms that are more difficult to culture, such as *Chlamydia trachomatis* and the genital mycoplasmas, have now been linked with disease in pregnancy and childbirth. In addition, pregnant women are subject to the same range of acute and chronic infections as non-pregnant women.

2 Urinary tract infection

Three to eight per cent of pregnant women harbour significant numbers of bacteria in their urine (bacteriuria), usually without exhibiting any symptoms, and 15–45 per cent of untreated women with symptomless bacteriuria will develop symptomatic infections of bladder or kidney (acute cystitis or pyelonephritis). Acute cystitis and acute pyelonephritis are found in approximately 1 per cent of pregnancies. Thus urinary tract infection is one of the most common medical complications of pregnancy.

Culture and colony count of a single voided specimen is the best currently available form of screening for bacteriuria. Other more economical methods of screening for infection, such as the detection of urinary nitrites or microscopic analysis of a clean catch spun urine, have been suggested. These may have a role in identifying which urines should be cultured in the interests of cost saving, but their sensitivity and specificity in pregnant women is not high enough to replace urine culture as an adequate screening test.

Recognition and treatment of asymptomatic bacteriuria in pregnancy will result in a substantially decreased risk of acute pyelonephritis and its short-term consequences to both mother and fetus. It appears to reduce the incidence of preterm delivery and low birthweight as well, although this relationship is somewhat more tenuous. The mechanism through which treatment of bacteriuria leads to a reduction in preterm delivery is not yet clear. Prevention of pyelonephritis may be a factor. Treatment of bacteriuria with antibiotics may also eradicate organisms colonizing the cervix and vagina. Antibiotic treatment of bacteriuria in pregnancy has not been shown to reduce the risk of subsequent infection in the long term, but the only trial with a follow-up is small.

The available evidence from controlled trials suggests that sulphonamides, nitrofurantoin, ampicillin, and the first-generation cephalosporins are equally effective in the treatment of asymptomatic bacteriuria when the bacteria are known to be susceptible.

The original approach to therapy for asymptomatic bacteriuria in pregnancy was continuous antibiotic treatment for the duration of pregnancy. Single-dose therapy for uncomplicated urinary tract infection in women who are not pregnant is well established, however, and trials suggest that this may be effective for pregnant women as well. It has obvious advantages in terms of compliance, minimization of adverse effects, and financial savings. No trials of other regimens have been reported.

Symptomatic lower-tract infection in pregnancy may also respond to single-dose treatment, but there are insufficient data for this treatment to be recommended. Regular follow-up urine cultures must be obtained. Failures, relapses, and recurrences must be treated appropriately, and when infection recurs consideration must be given to continuous treatment for the remainder of pregnancy. Recurrent infection during pregnancy may signify an underlying abnormality of the urinary tract, and these women should be evaluated radiographically after pregnancy.

Pyelonephritis is diagnosed clinically by the presence of fever, flank pain, and dysuria, together with a positive urine culture. Women should be hospitalized and started on adequate antibiotic therapy after blood and urine cultures have been taken. In the absence of frank septicaemia, oral therapy and intravenous therapy are associated with a similar

duration of maternal fever, and the same incidences of systemic complications of sepsis and of readmission for urinary tract infection.
Ampicillin or a first-generation cephalosporin is appropriate initial
treatment as infection is probably due to *Escherichia coli*. When there is
concern about antibiotic resistance or the woman is seriously ill, combination therapy with an aminoglycoside and ampicillin is appropriate.
Careful monitoring of serum aminoglycoside levels throughout therapy
is strongly recommended to minimize fetal exposure to the drug.

Following an episode of acute pyelonephritis, women are at risk of
relapse and recurrence of infection, but the only reported randomized
trial of suppressive therapy for the remainder of pregnancy failed to
detect any advantage over close surveillance with cultures.

3 Syphilis

Syphilis during pregnancy is particularly important because transmission of *Treponema pallidum* from mother to baby may result in congenital syphilis, with its tragic sequelae. Such transmission may cause
abortion, preterm birth, or perinatal death (20 per cent). Subclinical
congenital infection with resulting handicap is not uncommon.
Congenital syphilis can be largely prevented by identification and treatment of the infected mother during pregnancy. Transmission to the
fetus occurs particularly during the second trimester, although it may
also occur during the first trimester.

Most infected women are free of symptoms and can only be identified
by blood testing. A programme of screening and treating those women
found to be seropositive is cost effective, despite the rarity of syphilis in
pregnancy, because effective treatment is available and the consequences of untreated infection are so serious.

Treatment of mothers should consist of efficacious antibiotics, preferably a penicillin. Infants and the woman's sexual partner(s) should be
followed up, and treated if found to be infected.

The clinical diagnosis of congenital syphilis is difficult because the
presentation is variable, and many babies are free of symptoms.
Treatment of infants is recommended when the adequacy of treatment
of the mother is unknown, or if the mother received treatment for the
first time during the pregnancy with a drug other than penicillin.

4 Gonorrhoea

In some parts of the world, routine screening for gonorrhoea during
pregnancy is deemed worthwhile because of the severe effects of infection on both the mother and her baby. This can be accomplished by obtaining cervical or uterine swabs for culture at the first antenatal visit.

In women considered to be at particular risk, either on demographic grounds or because of a history of sexually transmitted disease, repeat cultures can be taken.

Culture and sensitivity testing remains the 'gold standard' for diagnosis of gonorrhoea. The Gram stain is not sensitive enough for specimens obtained from the female genital tract. Although the infection may be asymptomatic, pregnancy appears to increase the likelihood of both arthritis and systemic disease. Where the prevalence of penicillin resistant gonorrhoea is high, third generation cephalosporins are currently recommended for treatment.

The most common gonococcal infection in neonates is conjunctivitis. Gonococcal ophthalmia characteristically manifests itself early — two to five days after birth. If left untreated, this infection may lead to permanent corneal damage and even perforation of the eye. The ideal method of preventing neonatal ophthalmia is detection and early treatment of maternal disease.

There is no evidence that routine prophylactic medication is better than careful observation and prompt treatment of ophthalmia in the newborn in most populations. Prophylaxis is required when mandated by law (as is the case in some countries or states), or in populations with a high prevalence of gonorrhoea. When such prophylaxis is necessary, antibiotic regimens that are active against both gonorrhoea and chlamydia should be used. Cohort studies of tetracycline, erythromycin, and penicillin suggest that these agents are both less irritating and more effective prophylactic agents than silver nitrate, and they are also effective against chlamydia infection.

5 Rubella

Rubella is typically a mild childhood illness. Maternal infection, occurring early in pregnancy, can lead to fetal death, low birthweight for gestational age, deafness, cataracts, jaundice, purpura, hepatosplenomegaly, congenital heart disease, and mental retardation in the infant. The objective of rubella vaccination programmes is to prevent fetal infection and the congenital rubella syndrome.

The risk to the fetus of maternal infection decreases with increasing duration of pregnancy. In a prospective study, infants whose mothers had confirmed rubella at successive stages of pregnancy were followed for two years. No defects attributed to rubella were found in children infected after 16 weeks' gestation, while infants infected before week 11 had significant cardiac disease and deafness.

Two approaches to rubella vaccination have been used: universal vaccination and selective vaccination. Universal vaccination of young children to interrupt transmission has led to a significant decline in reported cases of rubella and the congenital rubella syndrome. A vaccination rate of close to 100 per cent will be needed if congenital infection is to be eliminated.

Following an attack of rubella, lifelong protection against the disease usually develops. Reinfections can occur, but the majority of these are asymptomatic and detected only by a booster response in rubella-specific antibodies. Vaccination produces an overall lower antibody response than natural infection, but protection against infection can be expected in almost all vaccinated women.

The diagnosis of rubella in a pregnant woman who has been exposed to or develops a rubella-like infection is often difficult. The laboratory must be provided with a detailed history, as routine screening tests are inadequate and additional testing is required. False-negative results can occur if the specimen is drawn too soon after exposure. The pattern of antibody response to acute infection and reinfection will vary according to the test method used, and expert consultation may be required for interpretation of data.

Pregnant women should not be given rubella vaccine, but there should be little concern if a pregnant woman is vaccinated unknowingly or if she becomes pregnant within three months of immunization. As rubella vaccine virus has been isolated from fetal tissue following induced abortion, the risk cannot be assumed to be zero, but receipt of rubella vaccine in pregnancy is not ordinarily an indication for termination of pregnancy. Available data suggest that the risk of teratogenicity from live rubella vaccine is virtually non-existent.

The most significant cost factors associated with rubella are related to the long-term consequences of congenital rubella. The costs of the congenital rubella syndrome far outweigh that of routine vaccination of all infants of both sexes, as well as teenage girls and postpartum rubella-seronegative women.

High immunization frequency must be achieved and maintained, and all susceptible women of childbearing age should be identified and vaccinated. Prenatal screening should be carried out on all pregnant women without documented immunity, and vaccination given following childbirth, miscarriage, or termination of pregnancy, when the probability of pregnancy occurring within the next 30 days is low. Almost all vaccinated women show seroconversion and side effects are mild. Where follow-up cannot be assured, rubella vaccination without prior serological testing may be preferable. A third to a half of current cases of the congenital rubella syndrome could be prevented if postpartum vaccination programmes were fully implemented.

Termination should be offered when maternal infection is diagnosed in the first 16 weeks of pregnancy. Routine use of immunoglobulins for post-exposure prophylaxis against rubella is not recommended, although it may have a role where maternal rubella occurs and termination of pregnancy is not an option.

6 Genital mycoplasmas

Genital mycoplasmas have been postulated as causative agents of recurrent abortion, chorioamnionitis, preterm birth, low birthweight, stillbirth, and postpartum fever. As these organisms are found in the vaginal secretions of 35–90 per cent of pregnant women, it is important to determine whether or not they are, in fact, pathogenic.

The suggestion that genital mycoplasmas might be a cause of pregnancy loss is largely based on case–control studies in which a higher rate of mycoplasma colonization was found among women with adverse pregnancy outcomes than among those without abnormal outcomes. As is true of all such studies, other factors, both known and unknown, may have caused the adverse outcomes, irrespective of the presence or absence of mycoplasma infection. For example, if such infections are simply markers for sexually transmitted disease in general, other organisms might be the actual causes.

In the light of these uncertainties, it also remains uncertain whether or not there is any value in screening for these organisms, or in instituting treatment if they are isolated. The organisms are sensitive to a number of antibiotics, but the effects of treatment have not been adequately evaluated. Trials of treatment of ureaplasma during pregnancy with erythromycin showed no effect on the incidence of prelabour rupture of the membranes, low birthweight, preterm birth, or perinatal mortality. One trial of a short course of erythromycin given between 26 and 30 weeks to inner-city women attending publicly supported antenatal clinics showed a reduction in the incidence of prelabour rupture of the membranes, but these results should not be extrapolated to all pregnant women.

7 Toxoplasmosis

Maternal infection with the protozoan parasite *Toxoplasma gondii* acquired during pregnancy may result in congenital infection of the infant, sometimes with serious sequelae. Individuals can be infected only once, and so a woman who is immune prior to pregnancy is not at risk of transmitting the organism to her infant.

Clinical manifestations of congenital toxoplasmosis, which consist of chorioretinitis, recurrent seizures, hydrocephalus, and intracranial calcifications, may be present at birth or appear later.

The vast majority of infections are asymptomatic in the mother, although lymphadenopathy may occur. Asymptomatic infection can be detected by a high antibody level to *Toxoplasma gondii* in the serum.

Studies of congenital toxoplasmosis have relied on serological results to identify maternal infection. Both the prevalence of seropositivity (indicative of past exposure and immunity) and the risk of acquisition during pregnancy vary among countries and even among different regions within the same country. This has been ascribed, at least in part, to different habits with respect to the handling and consumption of raw meat and the disposal of cat litter, both of which can be reservoirs of the organism. Routine screening for the condition is conducted in some countries (e.g. France, Belgium, and Luxembourg), but not in others (e.g. the United Kingdom or The Netherlands).

The risk of transmission of toxoplasmosis from mother to baby is dependent on the stage of pregnancy at which maternal infection occurs. In one major study the frequency of infection rose from 17 per cent in infants whose mothers were infected during the first trimester to 65 per cent in infants whose mothers acquired the infection in the third trimester. Although the incidence of transmission from mother to fetus, based on serology, was highest in third trimester infections, transmission in the first trimester was associated with more severe symptoms in the neonate. Severe disease occurred in 14 per cent of first trimester transmissions and in none of third trimester transmissions. Some of the early infections may result in spontaneous abortions. The high frequency of neonatal clinical disease after infections early in gestation has led French investigators to recommend termination of pregnancy where feasible, and an anti-protozoan drug, spiramycin, when termination is not possible.

When the incidence of new toxoplasmosis infection is low, the low pick-up would not justify the expense of a universal screening programme. A health education programme advising pregnant woman not to eat raw or undercooked meat, to wash their hands after its preparation, and to wear gloves when gardening or cleaning cat litter might be more cost effective. Although there is some evidence that knowledge of toxoplasmosis is increased following an educational session, there is no evidence that this knowledge is associated with changes in behaviour and a reduction in congenital toxoplasmosis.

Much work is still necessary to determine the true frequency of infection and the sequelae of congenital toxoplasmosis. Neither spiramycin nor pyrimethamine-sulpha are very efficacious against this parasite, and the latter agent is associated with a high frequency of side effects in pregnant women. Trials of new anti-protozoan drugs with potentially better efficacy and safety are needed. Such studies will require prolonged follow-up, as some of the sequelae of fetal infections may occur many years after birth.

8 *Chlamydia trachomatis*

Maternal infection with *Chlamydia trachomatis* is important primarily because of the potential adverse effects of infection on the newborn infant. The condition is often asymptomatic in the mother and may not be detected clinically, although some infected women may have a mucopurulent cervicitis, salpingitis, or an urethral syndrome.

The prevalence of *Chlamydia trachomatis* in pregnant women varies widely; estimates ranging from 2 to nearly 40 per cent have been reported. In the United States, high rates are found in young women, unmarried women, and black women, as well as in women from lower socio-economic groups and those attending inner city antenatal clinics.

The newborn infant can acquire chlamydial infection through contact with infected maternal genital secretions at birth. Inclusion conjunctivitis will develop in 18–50 per cent of infants born to infected mothers, making *Chlamydia trachomatis* the most common cause of neonatal conjunctivitis. The estimated risk that an infant born to an infected mother will develop chlamydial pneumonia ranges from 3 to 18 per cent.

The diagnosis of maternal *Chlamydia trachomatis* infection is made either by culture of the organism from an endocervical specimen or by the identification of chlamydial antigens directly in endocervical smears. Tissue culture is considered to be the gold standard for diagnosis. Commercial kits using immunofluorescent staining or enzyme immunoassay are available, and the sensitivity of these tests ranges from 70 to 100 per cent.

If the prevalence of maternal infection exceeds 6 per cent, it is cost effective to screen for chlamydia using cell-culture techniques and to treat the infected women. When less expensive diagnostic tests become available, screening may be justified at lower prevalence rates.

There have been no randomized controlled trials to guide care for women colonized with *Chlamydia trachomatis* during pregnancy. Tetracycline is contraindicated in pregnancy because of its hepatotoxicity and its effect on the development of bone and teeth in the fetus. Erythromycin is generally considered to be the drug of choice, but the optimal dose, duration, and timing of antibiotic therapy have not been established. A cohort study, using as controls women who refused therapy, showed a significantly lower prevalence of chlamydial infection in infants born to infected mothers treated with erythromycin 250 mg (base) four times daily for seven days at week 36 of pregnancy.

The natural history of *Chlamydia trachomatis* infections in pregnancy is inadequately known and the role of the organism in the adverse outcomes of pregnancy remains to be resolved. Well-designed trials are required to clarify the usefulness of screening for and treating this condition. As erythromycin is not completely effective in eradicating infection and some women do not tolerate the drug, trials of effective alternatives are needed.

9 Herpes simplex

Herpes simplex infection of the newborn, acquired from the mother, is a rare but potentially serious condition, occurring in between 1 in 2500 and 1 in 10 000 births. Its clinical presentation varies widely from asymptomatic, through involvement of only the skin, to involvement of the eye or nervous system, or widespread dissemination.

The risk of transmission from mother to baby at the time of birth is high in primary herpes infections, but the risk of infection from a mother with recurrent genital herpes is very low. Viral shedding is rare in the absence of a lesion. 'Prophylactic' caesarean section should not be offered to a woman with a history of herpes simplex infection who does not have clinically active herpes at the time of birth.

Clinical assessment remains the best criterion for identifying women who are shedding virus at the time of delivery. Asymptomatic shedding at delivery cannot be predicted on the basis of cultures during the pregnancy, and repeated cultures in asymptomatic women will not identify those women who are shedding asymptomatically at the time of delivery.

One controlled trial has studied the effects of acyclovir given to women with a history of recurrent genital herpes. In this study, pregnant women with recurrent genital herpes (mean of three symptomatic recurrences in previous six months) received acyclovir 200 mg four times daily beginning one week before the expected date of delivery. This treatment resulted in significant reductions in maternal viral shedding at the time of delivery, symptomatic recurrences within 10 days of delivery, and use of caesarean section for herpes. There were no cases of neonatal herpes in either the treatment or the control group, and no maternal or neonatal side-effects were seen. Although these results indicate that this drug can be effective in reducing maternal viral shedding and symptomatic recurrences, further confirmation is required before a firm conclusion about the role of antenatal acyclovir in recurrent genital herpes can be drawn.

There have been no randomized trials to evaluate care policies for women with herpes in pregnancy, and the evidence on which these are based is weak indeed. Current recommendations are that caesarean section should be carried out if there is clinical evidence of active disease, viral shedding cannot be ruled out, and the membranes have not been ruptured for more than four to six hours. An infant born to a woman with a history of genital herpes should be carefully monitored.

10 Group B streptococcus

Group B streptococcus has become the most frequent cause of overwhelming sepsis in neonates. The early, and most serious, form of infection is characterized by rapid onset of respiratory distress, sepsis, and shock. The likelihood of disease (approximately 2 per 1000 live births) is directly related to the density of colonization and the immaturity of the infant. Infants with birthweights of less than 2500 grams have a much higher overall infection rate than infants weighing 2500 grams or more. Prelabour rupture of the membranes and maternal fever are also associated with a higher incidence of infection.

Attempts to prevent disease by giving antibiotics either to all babies or to those considered to be at high risk have proved disappointing. Although the available data suggest that infant sepsis with group B streptococcus can be reduced by giving antibiotic prophylaxis to the baby, such prophylaxis may be accompanied by an increase in sepsis with penicillin-resistant organisms. This results in a higher rate of deaths from infection in the babies given antibiotic prophylaxis.

Since attempts at prophylaxis after the baby has been born may be too late, attention has focused on studies of the effectiveness of antepartum and intrapartum antibiotics. The available data show that a course of antibiotics given during pregnancy results in only a temporary eradication of group B streptococcal carriage, with no detectable effects on infant colonization or sepsis with group B streptococcus. Treatment during pregnancy, unless continued into labour, has only a transient effect on the vaginal flora and will not influence the rate of sepsis in the newborn.

There is no doubt that the use of intrapartum antibiotics reduces the transmission of group B streptococcus. The studies reviewed show a beneficial effect of treatment for women who are receiving comprehensive obstetrical care and are known to be colonized with group B streptococcus, but these results are not generalizable to all pregnant women. The optimal method of detecting colonization with group B streptococcus, whether by routine prenatal cultures or a rapid antigen test at the onset of labour in high risk patients, has not been determined. There is clear evidence that treatment should be given to high risk colonized women, but insufficient evidence to recommend routine screening of all women for group B streptococcus during pregnancy.

It seems reasonable that a pregnant woman in preterm labour, or a woman with either intrapartum fever or prolonged rupture of membranes, should receive intrapartum antibiotics if either a rapid diagnostic test is positive or not available, although there is no evidence from controlled trials to support these recommendations. An alternative option would be a screening culture for all women at 28 weeks, with treatment of women with a positive culture during labour if there are other risk factors such as preterm labour, prelabour rupture of membranes, or fever. In high prevalence populations it may be preferable to dispense with a preliminary screen, and to treat all women with the additional risk factors during labour.

As intrapartum prophylaxis of colonized pregnant women offers the possibility of reducing the incidence of infant sepsis, rapid methods for screening women in preterm labour are desirable. A number of such methods of rapid diagnosis of colonization are currently under development, but so far none are adequately sensitive.

11 HIV infection

Acquired immunodeficiency syndrome (AIDS) is now a major health problem. It is characterized by defects in the immune system with attendant susceptibility to infections by opportunistic micro-organisms and specific tumours. Despite the introduction of some specific antiretroviral chemotherapeutic agents such as azidothymidine and an aggressive approach to prophylaxis of opportunistic infections, the disease still progresses.

Paediatric AIDS was first recognized in 1982 and numerous case series have been reported since. The virus may be transmitted *in utero*. The condition may not be recognized at birth, and may present with recurrent bacterial infections and sepsis, persistent or recurrent thrush, and failure to thrive.

The organism is usually transmitted sexually in adults, but it is also transmitted through infected blood and blood products, or the sharing of needles among infected drug users. Intrauterine transmission from mother to infant may occur during pregnancy, at birth, or, more rarely, through breast milk. The advantages of breastfeeding will often outweigh the small risk of transmission through breast milk, particularly in less developed countries.

Both asymptomatic and symptomatic women may transmit the infection to their infants. Offering screening only to women who are considered to be at high risk will only detect a small proportion of infected women.

12 Conclusions

Screening pregnant women to detect asymptomatic bacteriuria and treating the condition with antibiotics is worthwhile. The practice will reduce the incidence of pyelonephritis and probably of preterm labour and low birthweight infants. The cost-effectiveness of this screening will depend on the prevalence of asymptomatic bacteriuria in the population.

Pyelonephritis in pregnancy requires intensive treatment with appropriate antibiotics, usually intravenously, but oral therapy may be appropriate for selected non-bacteremic women with acute pyelonephritis in pregnancy.

A high community level of rubella immunization should be obtained, and all rubella-susceptible women of childbearing age should be identified and vaccinated. Rubella vaccination in the early postpartum period is safe and effective. This opportunity for immunization should not be missed.

Congenital syphilis can be largely prevented by identification and treatment of the infected mother during pregnancy. Routine screening is justified by the simplicity of the test and the effectiveness of treatment with penicillin. The value of screening for other sexually transmitted diseases will depend on their prevalence in the community, and the effectiveness of the available treatments.

There is at present no evidence to warrant routine screening for ureaplasma or mycoplasma organisms in pregnancy, nor to warrant treatment if they are found.

The optimal method of detecting colonization with group B streptococcus, by routine prenatal cultures or by a rapid antigen test at the onset of labour in high risk patients, has not been determined. Antibiotics should be given to high risk colonized women during labour.

This chapter is derived from the chapter by Elaine Wang and Fiona Smaill (34) in EFFECTIVE CARE IN PREGNANCY AND CHILDBIRTH.

References to primary sources and more complete data for the statements made in this chapter can be found in the source chapter and/or in the following reviews from the *Cochrane pregnancy and childbirth database*:

Smaill, F.
— Antibiotic vs no treatment for asymptomatic bacteriuria. Review no. 03170.
— Single dose vs 4–7 day antibiotic for bacteriuria. Review no. 03171.
— Broad spectrum penicillin vs cephalosporin for bacteriuria in pregnancy. Review no. 03172.
— Nitrofurantoin vs broad spectrum antibiotic for bacteriuria. Review no. 03173.

— Single dose vs 7-day course of antibiotics for bacteriuria. Review no. 03176.
— Two week vs continuous antibiotic for bacteriuria. Review no. 05251.
— Oral vs intravenous antibiotics for acute pyelonephritis. Review no. 07204.
— Nitrofurantoin vs surveillance after pyelonephritis. Review no. 03166.
— Postpartum rubella vaccination. Review no. 06403.
— RA 27/3 vs Cendehill for postpartum rubella vaccination. Review no. 06404.
— Treatment with tetracycline during pregnancy. Review no. 07203.
— Erythromycin treatment at 26–30 weeks gestation. Review no. 07201.
— Prenatal education for congenital toxoplasmosis. Review no. 06434.
— Antenatal acyclovir for recurrent genital herpes. Review no. 06493.
— Antepartum antibiotics for Group B streptococcal colonization. Review no. 03005.
— Intrapartum antibiotics for Group B streptococcal colonization. Review no. 03006.

20 Diabetes in pregnancy

1 Introduction   2 Prepregnancy counselling and assessment
3 General care during pregnancy   4 Diabetic control
5 Obstetric care   6 Care in labour and delivery
7 Care after birth   8 Conclusions

1 Introduction

Perinatal mortality in pregnancy associated with diabetes has dropped tenfold in the last four decades, compared with a four- to fivefold drop in perinatal mortality overall. Nowadays, some specialist centres report perinatal mortality rates close to those of non-diabetic populations, but in general the risk of perinatal loss for the average diabetic woman remains higher than for non-diabetics, even after adjustment for fetal malformation. Population surveys suggest that the outcome of pregnancy overall is poorer than what would be inferred from the literature.

A number of factors have played a part in this remarkable improvement. These include, among others, increasing acceptance by physicians of the importance of tight control of diabetes, the introduction of programmes to achieve this control, the development of home glucose monitoring to facilitate such programmes, gradual trends towards prolongation of pregnancy, and advances in neonatal care.

Although there is general consensus on the adverse impact of overt diabetes in pregnancy, the significance and appropriate management of lesser degrees of hyperglycaemia are still widely debated. Diabetes is a disturbance of multiple metabolic pathways rather than of glucose metabolism alone, although the effects on carbohydrate metabolism are the most apparent. Generally accepted criteria for a diagnosis of diabetes in the presence of symptoms (polyuria, polydipsia, ketoacidosis) are random venous plasma glucose levels greater than 11 mmol/litre (200 mg/dl) or fasting levels of greater than 8 mmol/litre (140 mg/dl).

Normal values for plasma glucose values are defined as less than 8 mmol/litre on a random sample, and less than 6 mmol/litre fasting. Values between the normal and those of diabetes are considered to be 'equivocal', and evaluation with a glucose challenge is recommended (for example 75 grams taken orally after an overnight fast). Values over 11 mmol/litre two hours post challenge are taken to be diagnostic of diabetes, and those between 8 and 11 mmol/litre are termed 'impaired glucose tolerance'.

2 Prepregnancy counselling and assessment

Increasingly, women with diabetes wish to discuss the implications of pregnancy before they conceive, and this should be actively encouraged by all health care professionals before each pregnancy. While preconception clinics may fulfil an important role, the provision of adequate preconceptional care and advice does not necessarily depend on such specialized clinics. All who care for diabetic women should be aware of, and prepared to discuss, the importance of appropriate contraception and timing of pregnancy, the significance of pregnancy to the diabetic woman, the risks to the fetus and neonate as well as to herself, the importance of tight control of the diabetes just before and during pregnancy, and the need for accurate estimation of the date of conception.

Since diabetes is a chronic and progressive disease, the advice may need to include a discussion of the fact that postponement of pregnancy until a later age may worsen the prognosis.

The risks to the fetus are significant. Diabetes is associated with an increased incidence of congenital anomalies, up to three times as great as for the infants of non-diabetic mothers. Although no information is available from randomized trials, cohort studies suggest that tight control of the diabetes immediately before conception can reduce this risk significantly.

Macrosomia is still more common in the infants of diabetic mothers than in those of non-diabetic mothers, even with the best diabetic control currently available. Diabetes is not typically associated with intrauterine growth retardation unless the diabetes is complicated by microvascular disease. Women with vascular complications of diabetes will need particularly careful counselling.

Nephropathy without significant hypertension and a normal serum creatinine is not associated with a poor fetal outcome. The prognosis worsens in the presence of hypertension or impaired renal function. Renal disease that does not cause symptoms in non-pregnant individuals can jeopardize pregnancy outcome in some women. Although the majority of women with renal disease do not experience a deterioration of renal function during pregnancy, some women suffer significant deterioration that does not improve after delivery.

There is also concern about the effect of pregnancy on women with proliferative retinopathy. Pregnancy appears to be associated with a deterioration in the condition. However, cohort studies comparing pregnant and non-pregnant women with diabetic retinopathy show that visual acuity can be maintained with intensive laser treatment throughout pregnancy, and the prognosis is no worse than for the non-pregnant woman.

The risk of spontaneous preterm birth is not increased. The apparent excess of preterm births relates either to delivery for complications, particularly hypertensive disorders, or to policies advocating elective preterm delivery. Preterm births can be minimized by a reconsideration of such policies.

Diabetic women contemplating pregnancy will be reassured to learn that there is no good evidence of any long term adverse effects of their diabetes on the development or intelligence of their offspring, and that the risk of their children developing juvenile diabetes is of the order of 2 per cent or less.

Young diabetic women should have easy access to family planning services at all points of contact with the health care system, both before and between pregnancies.

3 General care during pregnancy

Diabetic pregnant women should be cared for by both obstetricians and physicians with special interest in this field. Ideally this might be carried out at joint clinical sessions, but other local arrangements can deliver quality integrated care. Such arrangements facilitate efficient deployment of other health care professionals such as dieticians and specialist liaison midwives or nurses.

The first few weeks of pregnancy are a period of readjustment, and many women require re-education about their diabetes and its control. Rotation of insulin injection sites, the interaction of diet and exercise, and the dietary requirements of pregnancy may be unfamiliar to many women. Specialist care in and for pregnancy should start as early as possible. Local organization should be such as to allow very rapid referral of diabetic women with suspected pregnancies. When the gestational age is in doubt, it should be estimated precisely by early ultrasound.

Hypoglycaemia can be troublesome at this stage, and control of the diabetes may be difficult to achieve because of poor motivation, nausea and vomiting, or changes in the hormonal milieu. Considerable education is needed to resist overtreatment of impending hypoglycaemic reactions. Glucose or sugar should be avoided; milk or a light snack, which can be repeated if necessary, are more appropriate. All diabetic women should be provided with glucagon for emergency situations.

In addition to routine prenatal assessment, obstetric care at this time should include assessment of renal function in diabetic women who have hypertension or proteinuria, and retinoscopy, particularly in women who have had diabetes for more than 10 years. This should be repeated once every trimester. Urine cultures should be repeated regularly in those with nephropathy. In view of the increased risk of malformation, detailed ultrasound is recommended. Additional expert fetal echocardiography is advisable given the high frequency of cardiac anomalies.

A diabetic woman without nephropathy or retinopathy, and with no other complications of pregnancy, usually experiences an uncomplicated second trimester. The educational and readjustment processes will hopefully be complete as far as possible, and unless there is a risk of compromised fetal growth or early pre-eclampsia there is little need for intensive obstetric supervision at this time.

Women with associated hypertension may need to be followed closely and, if necessary, treated with hypotensive drugs. Serial assessment of renal function can be particularly useful in following these pregnancies.

4 Diabetic control

In the non-pregnant diabetic, intensive control using continuous subcutaneous insulin, or regimens with or three or more injections daily, improves glycaemic control. Randomized trials have demonstrated that such regimens significantly reduce the risk of progression to nephropathy and the long term risk of retinopathy. In the short term, retinopathy may worsen and there is a trend towards more frequent severe hypoglycaemic episodes. The incidence of diabetic ketoacidosis is significantly higher using subcutaneous insulin regimens.

These findings may be relevant to pregnancy where intensive regimens are commonly used. The aim of diabetic control is to establish normoglycaemia, both fasting and before and after meals. Blood glucose levels can be monitored effectively and controlled by the woman at home, provided that she has readily available advice and support, predominantly through telephone contact. The use of home, instead of hospital, glucose monitoring can significantly reduce the time that the woman spends in hospital without affecting pregnancy outcomes.

The dose and type of insulin needed may require careful and frequent adjustment. Insulin may have to be given more frequently than before pregnancy, often requiring three and occasionally four injections a day. Pumps for continuous subcutaneous infusion of insulin are expensive and complex to use. Trials have shown no benefits for continuous infusion over conventional insulin administration in terms of metabolic control, use of caesarean section, or adverse pregnancy outcome. Temporary worsening of retinopathy may occur using such regimens. Despite the importance of the subject, to date the benefits and drawbacks of tight control of diabetes in pregnancy have been assessed in only one randomized trial. This trial compared the effects of very tight control (aiming to keep blood sugar levels below 5.6 mmol/litre), tight control (blood sugar levels between 5.6 and 6.7 mmol/litre), and moderate control (blood sugar levels between 6.7 and 8.9 mmol/litre).

Best results were obtained with tight, rather than either very tight or moderate, control. Very tight control was associated with episodes of hypoglycaemia, and conferred no benefits in other pregnancy outcomes compared to tight control. A policy of only moderate control, in which blood sugar levels were allowed to rise up to 8.9 mmol/litre, was associated with a higher incidence of macrosomia, urinary tract infection, and caesarean section, and a trend towards an increase in hypertension, preterm labour, respiratory distress syndrome, and perinatal mortality.

This evidence tends to confirm the conclusion from observational studies that keeping blood sugars within a well controlled range, between 5.6 and 6.7 mmol/litre, is better than either too strict or too lenient a regimen.

5 Obstetric care

As in any antenatal care programme, much attention is focused on the early detection of gestational hypertension and in the identification of problems with fetal growth.

The incidence of pregnancy-induced hypertension is increased in pregnancies of diabetic women and may occur earlier. Particularly at risk are those with pre-existing hypertension, those with nephropathy, and those with microvascular disease. Currently there are no good predictors of pregnancy-induced hypertension, and preventive therapy is untested in this population.

Problems with fetal growth divide into two groups — growth failure and macrosomia. Women with hypertension, nephropathy, or microvascular disease are at more risk of the complications of fetal growth failure. Women with moderate or poor diabetic control are at most risk of fetal macrosomia. Pregnancies in diabetic women should have additional ultrasound surveillance of growth in the late second and third trimester. Absolute size is not a good predictor of outcome and attention should be paid to growth velocity. As in the non-diabetic population, the precise value of screening for growth problems is not established.

Macrosomia can be assessed by ultrasound, but most formulae utilized to calculate fetal weight perform poorly in the larger fetus, and such estimates should be interpreted with caution.

Because of the increased perinatal mortality, most pregnancies are subjected to serial antepartum fetal assessment. Various regimens have been described, but to date there are no randomized trials addressing which is the most effective technique. Most studies are of a prospective interventional design, and claims of improved outcome for any particular regimen of antepartum fetal surveillance must be interpreted within these limits.

6 Care in labour and delivery

Elective preterm delivery has long been one of the classical management strategies applied in diabetic pregnancy, based on an observation in one influential study that the stillbirth rate rose above the neonatal death rate after 36 weeks of gestation. However, other cohort studies have shown that this conclusion was flawed and that pre-emptive delivery does not result in an improvement in perinatal mortality. There is no valid reason to terminate an otherwise uncomplicated pregnancy in a diabetic woman before the expected date of delivery.

Until recently assessment of pulmonary maturity has been considered a prerequisite for planned delivery of the diabetic woman. There is increasing evidence to suggest that in well-controlled diabetics, the lecithin/sphingomyelin ratio reflects the same degree of pulmonary surfactant production, and therefore the same risk of hyaline membrane disease, as in non-diabetics. The infants of poorly controlled diabetic mothers may have a disturbance in the development of pulmonary maturity. The problem becomes less important as delivery is delayed to a later gestational age and as elective caesarean section is used less frequently.

Although caesarean section in diabetic women, as in all women, should only be performed for obstetric indications, caesarean section rates tend to be much higher for diabetic women than for the general population. There seems to be little justification for this, except that the median birthweight for gestation is higher in diabetic pregnancies.

As blood glucose levels can be controlled confidently throughout labour with a glucose infusion with insulin added, the only increased hazard of vaginal delivery in diabetic women is that of birth trauma to the infant, particularly shoulder dystocia and brachial plexus injury secondary to macrosomia. In comparative studies these risks appear unexpectedly higher in infants of diabetic mothers compared with similarly sized infants of non-diabetic mothers. The management of the second stage of labour in a diabetic pregnancy with suspected macrosomia should reflect this knowledge and appropriately skilled senior staff should be readily available.

7 Care after birth

After birth and delivery of the placenta, insulin requirements should be recalibrated according to blood glucose levels. All women will require significantly less insulin than during pregnancy.

Breastfeeding should be encouraged, although there must be an awareness of the increased caloric intake sometimes needed to support this.

Family planning is an important consideration. There is no contraindication to the use of low-dose oral contraceptives, particularly in young non-obese non-hypertensive non-smoking diabetics. The development of headaches or hypertension is an indication to change to alternative methods.

Until recently intrauterine contraceptive devices have been a reasonable alternative, but concern over their effectiveness is making it less acceptable; barrier methods provide an acceptable second choice.

For all women, the risk of future pregnancy must be carefully weighed against the risk of the proposed method of contraception. Women with nephropathy or retinopathy should be counselled carefully about limiting family size.

Sterilization, if requested, is better not carried out at the time of caesarean section if this can be avoided, in view of the increased risk of still undiagnosed cardiac malformation in the newborn. The procedure can easily be performed by laparoscopy within the next few months.

8 Conclusions

Diabetic women embarking on pregnancy demonstrate a deep commitment to achieving a normal outcome. This involves much disruption of an already complex life-style. Their care must reflect this and should be individualized, so that disruption will be minimized and care tailored to meet the circumstances of each woman.

There is considerable evidence to suggest that pregnancy in diabetic women should be managed with fewer obstetric interventions than are currently practised. Specialized care and collaboration among various disciplines will achieve the best results, but good perinatal outcome is not confined to tertiary care centres. Many diabetic women can be treated as other pregnant women, with the one major addition of careful control of blood glucose levels. Tight, rather than either very tight or moderate, control is required.

Allowing pregnancy to continue at least until the expected date of delivery, associated with a decreased need to assess pulmonary maturity and more judicious use of caesarean section, may allow pregnant women with diabetes to feel more like their non-diabetic counterparts.

Much has been achieved in improving the care for, and the outcome of, pregnancies in diabetic women without resort to randomized clinical trials. The lack of controlled studies, however, has resulted in a blurring of the contributions made by the various components of care, and in doubts about the utility of some. The need continues for well-designed trials to assess the value of both current treatments and suggestions for their improvement. In addition, better survey data on the diabetic population are still needed, as data that refer only to women who attend specialist centres can be seriously misleading. Population based survey data may highlight deficiencies in the system of care and lead to improvements.

This chapter is derived from the chapter by David J. S. Hunter (36) in EFFECTIVE CARE IN PREGNANCY AND CHILDBIRTH.

References to primary sources and more complete data for the statements made in this chapter can be found in the source chapter and/or in the following reviews from the *Cochrane pregnancy and childbirth database:*

Walkinshaw, S.A.
— Continuous subcutaneous infusion vs conventional insulin treatment in diabetic pregnancy. Review no. 04064.
— Very tight vs tight control of diabetes in pregnancy. Review no. 04065.
— Tight vs moderate control of diabetes in pregnancy. Review no. 04066.
— Very tight vs moderate control of diabetes in pregnancy. Review no. 04067.
— Home vs intermittent clinic glucose monitoring of diabetes. Review no. 06647.
— Home vs hospital inpatient glucose monitoring of diabetes. Review no. 06651.

21 Bleeding in the latter half of pregnancy

1 Introduction   2 Placental abruption   2.1 *Clinical presentation*   2.2 *Treatment*   3 Placenta praevia
3.1 *Clinical presentation*   3.2 *Treatment*   3.3 *Delivery*
4 Bleeding of uncertain origin   5 Conclusions

1 Introduction

Bleeding in the second half of pregnancy is no longer a common cause of maternal death in the industrialized world, but it continues to be a major cause of perinatal mortality and of both maternal and infant morbidity. Approximately half the women who present with bleeding in the second half of pregnancy are eventually found to have either placental abruption or placenta praevia. No firm diagnosis can be made in the other half.

2 Placental abruption

Placental abruption, or retroplacental haemorrhage, is a major contributor to perinatal mortality among normally formed fetuses. Although maternal mortality is fortunately rare now, maternal morbidity in the forms of haemorrhage, shock, disseminated intravascular coagulation, and renal failure is sufficiently frequent to justify intensive treatment.

The frequency with which placental abruption is diagnosed will vary with the criteria used for the diagnosis. The perinatal mortality rate with confirmed abruption is high, often over 300 per 1000. More than half the perinatal losses are due to fetal death before the mother arrives in hospital. Neonatal deaths are principally related to the complications of preterm delivery. Among surviving infants, rates of respiratory distress, patent ductus arteriosus, low Apgar scores, and anaemia are more common than in unselected hospital series.

2.1 *Clinical presentation*

Abruption may occur at any stage of pregnancy. The diagnosis should be considered in any pregnant woman with abdominal pain, with or without bleeding. Mild cases may not be clinically obvious.

In severe abruption there may be heavy vaginal bleeding or evidence of increasing abdominal girth if the blood is retained within the uterus. Uterine hypertonus is a common physical sign in the more severe grades of abruption, particularly when the fetus has died. In these cases the woman is usually in severe pain, and may be shocked as a result of hypovolaemia. Absence of clotting may be obvious in vaginal blood. Other signs of clotting defects may be bleeding from the gums or venepuncture sites, or haematuria.

The amount of blood loss may not be obvious — some may have been lost before admission, and large volumes of blood may be retained in the uterus. Clinical signs of hypovolaemia may be masked by increased peripheral resistance. Cerebral and cardiac perfusion may be preserved, although renal blood flow is jeopardized. This will become manifest by diminished urine production.

Ultrasound examination of the uterus and its contents has an important role in the differential diagnosis of antepartum haemorrhage. Most important is its ability to localize the placenta. A low lying placenta brings placenta praevia into the differential diagnosis, while a posterior placenta might make the diagnosis of abruption more likely in a woman with back pain. The diagnosis of retroplacental haematoma by ultrasound is not straightforward.

2.2 *Treatment*

In suspected mild abruption, the symptoms may resolve, and if there has been bleeding this may cease, with the fetal condition apparently satisfactory. It may be impossible to confirm the diagnosis. In this case, the woman may safely be allowed home after a period of observation, as is the case for bleeding of unknown origin.

In moderate and severe abruption, maternal resuscitation and analgesia are priorities. Restoration of the circulating volume and emptying the uterus are the cornerstones of treatment. The use of whole blood has become traditional in volume replacement in these women. However, it is likely that a crystalloid infusion to precede the blood would be beneficial. It is probably best to use fresh-frozen plasma, at a rate of one unit for every four to six units of red cells transfused, to replenish clotting factors. No systematic attempts have been made to study alternatives to blood transfusion in this condition. Plasma substitutes such as plasma protein, dextran, gelatin, and starch may produce adverse reactions, and dextran in particular can interfere with platelet function *in vivo* and cross-matching *in vitro*.

A clotting defect should be sought, although defects of clinical significance are rare when there is a live fetus. The process of disseminated intravascular coagulation usually starts to resolve after delivery.

When the fetus is alive, a decision about delivery in its interest should be made in the light of its estimated maturity. In former years the usual policy was vaginal delivery at all reasonable costs because the prognosis of the newborn was so poor. More recently, policy has shifted to earlier resort to caesarean section in order to rescue the baby, and improved survival of the preterm newborn has often resulted. However, a recent series suggests that an attempt to deliver vaginally, inducing or augmenting labour with oxytocin when necessary and using continuous electronic fetal heart rate monitoring, may result in a 50 per cent reduction in the rate of caesarean section without significant difference in the risk of perinatal mortality.

In severe abruption, when the fetus is dead, vaginal delivery should be planned except where there is an obvious obstetrical indication for caesarean section such as transverse lie. Labour should be induced or augmented if needed, using oxytocin or prostaglandins if there is no satisfactory response to oxytocin. Caesarean section is required in the rare cases in which uterine contractions cannot be stimulated, or when clinical shock associated with haemorrhage has been uncontrollable. Coagulation defects are common in this situation, and the maternal risks are considerable.

When caesarean section in a woman with disseminated intravascular coagulation is judged to be inevitable, it should be undertaken after close consultation with the anaesthetist and a haematologist. Volume replacement and transfusions of whole blood, frozen plasma, and specific coagulation factors should be given before and during the surgical procedure.

A well-equipped hospital should be able to provide adequate emergency treatment for the mother, and will often be able to achieve safe delivery of a fetus that is alive on admission. Survival of such live-born infants depends to a large extent on the quality of neonatal care.

3 Placenta praevia

Placenta praevia is defined as a placenta which is situated wholly or partially over the lower pole of the uterus. The overall prevalence of the condition is slightly over 0.5 per cent. The major cause of both mortality and morbidity is haemorrhage. Prevention and effective treatment of haemorrhage has reduced the gravity of the condition. With modern care a perinatal mortality rate of 50–60 per 1000 is now attainable.

3.1 *Clinical presentation*

Although it is well recognized that a small proportion of women with placenta praevia do not bleed until the onset of labour, less than 2 per cent of cases of placenta praevia present in this way. Painless vaginal bleeding in the absence of labour is the most common presentation. Some form of fetal malpresentation (transverse, oblique or unstable lie, and breech presentation) is found in approximately one-third of cases. In cephalic presentation the presenting part is invariably high, and is often displaced slightly from the midline.

All placentae praeviae are asymptomatic before the first onset of bleeding. With routine ultrasound scanning in the early second trimester, approximately 5–6 per cent of placentae are found to be low lying. Over 90 per cent of asymptomatic placentae praeviae diagnosed by ultrasound in the early second trimester remain asymptomatic and become normally situated later. Women with low or cervical placental implantation found early in gestation should be rescanned between 30 and 32 weeks' gestation. Where an asymptomatic women is found on ultrasound after 32 weeks still to have a placenta that appears to cross the cervix, she should be considered as having placenta praevia for the purposes of subsequent care. Where there is a lesser degree of low placentation, the possibility remains that placenta praevia will not be present at the time of labour.

3.2 *Treatment*

A digital examination should never be performed when there is any possibility of placenta praevia, except in the operating room when termination of pregnancy is forced by bleeding or labour, or when the pregnancy has reached an adequate gestation for safe termination. The really dangerous haemorrhage is often the one that has been provoked by ill-advised obstetrical interference such as digital examination of the cervical canal at or very shortly after the time of the warning haemorrhage. Rectal examination is even more dangerous than vaginal examination.

If bleeding is less severe or has stopped, confirmation of the diagnosis by ultrasound should be performed at the earliest opportunity. The early and accurate diagnosis of placenta praevia is imperative to spare women with a normally implanted placenta the economic, emotional, and social expense of long-term hospitalization.

The object of expectant management is to reduce the number of preterm births by allowing the pregnancy to continue until the baby has grown to a size and age that will give it a reasonable chance of survival. This form of care usually requires that the woman must remain in a fully equipped and staffed maternity hospital from the time of diagnosis until delivery because of the risks to both mother and fetus from further major haemorrhage. Some clinicians have adopted a policy of permitting selected women to return home as part of expectant management. Many women sent home require readmission to hospital for significant maternal bleeding, but no maternal deaths and no significant differences in perinatal outcome compared with those kept in hospital have been reported.

Preterm birth continues to be a major problem even when expectant management is used. Maternal and fetal well-being should be monitored. The mother should not be allowed to become anaemic, and her haemoglobin should be maintained at a normal level by haematinics or, if necessary, by transfusion.

With every episode of bleeding, the rhesus-negative woman should have a Kleihauer test performed for the presence of fetal cells, and be given prophylactic anti-D immunoglobulin.

The optimal timing for delivery remains controversial. Although expectant management until 37 weeks is most generally accepted, some clinicians have recommended elective preterm delivery after 34 weeks when amniocentesis has confirmed pulmonary maturity. No controlled trials to evaluate either approach have been reported.

3.3 Delivery

There is almost no indication for vaginal delivery for women with even marginal placenta praevia whose babies have attained a viable age. The hazards of vaginal delivery include profuse maternal haemorrhage, malpresentation, cord accidents, placental separation, fetal haemorrhage, and dystocia resulting from a posterior placental implantation. If the fetus is previable, malformed, or dead, vaginal delivery may occasionally be appropriate.

With improved ultrasound diagnosis, many authorities suggest elective caesarean section without prior digital confirmation. This approach has considerable merit, since digital examination may cause serious haemorrhage. Digital examination in the operating theatre does have a place when ultrasound is not available, the ultrasound appearances are equivocal, or the clinical signs of placenta praevia are not confirmed by ultrasound. If digital examination is indicated, it should be carried out only in the operating room, with the staff scrubbed and prepared for immediate caesarean section should catastrophic haemorrhage be provoked.

4 Bleeding of uncertain origin

Haemorrhage of undetermined or uncertain origin is the most common type of antepartum haemorrhage. Although in some cases the cause of bleeding later becomes clear, in the majority no cause can be demonstrated. The importance of this subgroup lies in its frequency, in the clinical problems that it presents in diagnosis and management, and in the associated high fetal loss.

Haemorrhage of uncertain origin is a collective clinical category and must include minor but unrecognized cases of all specific types of antepartum haemorrhage — localized abruption, marginal haemorrhage, cervical and vaginal lesions, and excessive show.

The clinical presentation of bleeding of unknown origin is painless antepartum haemorrhage without evidence of placenta praevia. In the majority of cases the blood loss is not great enough to cause serious concern and usually settles spontaneously. The most serious threat to the fetus is preterm labour and delivery.

The management of painless antepartum haemorrhage depends primarily on the gestational age of the fetus at the time of the initial bleed. An ultrasound examination for placental localization should be performed as soon as possible, and if placenta praevia is diagnosed or cannot be excluded, then management should follow the plan already discussed for placenta praevia. When the placenta is clearly defined in the upper segment, the woman should be allowed home after a period of rest and observation in hospital, provided that she has no recurrence of bleeding. The risk to the fetus in such cases is preterm delivery, and the great majority of perinatal deaths are due to preterm delivery occurring within seven to ten days of the initial haemorrhage. After the bleeding has settled, and before discharge from hospital, both a speculum examination to exclude a local cause for the bleeding and a digital examination to exclude advanced cervical dilatation should be performed.

If a policy of expectant management is adopted, fetal well-being should be monitored. Although frequently recommended, routine induction of labour at 38 weeks should not be performed and women should be allowed to go into spontaneous labour.

5 Conclusions

Bleeding in the second half of pregnancy constitutes a possibly life-threatening condition. All professionals who care for women during pregnancy and childbirth must be aware of the causes and prognosis of such bleeding, and have a clear plan in mind for its differential diagnosis and management.

This chapter is derived from the chapter by Robert Fraser and Robert Watson (37) in EFFECTIVE CARE IN PREGNANCY AND CHILDBIRTH.

References to primary sources and more complete data for statements made in this chapter can be found in the source chapter.

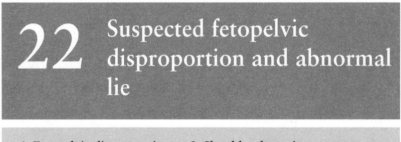

22 Suspected fetopelvic disproportion and abnormal lie

1 Fetopelvic disproportion   2 Shoulder dystocia
3 Breech presentation   4 External cephalic version for breech
presentation   5 Oblique and transverse lie   6 Conclusions

1 Fetopelvic disproportion

Fetopelvic disproportion exists when the capacity of the woman's pelvis is insufficient for the safe vaginal delivery of the baby. In the past, antenatal strategies to diagnose fetopelvic disproportion received considerable attention, with the objective of early induction of labour or delivery by planned caesarean section if disproportion was diagnosed. Attempts to predict the occurrence of fetopelvic disproportion have included measurement of maternal height and shoe size, and clinical and X-ray pelvimetry.

Maternal height has limited value for predicting fetopelvic disproportion, although short women tend to have a higher rate of caesarean section. The statistical correlations between maternal height or shoe size and cephalopelvic disproportion are of limited clinical use because of the large overlap in obstetrical outcome between small and large women. There is a reasonable correlation between clinical and radiological assessment of pelvic dimensions, but neither are particularly accurate in predicting the outcome of labour. The effects of clinical pelvimetry have not been evaluated by randomized studies.

The meaning of non-engagement of the fetal head near term as an indicator of cephalopelvic disproportion is not well established. In black women having their first babies, in whom such non-engagement commonly occurs, it is associated with longer labours but not with operative delivery or increased maternal or fetal morbidity. The most reliable predictor of pelvic adequacy remains the history of the uncomplicated birth of a baby of similar or greater birthweight than that estimated for the current pregnancy.

The formerly widespread use of X-ray pelvimetry to predict cephalopelvic disproportion has come under critical scrutiny since the reports of an association between prenatal irradiation and childhood leukaemia. As a general principle, irradiation that is not likely to benefit mother or baby should be avoided. The use of ultrasound for pelvimetry has been reported, but has not been widely adopted.

There is wide variation among individuals and institutions in the use of X-ray pelvimetry. Its utility has been questioned because of its poor predictive value as a screening test for cephalopelvic disproportion and the infrequency with which the results influence management. Two prospective randomized studies of X-ray pelvimetry failed to detect any clear benefit to either mother or baby. Both trials, and their combined results, show a substantial increase in the rate of caesarean section with the use of X-ray pelvimetry.

Computed tomographic pelvimetry, which greatly reduces the radiation exposure to the fetus, has been found in two small studies to be easier to perform and, in the measurement of a model pelvis, probably more accurate than conventional X-ray pelvimetry.

Neither X-ray nor clinical pelvimetry have been shown to predict cephalopelvic disproportion with sufficient accuracy to justify elective caesarean section with cephalic presentations. Cephalopelvic disproportion is best diagnosed by a carefully monitored trial of labour, and X-ray pelvimetry should seldom, if ever, be necessary. It is important to distinguish between the risk of disproportion related to the presenting fetal part and that related to a non-leading part of the fetus because of the special hazards of obstructed labour to the partly born fetus, as in shoulder dystocia, or obstruction to the after-coming head in a breech presentation.

2 Shoulder dystocia

Shoulder dystocia (where the baby's head is delivered, but the shoulders remain stuck) is an obstetric emergency associated with considerable risk of fetal trauma. If a high likelihood of the condition could be predicted antenatally it could be prevented by caesarean section. Caesarean section has been suggested for diabetic women with a fetal weight estimated to be above 4000 grams and for non-diabetic women with an estimated fetal weight above 4500 grams and slow progress of labour. Such a policy has not been subjected to prospective evaluation. Even if fetal weight estimation were accurate, this would prevent only a small proportion of potential cases of shoulder dystocia, while the majority of the caesarean sections would be unnecessary. Almost half of the cases of shoulder dystocia occur in infants weighing less than 4000 grams.

There is at present no reliable method of antenatal prediction of shoulder dystocia, and efforts to reduce the problem should be directed towards ensuring that birth attendants are skilled in the management of this condition when it occurs. The woman should be placed in the lithotomy position with exaggerated flexion of her thighs and her knees pushed against her chest. A large episiotomy should be done. Attempts to deliver the baby with downward traction with or without suprapubic pressure are then often successful. Rarely, deliberate fracture of the baby's clavicles may be necessary. Recent reports of novel methods of management require further evaluation. These include squatting, symphysiotomy, and even pushing the head back up followed by caesarean section.

3 Breech presentation

Breech presentation (where the baby's bottom, foot, or feet present instead of the head), although sometimes associated with uterine or fetal abnormalities, is usually simply an error of orientation that places a healthy mother and a healthy baby at risk. The prevalence of breech presentation decreases from about 15 per cent at 29–32 weeks' gestation to between 3 and 4 per cent at term. Spontaneous changes from breech to cephalic presentation occur with decreasing frequency as gestational age advances in the third trimester.

The risk to the mother is an increased likelihood of caesarean section, and the risk to the baby is the hazard of breech delivery. There is little doubt that the perinatal outcome following breech birth is less favourable than for cephalic presentations. The widespread use of caesarean section for term breech presentation has been based on the unproven assumption that this poorer outcome is at least in part the result of damage sustained during vaginal breech delivery, rather than a reflection of an intrinsically poorer outlook for babies presenting by the breech irrespective of the method of delivery. Because of the considerable contribution of elective caesarean section for breech presentation to caesarean section rates in many areas, it is important to determine whether caesarean section for term breech presentation reduces important measures of adverse perinatal outcome.

Cephalopelvic disproportion, if present, is perilous in a breech presentation because of the risk of entrapment of the after-coming head. Several recent studies have confirmed the safety of vaginal breech delivery in women carefully selected to exclude cephalopelvic disproportion using X-ray pelvimetry and usually estimation of fetal weight, assessment of neck flexion, and exclusion of double footling presentation. Such studies in no way establish that X-ray pelvimetry contributes significantly to the selection of women for either vaginal or abdominal delivery. For this, a randomized trial comparing women with and without pelvimetry would be required. No such trial has been carried out.

There is no reason to expect that X-ray pelvimetry should be more accurately predictive of cephalopelvic disproportion for breech than it is for cephalic presentations. In breech presentations, however, it is usual to avoid vaginal delivery when pelvic dimensions are even slightly reduced, in an attempt to ensure exclusion of all potential cases of cephalopelvic disproportion. Careful clinical assessment of fetopelvic relationships should be possible even in advanced labour, with the use of betamimetics to temporarily delay the progress of labour if necessary. If the frank or complete breech passes easily through the pelvis, the head can be expected to follow without difficulty.

The place of elective caesarean section for breech presentation at term is still unclear. The two small randomized trials that have been performed suggest that a policy of elective caesarean section for term breech presentation is associated with some decrease in short-term neonatal morbidity and an increase in maternal morbidity. The numbers studied are far too small to address the important issues of perinatal mortality or even low Apgar scores, nor has the question of long-term neonatal morbidity been addressed. A key factor is the experience of the attendant. The use of caesarean section for breech delivery in the belief that it is safer may become a self-fulfilling prophecy as attendants become less skilled at breech delivery.

The rare condition of deflexion of the head with a breech presentation may add additional risk to vaginal delivery. Ultrasound examination in early labour can lead to rapid diagnosis or exclusion of the condition.

4 External cephalic version for breech presentation

Over the years, many postural techniques have been used by midwives, doctors, and traditional birth attendants to turn the baby so that the head presents (cephalic version). However, little has appeared in the medical literature on this subject. A midwifery technique claimed to be effective is to ask the mother to lie on her back, with hips slightly elevated and hips and knees flexed, and to gently roll through 180° from side to side for ten minutes, repeating this three times a day. Another example of traditional practice is to attempt to correct abnormal presentations during labour by manually shaking the uterus while the mother is in the knee–elbow position on the floor.

Uncontrolled studies have suggested that use of the knee–chest position is associated with a high rate of spontaneous version and normal vaginal birth. The woman is instructed to kneel with her hips flexed slightly more than 90°, but with her thighs not pressing against the abdomen, while her head, shoulders, and upper chest lie flat on the mattress. This is done for 15 minutes every two waking hours for five days. The small reduction in non-cephalic births found in controlled studies, and the marginally lower overall caesarean section rate and rate of low Apgar score at 1 minute, could all be the result of chance. They are consistent with anything between a large positive and a large negative effect. Therefore the studies must be regarded as too small to establish whether or not postural management is effective.

There are fundamental differences between external cephalic version (where the attendant attempts to turn the baby by rotating its body through the mother's abdominal wall) attempted before term and at term, and these two approaches must be considered separately.

While external cephalic version before term can be readily accomplished, it serves little purpose. A high proportion of fetuses revert to breech presentation. Three randomized controlled trials failed to demonstrate any effect of external cephalic version before term on the incidence of breech birth, caesarean section, or perinatal outcome. The close to 1 per cent fetal mortality rate associated with the procedure led to a decline in its use.

Interest in external cephalic version for breech presentation was revived by a 1975 report that cephalic version could be achieved in the majority of cases after 37 weeks' gestation, provided that the uterus was relaxed with betamimetic agents when necessary. Delay of external cephalic version attempts until term allows time for a maximal number of spontaneous versions to take place and for any obstetrical complications that may require delivery by caesarean section to become apparent. Thus, by waiting until term, fewer unnecessary attempts at external cephalic version are required. At term, complications of the version can be readily managed by prompt abdominal delivery of the mature infant. Following successful external cephalic version, fewer reversions to breech presentation occur. The only disadvantage of delaying external cephalic version until term is that the opportunity to attempt external cephalic version may be missed in women whose membranes rupture or whose labour commences before term.

In contrast with external cephalic version before term, external cephalic version at term shows far better results. The procedure, with or without tocolysis, results in a reduction of over 80 per cent in the odds of a non-cephalic presentation at birth, and a reduction of over 50 per cent in the rate of caesarean section. Whether or not routine tocolysis should be employed is not clear. In studies that have made a direct comparison of version with and without tocolysis, little difference was found in either the immediate success of the attempted version or the prevalence of non-cephalic presentation at birth. Based on the limited evidence available at present, it would seem reasonable to restrict the use of tocolysis for external cephalic version at term to those women in whom an initial attempt without tocolysis is unsuccessful. The relative benefits of routine tocolysis in terms of possibly reducing the force required for success, and the possible risks of maternal cardiovascular side-effects, need to be addressed.

External cephalic version in early labour is worthy of consideration as an extension of the trend towards version later in pregnancy. Earlier reports have included occasional references to attempted external cephalic versions during labour and success rates up to almost 75 per cent have been reported.

In theory this approach has several advantages. Maximum time would be allowed for spontaneous version to take place and for possible contraindications to external cephalic version to appear, thus limiting the number of versions that would be necessary. The risks of external cephalic version may be reduced further by performing the procedure in the labour ward, with continuous monitoring of the fetal condition until delivery. In cases assessed as unsuitable for vaginal breech birth, the external cephalic version may be attempted in the operating theatre, and in the event of failure followed immediately by caesarean section.

Because waiting for the onset of labour would involve the inconvenience of performing version as an emergency rather than as an elective procedure, and because the breech may be well engaged in some cases, it is unlikely that external cephalic version during labour will become a first-line approach for breech presentation. However, when breech presentation is encountered during labour while the membranes are still intact, the limited data available suggest that external cephalic version with tocolysis is a reasonable procedure to consider.

The risk of attempted external cephalic version to the mother is exceedingly small. It consists of the possibility of adverse effects from any of the drugs used to facilitate version and the hazards of placental abruption, a rare but recognized complication.

Attempted external cephalic version must be recognized as an invasive procedure that involves some risk to the fetus. Fetal risks are related to gestational age and to the methods employed. The complication rate is greater when external cephalic version is attempted before 37 weeks' gestation, when general anaesthesia is employed, and when the placenta is situated anteriorly. Provided that fetal well-being is confirmed and monitored, and provided that appropriate precautions are observed (including administration of anti-D immunoglobulin to rhesus-negative women), the risk to the mature fetus appears to be small.

5 Oblique and transverse lie

Breech presentation is an error of polarity in which the forces that maintain a longitudinal lie of the fetus are intact. Oblique and transverse lies are the result of an entirely different situation in that the fetus fails to adopt a longitudinal lie. They are associated with multiparity, abdominal laxity, uterine and fetal anomalies, shortening of the longitudinal axis of the uterus by fundal or low-lying placenta, and conditions that prevent the engagement of the presenting part, such as pelvic tumours and a small pelvic inlet. Abnormal lie, particularly oblique lie, may be transitory and related to maternal position.

When non-longitudinal lie is encountered after 32 weeks' gestation, underlying abnormalities should be sought. Further care options include antenatal attempts at external version, version at term followed by induction of labour, or expectant care with or without intrapartum attempts at version, if the abnormal lie persists.

The role of external version in the management of oblique and transverse lie has not been assessed in a randomized trial. A number of descriptive case series have been reported. With expectant care most cases of abnormal lie will revert to the longitudinal lie by the time of delivery. Less than 20 per cent of transverse lies observed after 37 weeks' gestation persist to delivery. Given the high spontaneous version rate and the unstable nature of the non-longitudinal lie, with a high probability of reversion following external version, there is no firm case for external version prior to labour or planned delivery.

Care of the woman with a transverse or oblique lie encountered during labour is more straightforward, as the choice lies between caesarean section and external version. A prospective but uncontrolled study of external version for transverse lie in labour showed a modest rate of success. There were no fetal or maternal complications associated with the procedure, although larger studies would be needed to evaluate potential risks.

In the absence of controlled trials, the timing of intervention in pregnancies complicated by non-longitudinal lie must remain a matter of clinical judgement. The advantage of gaining fetal maturity, allowing time for spontaneous version to take place and allowing labour to begin spontaneously, must be weighed against the risk of membrane rupture or cord prolapse before the version can be attempted. Once labour has begun or the decision has been made to deliver the baby, attempted version with immediate recourse to caesarean section, if necessary, is a reasonable option.

6 Conclusions

No reliable methods are available for accurate prediction of fetopelvic disproportion before labour. Labour is the best test of pelvic adequacy in cephalic presentations.

At present there is inadequate evidence either to substantiate or refute the effectiveness of caesarean section for term breech presentation in improving perinatal outcome. The place of pelvimetry in breech presentation has not been established and a trial determining its benefits, if any, is warranted if this common practice is continued.

The controlled trials reported to date are too small to support or refute the alleged benefits of postural management for breech presentation. External cephalic version for breech presentation before term is not warranted. However, external cephalic version at term substantially reduces the incidence of breech presentation at birth and of caesarean section.

This chapter is derived from two chapters by G. Justus Hofmeyr (31, 42) in EFFECTIVE CARE IN PREGNANCY AND CHILDBIRTH.

References to primary sources and more complete data for statements made in this chapter can be found in the source chapters and/or in the following reviews from the *Cochrane pregnancy and childbirth database*:

Fraser, W.D.
— X-ray pelvimetry in cephalic presentations. Review no. 03229.

Hofmeyr, G.J.
— Cephalic version by postural management. Review no. 03393.
— External cephalic version before term. Review no. 02877.
— External cephalic version at term. Review no. 03087.
— Routine tocolysis for external cephalic version at term. Review no. 04026.
— Planned elective caesarean section for term breech presentation. Review no. 05287.

23 Prelabour rupture of the membranes

1 Introduction

Prelabour rupture of the membranes is defined as spontaneous rupture of the membranes before the onset of regular uterine contractions. It is often referred to as 'premature rupture of the membranes', but 'prelabour rupture' is a more precise and appropriate expression because of the diverse connotations of the word 'premature'. When this condition occurs before 37 weeks it is referred to as 'prelabour rupture of the membranes preterm'; at or after 37 weeks, as 'prelabour rupture at term'. The distinction, while arbitrary, is important for both prognosis and care.

2 Diagnosis

2.1 *Ruptured membranes*

The diagnosis of rupture of the membranes is in many instances obvious from the sudden gush of clear amniotic fluid from the vagina and its continued dribbling thereafter. If the rupture has occurred recently, it will often be possible to collect some fluid by asking the woman to sit on a suitable receptacle. Alternatively, it may be possible to obtain a sample from a pool of amniotic fluid in the posterior fornix on speculum examination.

The nitrazine test is probably the most widely used test for differentiating liquor amnii from other body fluids, but it has a false positive rate of about 15 per cent. For this reason an additional test, usually microscopic observation of ferning, is worthwhile. The fern test is less likely to produce false positive results, although it has a higher rate of false negatives.

If the rupture has occurred some hours previously and most of the fluid has escaped from the vagina, it may be difficult or impossible to establish or confirm the diagnosis with any degree of confidence. In these circumstances, much depends on taking a careful history from the woman. Information can be obtained as to when and how the gush of fluid occurred, whether anything like it has ever happened before, approximately how much fluid was lost, what colour it was, whether it smelled of anything, and whether there was anything else remarkable. The latter question may elicit a comment on the presence of white or greasy particles. The ultrasound finding of oligohydramnios is strong confirmatory evidence of prelabour rupture of membranes when there is a history of sudden release of fluid.

It is not clear whether or not high (hindwater) rupture of the membranes should be considered as clinically distinct from low rupture. There is little information about how these two types of rupture can be differentiated, or whether or not they warrant different forms of care. In the absence of such data, the only practical approach is to consider them as equivalent.

2.2 Vaginal examination

It is likely that vaginal examinations can introduce or increase the risk of intrauterine infection, although no controlled comparisons have been conducted to substantiate or refute this belief. The only reason to perform a vaginal examination would be to obtain information that would be useful in determining further care and that cannot be obtained in a less invasive way. There is probably little benefit to be derived from performing both digital and speculum examinations. The information that can be obtained by speculum examination, which may include visualization of amniotic fluid 'pooling' in the posterior fornix, collection of some of that fluid for nitrazine test, microscopic examination for ferning, or phosphatidylglycerol determination, and the collection of material for culture or screening for group B streptococci, is likely to be superior to that obtained by digital examination. However, speculum examination is likely to be more unpleasant for the mother than digital examination and is unlikely to provide much useful information if the membranes have been ruptured for some time. Unfortunately, no controlled comparisons have been conducted to establish the benefit, if any, of either digital or speculum examination.

2.3 Assessing the risk of infection

Any woman with prelabour rupture of the membranes should be assessed for signs of intrauterine infection. These include fever and maternal or fetal tachycardia. If any one of these is accompanied by a tender uterus and foul-smelling liquor there will be no doubt about the diagnosis.Uterine tenderness and fetid discharge, however, are late signs of intrauterine infection.

The earliest clinical signs of intra-amniotic infection are fetal tachycardia and a slight elevation of maternal temperature, but both these signs are rather non-specific. A few years ago, there were hopes that the estimation of C-reactive protein in the maternal circulation might be reasonably reliable as an early sign of intrauterine infection, but these hopes have not materialized, nor has the value (if any) of using C-reactive protein estimation ever been assessed in controlled comparisons.

Although intrauterine infection may on occasion precede rupture of the membranes, the main risk is infection ascending from the vagina into the uterine cavity. Therefore information on the presence of pathogens may be useful, particularly about those organisms that are responsible for the majority of fetal infections, such as the group B streptococci, *Escherichia coli*, and *Bacteroides*. Intrapartum antibiotic treatment of mothers carrying group B streptococci in the vagina reduces the incidence of neonatal sepsis and neonatal death from infection.

In populations with a high prevalence of group B streptococci carriers, either screening for the organism or routine antibiotic treatment should be adopted as standard care. In all other populations, an initial culture should be part of the care provided after prelabour rupture of the membranes preterm.

Amniocentesis has been advocated to assess the risk of infection, particularly in the preterm period. Since the main risk is ascending infection, amniocentesis in order to obtain culture data directly from within the uterus would seem to be a pointless exercise. Additional problems with amniocentesis include failure to obtain amniotic fluid in a significant proportion of cases, the invasiveness and risks associated with the procedure itself, and, most importantly, the poor correlation between the results of diagnostic tests applied to the liquor and the development of fetal infection. Bacteria are not found in all women with clinical signs of intra-amniotic infection, nor are they always absent in women without signs of infection. Some studies have suggested that white cells in the amniotic fluid are more predictive of infectious morbidity than bacteria, but this has not been confirmed by others.

There has been only one controlled trial to assess the value of amniocentesis for detection of intrauterine infection by Gram stain and culture after prelabour rupture of the membranes preterm. The use of amniocentesis did not reduce the proportion of women delivered because of clinical amnionitis or the number of perinatal deaths, but it did result in a reduced frequency of abnormal fetal heart patterns in labour and a reduction in the average number of days that the infants remained in hospital after the mother had been discharged. On the whole, there is inadequate evidence to judge whether the use of amniocentesis in women with prelabour rupture of the membranes preterm confers more benefit than harm to mother and baby.

A number of observational studies suggest that fetal breathing movements and gross body movements cease when intra-amniotic infection develops. These changes in fetal behaviour may prove to be as reliable for the detection of intrauterine infection as the more invasive technique of amniocentesis, but they too require further assessment.

2.4 *Assessing the risk of fetal immaturity*

Assessing the risk of fetal immaturity largely depends on a careful assessment of gestational age and on ascertainment of any such assessments made earlier in pregnancy. There will usually be little difficulty in identifying fetuses who are either profoundly immature or who are clearly close to term. Between these two categories, and most typically between 26 and 34 weeks of gestation, weighing the risk of relative immaturity is considerably more difficult.

Pulmonary maturity, as determined by analysis of amniotic fluid, is associated with a decreased risk of mortality and morbidity from respiratory disorders, but does not necessarily imply that other hazards associated with preterm birth, most notably periventricular haemorrhage, will be avoided.

3 Prelabour rupture of the membranes preterm

3.1 *Risks*

The most serious and most common consequence of prelabour rupture of the membranes preterm is preterm delivery. The risk associated with this is directly related to the gestational age and maturity of the fetus. At the extreme lower end of the gestational age range, improvement in perinatal outcome will depend entirely on maintaining the pregnancy. At the other end of the preterm gestational age range, care policies should differ little, if at all, from those that apply after rupture of the membranes at term. Prelabour rupture of the membranes at gestational ages between these two extremes (roughly between 26 and 34 weeks) presents one of the greatest dilemmas in obstetrical care.

Infectious morbidity, mostly due to ascending intrauterine infection, is the second most important hazard for the baby. This risk is also greater at lower gestational ages, possibly because of the relative immaturity of antibacterial defence mechanisms as well as underdeveloped bacteriostatic properties of amniotic fluid at early gestational age.

In addition to these two main risks, other hazards of prelabour rupture of the membranes before term include pulmonary hypoplasia and various deformities associated with persistent oligohydramnios; placental abruption; umbilical cord complications, either immediately or when labour supervenes; and the mechanical difficulties (if caesarean section becomes necessary) of delivering a baby from a uterus that contains little, if any, amniotic fluid and has a poorly developed lower segment.

3.2 Care before the onset of labour

The first decision in the care of a woman with prelabour rupture of the membranes preterm is whether the primary objective should be to effect delivery or to prolong pregnancy. Most women with prelabour rupture of the membranes in the preterm period will deliver within a week of the rupture.

Preterm delivery is *the* main consequence of prelabour rupture of the membranes preterm. Where adequate facilities for intensive perinatal and neonatal care are lacking, the most effective form of care is referral to a centre where such facilities are readily available. It is possible that leakage will stop and that amniotic fluid will accumulate again; however, this is the exception rather than the rule.

Four forms of care for women whose membranes ruptured preterm, and in whom there is no evidence of either uterine contractions or infection, have been evaluated in controlled comparisons. These are prophylactic antibiotics, prophylactic tocolytics, corticosteroids to promote fetal pulmonary maturity, and induction of labour. Some of the controlled comparisons have evaluated combinations of these interventions.

3.2.1 Prophylactic antibiotics for prelabour rupture of the membranes preterm

Treatment with prophylactic antibiotics is associated with statistically significant reductions in the risk of preterm delivery occurring within one week, of infection in the mother before delivery, and of infection in the baby. One would expect that this reduction in neonatal infection, coupled with the improved fetal maturity achieved by prolonging the time to delivery, would increase neonatal survival.

However, controlled trials have so far failed to show an effect on perinatal mortality.

3.2.2 Prophylactic tocolytics

The question of whether the administration of tocolytic drugs might improve outcome in women not in labour after prelabour rupture of the membranes preterm has been addressed in two small studies of the use of oral ritodrine. There was no difference in the proportion of women who delivered within ten days and the number of women involved was too small to assess the influence, if any, on neonatal infection, respiratory distress syndrome, or perinatal mortality.

These data, and data from other placebo controlled trials on the prophylactic use of betamimetic drugs in women without ruptured membranes, offer no support for suggestions that prophylactic tocolysis before the onset of uterine contractions is worthwhile in women with prelabour rupture of the membranes.

3.2.3 Corticosteroid administration

The theoretical concerns that corticosteroids might be both superfluous and hazardous in women with prelabour rupture of the membranes have not been confirmed by the results of controlled trials. These concerns were based on the observation that rupture of the membranes *per se* may enhance fetal pulmonary maturity, and thus make the use of other agents to promote maturity unnecessary. In addition, there were fears that the immunosuppressive effects of corticosteroids might both increase susceptibility to intrauterine infection and mask early signs of infection in women with prelabour rupture of the membranes.

A systematic review of the randomized trials of corticosteroid administration to women with prelabour rupture of the membranes preterm provides a clear response to these concerns. Irrespective of any effects that prelabour rupture of the membranes itself may have on fetal pulmonary maturity, the incidence of respiratory distress syndrome is further reduced to an important extent by corticosteroid administration. The incidence of neonatal infection was not statistically significantly higher in the corticosteroid treated group than in the control group among the seven trials that reported this outcome, either individually or combined.

Although there is no statistically significant increase in neonatal infection following corticosteroid therapy, this remains a possible risk. The concomitant use of antibiotics may be appropriate.

3.2.4 Induction of labour

Controlled comparisons between an 'active' policy intended to effect delivery after prelabour rupture of the membranes preterm and a control policy in which no such measures were taken do not suggest any protective effect of the more active policy in respect of either maternal sepsis or infant outcome measures, such as neonatal infection, neonatal sepsis, respiratory distress syndrome, intracranial haemorrhage, and perinatal death from causes other than congenital malformations. Indeed, there is a tendency for outcomes to be less favourable in the actively managed group. Even when corticosteroids to improve pulmonary maturity are used together with induction of labour, the active policy confers no benefit on the infant. The beneficial effects of antenatal corticosteroids in prelabour rupture of the membranes preterm appear to be offset by curtailing the duration of pregnancy with elective delivery.

3.2.5 *Surveillance* Most women will go into labour within hours or a few days after rupture of the membranes. In some women, however, labour will be delayed much longer. Among these will be some in whom the diagnosis of ruptured membranes was made erroneously, and some in whom a high leak may have sealed over.

Provided that mother and fetus are well at the initial assessments, the main concerns in the first few days after rupture of the membranes should centre on detecting the onset of infection or of uterine contractions. This can be accomplished by regular assessments of maternal temperature and pulse, uterine contractility, and fetal heart rate. It is not clear whether regular determinations of leucocyte counts add anything to this surveillance. Variation in leucocyte counts can be quite large, particularly when the influences of labour or corticosteroid administration are added.

Other elements of surveillance are guided by the other complications which are known to occur more frequently after prelabour rupture of the membranes. These are primarily related to compression or prolapse of the umbilical cord, placental abruption, and the development of fetal deformities and lung hypoplasia.

Prolapse of the umbilical cord is a well-recognized complication of prelabour rupture of the membranes. It may occur either at the time of membrane rupture or later with the onset of labour. Any change in an apparently stable situation, such as a resuming loss of liquor or the onset of uterine contractions, should alert the caregiver to this possibility. Frequent assessments of the fetal heart rate, either by auscultation or cardiotocography, and a careful ultrasound examination may be useful in these circumstances.

Compression of the umbilical cord may occur owing to the loss of the protective effect of amniotic fluid. The risk of local increase in pressure escalates with the onset of uterine contractions, and the incidence of severe fetal heart rate decelerations is directly related to the degree of oligohydramnios.

The risk of placental abruption should be considered whenever blood loss or abdominal pain occur in a woman with ruptured membranes preterm.

Prolonged rupture of the membranes with oligohydramnios for several weeks may lead to the development of a spectrum of fetal postural and compression abnormalities. Pulmonary hypoplasia, the most dreaded of these complications, becomes the rule rather than the exception when oligohydramnios has been severe enough and of long enough duration to cause positional deformities. It is not clear whether the presence of fetal breathing movements indicate that lung growth is preserved; the reported observational studies have yielded conflicting results.

Renewed accumulation of amniotic fluid may imply that the woman can return home with a reasonable degree of safety, although this has never been assessed in a controlled comparison.

3.3 Care *after the onset of labour*

The onset of uterine contractions may be the result of intrauterine infection. If this is the case, labour should be allowed to proceed to delivery as swiftly as possible in the interests of both mother and fetus, irrespective of the latter's degree of maturity. Decisions about the method of delivery and whether caesarean section will be necessary should differ little from those for other preterm deliveries.

3.3.1 *Antibiotics*

Whether or not antibiotics should be administered at once in these circumstances has not been addressed by controlled studies. Both clinical common sense and data from studies that have compared intrapartum with immediate postpartum antibiotic treatment for intra-amniotic infection tend to support the use of antibiotic treatment as soon as a clinical diagnosis of intrauterine infection is reached.

Whether or not the fetus is mature, preterm labour starting after the membranes have been ruptured for some time should probably be allowed to proceed to preterm delivery. Considerations such as the anticipated duration of neonatal hospitalization, infant–mother separation, and the costs of neonatal care should be less influential than the probability that the risk of infection is likely to be increased if other care options are chosen. When the fetus is so immature as to have no chance of extrauterine survival, attempts to prolong pregnancy should depend not only on what might be gained in terms of infant outcome, but even more on the maternal risks and the opinions of the parents.

3.3.2 *Tocolysis*

The small controlled trials that have compared tocolysis with no tocolysis in preterm labour following prelabour rupture of the membranes show no statistically significant differences in any of the outcome measures examined. These include delay of delivery, recurrence of preterm labour, preterm delivery, birthweight, mortality, and respiratory morbidity.

There is no evidence that tocolytic agents *per se* improve perinatal outcome, and they are not innocuous. They should be used only when the inhibition of labour permits the implementation of other measures that are known to be effective in improving outcome for the preterm infant, such as administration of corticosteroids or transfer of the mother to a centre with adequate facilities for preterm delivery and care of the preterm infant.

3.3.3 *Amnioinfusion* Decelerations of the fetal heart rate during preterm labour occur more frequently if the membranes have ruptured before the onset of labour. Many of the abnormal fetal heart rate patterns are suggestive of umbilical cord compression and may well be due to the loss of the protective effect of amniotic fluid. The relative merits and hazards of using amnioinfusion during labour in women with prelabour rupture of the membranes preterm have not yet been thoroughly assessed. Thus far, available data suggest a statistically significant reduction in the number of mild, moderate, and severe variable decelerations per hour with amnioinfusion. The merits (and hazards) of this approach warrant further evaluation.

4 Prelabour rupture of the membranes at term

Most women with prelabour rupture of membranes at term will go into labour soon after the membranes rupture. Almost 70 per cent of these women will deliver within 24 hours, and almost 90 per cent will do so within 48 hours. A remarkably constant 2–5 per cent will be undelivered after 72 hours, and almost the same proportion will remain undelivered after seven days. It is possible that these women may have a deficiency in prostaglandin production or in their prostanoid biosynthesis pathway, and that this is responsible not only for their failure to go into spontaneous labour but also for the frequently observed poor progress in cervical dilatation when labour is induced with oxytocin.

The main concerns related to prelabour rupture of membranes occurring at term are maternal and neonatal infection and an increased incidence of caesarean section. The increased caesarean section rate may be because induction of labour fails to achieve sufficient progress in cervical dilatation, because of decelerations of the fetal heart rate (possibly related to lack of liquor and cord compression), or because of underlying dystocia (which may have been the cause of the prelabour rupture of the membranes).

Infection, both maternal and neonatal, has been the main complication to worry obstetricians. Reports in the 1950s showed that prelabour rupture of the membranes at term was associated with maternal and fetal or neonatal infection and a high risk of maternal and perinatal mortality. A controlled trial conducted at that time showed that antibiotics in labour resulted in a statistically significant reduction in perinatal mortality, but two subsequent trials conducted in the 1960s failed to show this effect.

The prognosis of prelabour rupture of the membranes at term has changed considerably in the second half of this century. Data collected over long periods, some of which include women who delivered more than twenty years ago, are of questionable relevance to current obstetric practice. The maternal deaths noted at that time mostly occurred in women with prolonged severe intrauterine infection who often had inadequate antibiotic therapy judged by today's standards. Some received no care at all before becoming moribund. Maternal mortality is almost never found in more recently conducted studies of prelabour rupture of the membranes at term, and perinatal death from infection has also become a rarity.

4.1 *Induction of labour or expectant care for prelabour rupture of the membranes at term*

The widespread practice of early induction of labour after a diagnosis of prelabour rupture of membranes at term arose from concern about maternal and fetal infection.

Policies of induction of labour with oxytocin at 34+ weeks (when the main dangers of pulmonary immaturity have been overcome), compared with an expectant policy, are associated with a higher caesarean section rate but a lower risk of neonatal infection. The possibility of bias in ascertainment of infectious outcomes leaves some uncertainty about the nature and extent of these effects, and further research is needed to clarify them. Trials of inductions involving the use of prostaglandins with or without oxytocin (as opposed to inductions using oxytocin alone) suggest that with prostaglandins the caesarean section rate is similar to that for expectant management, although the reduction in neonatal infection noted is not statistically significant. Trials of induction after prelabour rupture of the membranes at 37+ weeks show similar differential effects as found in the trials of induction after 34 weeks.

Thus active management may have different effects if prostaglandins are used than if they are not. Direct comparisons of induction with prostaglandin versus induction with oxytocin show that the use of prostaglandins results in a lower rate of caesarean section than does the use of oxytocin alone. It does not appear to be associated with higher or lower risks of other adverse maternal or fetal outcomes, although the studies carried out have not been large enough to excluded differences in either direction.

Questions about the length of labour, analgesia, and the degree of discomfort experienced by the woman have not been adequately addressed, although one trial reported a much higher use of analgesia with the active policy. For some trials the duration of labour could be estimated from the data given. The interval from time of membrane rupture to delivery was reduced from an average of about 30 hours to about 24 hours by a policy of induction, but the duration of labour averaged 15 hours in the induced group versus 6 hours in the group with expectant care.

The evidence thus suggests that a policy of induction of labour with oxytocin for prelabour rupture of the membranes at term exposes the mother to a longer and probably less comfortable labour, and to a higher risk of operative delivery and caesarean section. This risk is less if prostaglandin is chosen as the method of induction. Induction for prelabour rupture of the membranes at term appears to reduces the rate of neonatal infection, but further data are required to determine whether this effect is real or an artefact of biased ascertainment.

4.2 Prophylactic antibiotics for prelabour rupture of the membranes at term

The two controlled trials of prophylactic antibiotics conducted in the 1960s involved a total of more than 2000 women with prelabour rupture of the membranes at variable gestational ages, most of whom were at term. These trials utilized antibiotic treatments that are no longer used. Although they failed to show any effect on the incidence of fetal and neonatal infection, a statistically significant reduction in the incidence of infectious maternal morbidity postpartum was observed after antibiotic prophylaxis. In some centres, this has led to the adoption of policies involving routine administration of antibiotics *after* delivery to women with prolonged rupture of the membranes at term. Although the utility of this approach has not been addressed in randomized comparisons, it would be worth assessing which women, without overt signs of infection, might benefit from postpartum antibiotic prophylaxis.

One small randomized trial has addressed the question of whether prophylactic antibiotics should be given to the baby after birth in women with prolonged prelabour rupture of membranes, some of whom were at term. There was some support for the idea that prophylactic antibiotics reduced the risk of infection, but the trial requires replication on a larger sample with blind assessment of infant outcomes.

5 Conclusions

Any woman with a history suggestive of prelabour rupture of the membranes should be assessed as soon as possible. Attention should be directed to whether the membranes are indeed ruptured, to a careful review of the menstrual history and assessment of gestational age, to possible signs of incipient or established infection, to signs of fetal distress due to cord compression or prolapse, and to signs of uterine contractions.

For the woman with preterm prelabour rupture of the membranes who is not in labour, is not infected, and shows no evidence of fetal distress or other fetal or maternal pathology, continuation of the pregnancy is more likely to be beneficial than harmful. Administration of antibiotics prophylactically to women with prelabour rupture of the membranes preterm delays delivery and reduces maternal and neonatal infection. No effect has yet been demonstrated on perinatal mortality. In women known to be carrying group B streptococci, intrapartum antibiotics should be adopted as standard care (see Chapter 19).

There is no evidence that the prophylactic use of betamimetic agents before uterine contractions begin is of value in preventing the onset of preterm labour.

Because preterm delivery frequently follows prelabour rupture of the membranes in the preterm period, corticosteroids should be administered if there is no evidence of sufficient pulmonary maturity. The theoretical increase in the risk of neonatal infection is more than offset by the real benefit of reduction in respiratory distress. Although there is no evidence from controlled trials, combining antibiotics with corticosteroids is likely to do more good than harm.

The routine use of measures to effect early delivery after prelabour rupture of the membranes preterm is more harmful than beneficial. There is no evidence that this policy will reduce infectious morbidity. However, if there are signs of intrauterine infection antibiotic treatment should be started and delivery effected. A skilled neonatologist should be present at birth.

Women with prelabour rupture of the membranes at gestational ages of 34 weeks or more should be informed that their risk of having a caesarean section is higher with oxytocin induction of labour than with expectant care. Because of the potential for biased assessment in the trials that have been carried out, there is still no clear evidence as to whether or not a policy of induction will reduce the risk of neonatal infection. Further trials that are currently being conducted may address some of these uncertainties.

This chapter is derived from the chapters by Marc J.N.C. Keirse, Arne Ohlsson, Pieter E. Treffers, and Humphrey H.H. Kanhai (43), and by John Grant and

Marc J.N.C. Keirse (64) in EFFECTIVE CARE IN PREGNANCY AND CHILDBIRTH

References to primary sources and more complete data for the statements made in this chapter can be found in the source chapters and/or in the following reviews from the *Cochrane pregnancy and childbirth database*:

Crowley, P.
— Antibiotics for preterm prelabour rupture of membranes. Review no. 04391.
— Corticosteroids after preterm prelabour rupture of membranes. Review no. 04395.
— Elective delivery after preterm prelabour rupture of membranes. Review no. 04473.
— Corticosteroids plus induction of labour after PROM preterm. Review no. 06871.

Hannah, M.E.
— Active management of prelabour rupture of membranes at 34+ weeks. Review no. 07126.
— Oxytocin for prelabour rupture of membranes at 34+ weeks. Review no. 05208.
— Prostaglandins for prelabour rupture of membranes at 34+ weeks. Review no. 07154.
— Prostaglandins vs oxytocin for prelabour rupture of membranes at 34+ weeks. Review no. 07152.
— Active management of prelabour rupture of membranes at 37+ weeks. Review no. 03272.
— Prostaglandins vs oxytocin for prelabour rupture of membranes at 37+ weeks. Review no. 03273.

Hofmeyr, G.J.
— Amnioinfusion for preterm prelabour rupture of membranes. Review no. 04390.

Keirse, M.J.N.C.
— Betamimetics after preterm prelabour rupture of membranes. Review no. 04396.

24 Preterm labour

1 Introduction

Preterm birth is associated with an inordinate amount of mortality and morbidity among normally formed babies. Prevention and treatment of preterm labour is important, not as an end in itself, but as a means of preventing preterm birth.

Prevention of preterm birth is not always wise; many preterm births occur as a result of conditions such as prelabour rupture of the membranes with its inherent risk of amnionitis, or of obstetric intervention aimed at ending pregnancy before term because of problems with fetal well-being or growth. These situations are discussed in other chapters.

Preterm labour differs little physiologically from labour at term, except that it occurs too early. However, it will be accompanied by increased anxiety for the mother and her partner. It is not easy to tell whether preterm labour really has or has not commenced. In many instances apparently progressive preterm labour stops, irrespective of whether any treatment is instituted. An overly expectant attitude while watching for signs of progress can be dangerous, however, as more advanced preterm labour is more difficult to stop.

Successful suppression of uterine contractions does not necessarily improve the outcome for the infant. Birth may not be postponed to a clinically useful extent, while any treatment that is powerful enough to suppress uterine contractions may have other effects on the mother or the baby. Some of these may be undesirable or dangerous.

2 Prevention

2.1 Social interventions

There is a strong association between a woman's social and economic circumstances and her risk of preterm delivery. This association has prompted a number of social programmes aimed at reducing that risk. However well intentioned these interventions are, controlled evaluation (discussed more fully in Chapter 3) has not detected any effects on the rate of preterm birth.

2.2 Physical measures

2.2.1 *Home uterine activity monitoring* Several trials have addressed the question of whether electronic monitoring of uterine activity at home, with daily transmission to a monitoring centre by telephone, can reduce the frequency of preterm birth by early identification of women who are developing preterm labour. Unfortunately, there is enormous potential for bias in most of the trials reported, and this approach cannot be evaluated properly from the available data.

Considering the enormous cost of the intervention to the women concerned (or their medical insurers) and the notable lack of data on infant outcome, this intervention should not be used outside the context of well-controlled studies.

2.2.2 *Bed-rest* Bed-rest in the hope of reducing the incidence of preterm birth has been used predominantly in multiple pregnancies. The intervention has not been demonstrated to be effective for this purpose. The problem is addressed in more detail in Chapter 17.

2.2.3 *Cervical cerclage* Cervical cerclage is useful for preventing birth too early in pregnancy in a small proportion of women. Unfortunately, no satisfactory methods of identifying the women who are likely to benefit from this intervention have been found. Benefits are more likely to occur in women who have had two or more past pregnancies which ended too early.

The intervention should be avoided in women who are unlikely to benefit because of the potential hazards associated with its use, including those attendant on any surgery plus the additional risk of stimulating uterine contractions.

2.2.4 *Cervical assessment*   A few small trials have addressed the question of whether vaginal examination or ultrasound assessment of cervical length may help to recognize women who are likely to deliver too early in time to institute useful preventive measures. Most of these trials have been small and offer no guidance for clinical practice at this time. While some of these approaches may be promising, there are also distinct disadvantages both to the procedures themselves and to the interventions that may be precipitated by their results.

2.3 Pharmacological approaches

2.3.1 *Betamimetic drugs*   Many clinicians prescribe betamimetic drugs to prevent uterine contractions in women who, for one reason or another, are considered to be at increased risk of preterm labour. Trials of prophylactic betamimetics, both in multiple pregnancy and in women with singleton pregnancies believed to be at high risk of preterm birth, have failed to detect any reduction in the risk of preterm birth, low birthweight, or perinatal mortality.

2.3.2 *Magnesium*   The effects of routine magnesium supplementation on a number of adverse pregnancy outcomes, including preterm labour, have been addressed in a few trials. Unfortunately, there is considerable potential for bias in all of these and interpretation of the results must be guarded. While they suggest a reduction in the incidence of preterm birth, far better evidence is required before this can be accepted as a true effect. No statistically significant differences were observed in any of the infant outcomes for which the trials provide data.

2.3.3 *Calcium*   On the basis of the few data that are available, calcium supplementation during pregnancy appears to reduce the risk of preterm birth substantially. Controlled evaluations have shown beneficial effects of calcium supplementation on the incidence of hypertension and pre-eclampsia. There is still insufficient evidence for any reliable conclusions about other important outcomes such as perinatal death, failure of intrauterine growth, or caesarean section. The role, if any, for routine calcium supplementation in pregnancy is not yet clear.

2.3.4 *Progestogens*   Regular intramuscular injections of 17α-hydroxyprogesterone caproate may reduce the incidence of preterm labour and preterm birth in women considered to be at high risk of preterm labour, but they have not been shown to decrease perinatal mortality or morbidity. The findings are encouraging enough to warrant further evaluation, preferably using less invasive forms of progestogen administration.

3 Treatment of active preterm labour

3.1 *Hydration*

Hydration with intravenous fluid, with or without sedation, is frequently used as a primary approach to stopping preterm labour, particularly in North America. This approach has not been evaluated properly and the data that are available do not show it to be more useful than no treatment at all. It entails an increased risk of pulmonary oedema if labour-inhibiting drugs are subsequently used. This practice should be abandoned unless evidence is brought forward to substantiate it.

3.2 *Betamimetic drugs*

Betamimetics are used more extensively than any of the other labour-inhibiting agents that are employed to suppress uterine contractions preterm. A variety of betamimetics have been introduced in the hope of developing agents that would have a maximum effect on uterine relaxation with a minimal effect on the heart or other body organs.

Only three of the many betamimetic agents available have ever been compared with a placebo or no-treatment control group for inhibition of preterm labour. Some of the drugs that are widely used, such as salbutamol or fenoterol, have never been tested. The majority of the controlled trials dealt with ritodrine.

Data from these trials show that betamimetics reduce the proportion of deliveries that occur within the first 24 hours and within 48 hours after beginning treatment. They also reduce the incidence of delivery too early in pregnancy. No decrease in perinatal mortality or serious morbidity, such as respiratory distress syndrome, has been detected.

At least three factors may contribute to this lack of effect on important adverse outcomes. First, the trials may have included too many women who were already sufficiently advanced in gestation, so that postponement of birth and prolongation of pregnancy were unlikely to confer any substantial benefit to the baby. Second, the time gained by betamimetic drug treatment may not have been used to implement measures with direct beneficial effects such as promoting fetal lung maturity or transfer to a centre with adequate perinatal care facilities. Third, there may also be direct or indirect adverse effects of the drug treatment (including prolongation of pregnancy when this is contrary to the best interests of the baby) which counteract their potential gain.

None of the studies comparing one betamimetic drug with another has been large enough to have had a chance of detecting or excluding important differences in the outcomes that really matter. Nor have any of the trials shown any clear differences in serious maternal outcomes, such as pulmonary oedema. Taken together, the trials comparing different betamimetic agents show no reason to prefer one agent over another.

The placebo-controlled trials do not suggest that betamimetic drug treatment frequently poses great hazards to either mother or baby, but other data in the literature show that these drugs are not harmless. The most frequently observed symptoms associated with betamimetic use are palpitations, tremor, nausea, and vomiting. Headache, vague uneasiness, thirst, nervousness, and restlessness may occur.

The most common, and dose-related, side effect observed in all betamimetic treated women is an increase in heart rate. Only rarely will effective labour inhibition be achieved with maternal heart rates below 100 beats/minute. Heart rates of 130–140 beats/minute, however, should preclude further increases in the dose of betamimetics administered. Chest discomfort and shortness of breath should alert those providing care to the possibility of pulmonary congestion.

Pulmonary oedema is a well-recognized complication of betamimetics used in preterm labour. Most cases are due to aggressive intravenous hydration and to the neglect of signs of a fluid accumulation. It is safer to administer betamimetic drugs in a small volume of fluid with the use of an infusion pump than to rely on intravenous infusion of dilute solutions of the drug. Many of the cases of pulmonary oedema have been observed in women with twin pregnancies. Plasma volume expansion is larger in women with multiple pregnancies, and these women are at greater risk of developing pulmonary oedema during treatment with betamimetics than women with singleton pregnancies.

Myocardial ischaemia has been described as the other serious, although rare, complication of betamimetic drug treatment. Betamimetic drug administration in pregnancy results in a marked increase in cardiac output, of the same order as that observed in moderate exercise. The additional work imposed on the myocardium may be too much for women with pre-existing cardiac disease. These women should not be given betamimetic drug treatment, as the hazards for them are likely to be greater than any possible benefits that might be derived.

All betamimetic agents show a clear tendency to lower diastolic blood pressure. This is usually accompanied by an increase in systolic blood pressure, with the effect of a net increase in pulse pressure. Clinically significant hypotension is less frequently encountered with currently used betamimetic drugs, such as ritodrine and terbutaline, than with earlier agents, such as isoxsuprine, but the problem has not been eliminated.

Other drugs, including calcium antagonists (verapamil) and beta$_1$-blockers (atenolol, metoprolol) have been tried as adjuncts to betamimetics in attempts to reduce the cardiovascular side effects. The use of these agents has not been shown to achieve the desired effects, and the available data do not justify their use.

All betamimetic agents influence carbohydrate metabolism. Blood sugar levels increase by about 40 per cent, and there is an increase in insulin secretion. In women with diabetes the rise in glucose levels is even more pronounced. Thus a woman with well-controlled diabetes is likely to become deregulated when betamimetics are administered. This applies even more forcibly when betamimetics are combined with corticosteroids, which also have diabetogenic effects.

There is no doubt that betamimetic agents cross the placenta. Stimulation of beta receptors in the fetus evokes roughly the same effects as it does in the mother. The cardiovascular effects result in fetal tachycardia, although this is usually less pronounced than it is in the mother. Since the metabolic effects in mother and fetus may result in hypoglycaemia and hyperinsulinism after birth, assessment of blood sugar levels is advisable in infants born during or shortly after use of betamimetics to inhibit labour.

A few studies have compared long-term outcomes between infants whose mothers had received betamimetic drugs and infants whose mothers had not received such treatment. All these studies have been rather small, and the control groups have been variously constructed. No long-term ill effects have been observed as yet.

3.3 Inhibitors of prostaglandin synthesis

As prostaglandins are of crucial importance in the initiation and maintenance of human labour, suppression of prostaglandin synthesis is a logical approach to the inhibition of preterm labour. Several agents with widely different chemical structures inhibit prostaglandin synthesis. Those that have been used to treat preterm labour include naproxen, flufenamic acid, and aspirin, but the most widely used has been indomethacin.

All these drugs act by inhibiting the activity of the cyclo-oxygenase enzyme necessary for the synthesis of prostaglandins, prostacyclin, and thromboxane, but the mechanisms of inhibition may be different. Aspirin, for example, causes an irreversible inhibition of the enzyme, whereas indomethacin results in a competitive and reversible inhibition.

All prostaglandin synthesis inhibitors are effective inhibitors of myometrial contractility both during and outside pregnancy. They are more effective in this respect than any of the betamimetic drugs. No case has been reported in which a betamimetic drug resulted in suppression of uterine contractility after inhibition of prostaglandin synthesis had failed; the reverse has been observed repeatedly. Trials of indomethacin, although of a heterogeneous nature, show that this drug reduces the frequency of delivery within 48 hours, and within seven to ten days, of beginning treatment. It also reduces the incidence of preterm birth and low birthweight. There is a trend, not statistically significant, towards a reduction in the incidence of perinatal death and respiratory distress syndrome.

Only a few reports on the use of naproxen, flufenamic acid, and aspirin have appeared in the literature. These drugs have not been as widely used as indomethacin, and there have been no controlled trials of their use.

Inhibitors of prostaglandin synthesis are not innocuous. The most serious potential maternal side-effects are peptic ulceration, gastrointestinal and other bleeding, thrombocytopenia and allergic reactions. Nausea, vomiting, dyspepsia, diarrhoea, and allergic rashes have all been observed in women treated, even briefly, with prostaglandin synthesis inhibitors in preterm labour. Headache and dizziness may occur at the very start of treatment.

Gastrointestinal irritation is common with the use of prostaglandin synthesis inhibitors, and it can occur irrespective of the route of administration. With indomethacin it is less frequent with rectal than with oral administration; as the drug is equally well absorbed with both routes of administration, the rectal route offers some advantage.

Signs of infection may be masked by administration of prostaglandin synthesis inhibitors, and this could hamper or postpone the diagnosis of incipient intrauterine infection. The prolongation of bleeding time seen with prostaglandin synthesis inhibitors may be important, particularly when epidural anaesthesia is considered.

Prostaglandin synthesis inhibitors cross from the mother to the fetus and may influence several fetal functions. Apart from a prolonged bleeding time, which is a constant feature in infants born with detectable levels of such drugs, knowledge of effects in human fetuses and neonates is based more on case reports than on controlled trials.

The areas of major concern relate to cardiopulmonary circulation, renal function, and coagulation. The main worry about the use of these drugs for the inhibition of preterm labour relates to constriction of the ductus arteriosus. This probably has little effect on fetal oxygenation in the short term, but with prolonged treatment may result in changes similar to those seen in persistent pulmonary hypertension in the newborn. Several reports have linked persistent pulmonary hypertension in the neonate to the prenatal use of prostaglandin synthesis inhibitors.

Indomethacin treatment may reduce both fetal and neonatal renal function. The effect is dose related and appears to be transient. Several reports have noted impaired renal function in fetuses and in neonates at birth following administration of prostaglandin synthesis inhibitors to the mother. Long-term maternal treatment may influence fetal urine output enough to alter amniotic fluid volume, although other mechanisms may also be involved in the reduction of amniotic fluid volume that can be seen during indomethacin treatment. However, there is no evidence that the use of this drug in preterm labour leads to permanent impairment of renal function in the infant.

Inhibitors of prostaglandin synthesis all inhibit platelet aggregation and prolong bleeding time. They do so in the mother, in the fetus, and in the neonate at birth. Since neonates, and particularly preterm neonates, eliminate these drugs far less efficiently than their mothers, these effects will be of longer duration in the baby than in the mother.

Indomethacin, like betamimetics, may be a useful drug for obtaining sufficient delay of delivery to improve infant outcome if used properly. More, and better-controlled, data will be needed before an adequate assessment of its usefulness in care for preterm labour can be made.

The lasting effect of salicylates on platelet function and the large doses required to arrest uterine contractions preclude the use of these drugs for preterm labour.

3.4 *Ethanol*

Ethanol, for a long time one of the main labour-inhibiting drugs, is now only of historical interest. It is less efficacious than other drug treatments and has serious side-effects in both mothers and babies.

3.5 *Progestogens*

The small amount of controlled research on the use of progesterone in established preterm labour has not demonstrated any useful labour-inhibiting effects.

3.6 *Magnesium sulphate*

Magnesium sulphate has also been used for inhibition of preterm labour, although the placebo controlled trials have not shown it to be effective in reducing the frequency of any adverse outcomes. Comparative studies of magnesium sulphate and betamimetics have not provided any useful information. The combination of ritodrine and magnesium sulphate results in a higher incidence of serious side-effects than ritodrine alone.

Pulmonary oedema has been reported in association with magnesium sulphate and corticosteroid administration in preterm labour. As magnesium is primarily excreted by the kidney, hypermagnesaemia can occur if renal function is impaired. This may lead to impaired reflexes, respiratory depression, alteration in myocardial conduction, cardiac arrest, and death. Regular examination of the knee jerk reflex is said to offer protection against such complications, since these reflexes disappear at less elevated magnesium levels than those that cause respiratory depression and cardiac conduction defects.

Magnesium levels in the fetus closely parallel those in the mother. Infants born during or shortly after treatment are reported to be drowsy; they have reduced muscle tone and low calcium levels, and may need three or four days to eliminate the excess magnesium.

Although magnesium sulphate may be efficacious for arresting uterine contractions in women who are not actually in preterm labour, its place in established preterm labour has not been demonstrated and it can have serious side-effects.

3.7 Other drug treatments

3.7.1 *Calcium antagonists* 'Calcium-channel blockers' or 'calcium antagonists' include a wide range of different and apparently unrelated compounds, some of which, such as verapamil and nifedipine, have been used in the treatment of ischaemic heart disease and arterial hypertension and have also been used for the treatment of hypertension in pregnancy.

Apart from trials in which calcium antagonists were used mainly to supplement labour-inhibiting treatment with betamimetic drugs, there have been few attempts to evaluate these agents in preterm labour. There are not enough data on any of these agents to justify their use outside the context of well-designed and carefully monitored randomized trials.

3.7.2 *Diazoxide* Diazoxide is a powerful antihypertensive agent which also inhibits uterine contractions. It shares many of the properties of the betamimetic drugs with respect to both the cardiovascular system and carbohydrate metabolism.

No controlled trials of this drug in preterm labour have been reported, although it is said to be the principal tocolytic agent in at least a few centres in North America. The available evidence does not justify its use in pregnancy and certainly not for the inhibition of preterm labour.

3.7.3 *Antimicrobial agents* Subclinical infection and bacterial colonization may cause preterm labour with or without prior rupture of the membranes. A wealth of data in support of these suggestions has been described for many years in various epidemiological, microbiological, and histological associations between preterm birth and infections of the reproductive tract. Thus the hypothesis that antibiotic therapy might be of benefit in the care of women in preterm labour is attractive. Although the hypothesis has recently been assessed in a few trials, particularly using antibiotics in addition to labour-inhibiting drugs, no confident conclusion for practice can be derived from these data as yet.

3.7.4 *Oxytocin antagonists*   Several oxytocin analogue antagonists are currently under investigation. Results to date do not provide any guide to clinical treatment.

4 Maintenance of labour inhibition

Successful arrest of preterm labour does not imply that the problem may not recur before adequate fetal maturity has been achieved. Thus attention has been devoted to efforts to detect recurrences or to maintain labour inhibition for as long as necessary.

Home uterine activity monitoring has been used for early detection of recurrences in women in whom contractions were said to have been effectively stopped by treatment of preterm labour.

Betamimetics given orally to maintain labour inhibition after uterine contractions had been arrested by intravenous therapy will reduce the risk of recurrent preterm labour, but they have not been shown to reduce the incidence of preterm birth. The few reported trials of oral maintenance of labour inhibition failed to detect any effect on the incidence of respiratory distress syndrome or perinatal death.

5 Conclusions

Social and physical interventions have proved to be disappointing in their lack of effect in preventing preterm labour. Enhanced social support, despite its promise, has been shown to be ineffective for reducing the risk of preterm labour and birth. Home uterine activity monitoring is an expensive and invasive intervention which has not so far been demonstrated to result in any substantive benefit. Bed-rest, which has been evaluated mainly with multiple pregnancy, does not reduce the risk of preterm delivery. No benefits (or hazards) have been shown for repeated vaginal examinations or ultrasound assessment of cervical length.

There is no evidence that the prophylactic use of oral betamimetic agents does more good than harm. Because long-term treatment with these agents cannot be assumed to be free from adverse effects on the baby, they should not be used outside the context of controlled trials. However, there is reasonable evidence that oral maintenance treatment after inhibition of active preterm labour with intravenous betamimetics reduces the frequency of recurrent preterm labour, and the need for repeated hospitalization and intravenous treatment with betamimetic agents. This use of oral betamimetics would thus seem to be worthwhile.

At present, only two categories of drugs merit consideration for the inhibition of preterm labour: betamimetic agents and inhibitors of prostaglandin synthesis. All the others are either obsolete, excessively hazardous, or still at an experimental stage. There is no longer a place for ethanol or progesterone in the treatment of preterm labour. Oxytocin analogues and calcium antagonists have been insufficiently studied to assess whether they have any beneficial effect. Magnesium sulphate, although widely used in some centres, has never been adequately evaluated. Other drugs, such as diazoxide, should not be used in attempts to inhibit preterm labour because of their potential for serious side-effects.

The rejection of other agents does not imply strong endorsement of either betamimetic agents or the inhibitors of prostaglandin synthesis. Although both are effective in temporarily postponing delivery, there is no evidence that the use of these drugs *per se* reduces infant morbidity. They can be useful when the time that is gained before delivery is used to implement effective measures such as transfer of the mother to a centre with adequate facilities for intensive perinatal and neonatal care, the administration of corticosteroids to reduce neonatal morbidity, or judicious use of 'expectant management' in the period of gestation in which the infants chances of intact survival are very poor. Treatment with these powerful drugs may be dangerous for the mother and can occasionally result in maternal death.

The potential benefits of betamimetics, weighed against the risk of adverse effects, do not justify their use in women with heart disease, hyperthyroidism, or diabetes. If labour needs to be inhibited in these women, prostaglandin synthesis inhibitors are the logical choice. For other women who require labour inhibition, betamimetic drugs are currently the drugs of choice.

This chapter is derived from the chapter by Marc J.N.C. Keirse, Adrian Grant, and James F. King (44) in EFFECTIVE CARE IN PREGNANCY AND CHILDBIRTH.

References to primary sources and more complete data for statements made in this chapter can be found in the source chapter and/or in the following reviews from the *Cochrane pregnancy and childbirth database*:

Duley, L.
— Routine calcium supplementation in pregnancy. Review no. 05938.
Grant, A.M.
— Cervical cerclage (all trials). Review no. 04135.
— Cervical cerclage for moderate risk of early delivery. Review no. 03281.
— Stutz pessary vs cervical cerclage. Review no. 03282.
— Cervical cerclage for high risk of early delivery. Review no. 03279.
— Betamimetic infusion following cervical cerclage. Review no. 06596.

— Antibiotic prophylaxis following cervical cerclage. Review no. 06595.

Hodnett, E.D.
— Support from caregivers during at-risk pregnancy. Review no. 04169.

Kaufman, K.
— Weekly vaginal examinations. Review no. 06818.

Keirse, M.J.N.C.
— Home monitoring in prevention of recurrence of preterm labour. Review no. 06463.
— Home uterine activity monitoring for preventing preterm delivery. Review no. 06656.
— Home uterine activity monitoring in twin pregnancies. Review no. 06661.
— Ultrasound vs pelvic examination for prevention of preterm delivery. Review no. 06817.
— Betamimetic tocolytics in preterm labour. Review no. 03237.
— Prophylactic oral betamimetics in pregnancy. Review no. 04401.
— Prophylactic oral betamimetics in singleton pregnancies. Review no. 03461.
— Oral betamimetics for maintenance after preterm labour. Review no. 04380.
— Routine magnesium supplementation in pregnancy. Review no. 04008.
— Magnesium for maintenance after preterm labour. Review no. 06819.
— Antibiotics in preterm labour with intact membranes. Review no. 05531.
— Tocolytic treatment during preterm labour after PROM. Review no. 04397.
— Hydration in preterm labour. Review no. 06189.
— Prophylactic oral betamimetics in twin pregnancies. Review no. 03462.
— Solvents for betamimetic drugs in preterm labour. Review no. 06191.
— Terbutaline vs ritodrine in preterm labour. Review no. 06190.
— Ethanol tocolysis in preterm labour. Review no. 04377.
— Betamimetics vs ethanol in preterm labour. Review no. 04378.
— Magnesium sulphate vs ethanol for tocolysis in preterm labour. Review no. 04379.
— Progesterone in active preterm labour. Review no. 04381.
— Magnesium sulphate vs betamimetics for maintenance after preterm labour. Review no. 06822.

— Magnesium sulphate plus betamimetics for tocolysis in preterm labour. Review no. 05558.
— Magnesium sulphate in preterm labour. Review no. 06192.
— Magnesium sulphate vs betamimetics for tocolysis in preterm labour. Review no. 06194.
— Calcium antagonists vs betamimetics in preterm labour. Review no. 06195.
— Indomethacin tocolysis in preterm labour. Review no. 04383.

Prendiville, W.J.
— 17alpha-hydroxyprogesterone caproate in pregnancy. Review no. 04399.

25 Promoting pulmonary maturity

1 Introduction   2 Benefits of antenatal corticosteroid administration   2.1 *Respiratory distress syndrome*
2.2 *Other neonatal morbidity and mortality*   3 Potential risks of antenatal corticosteroid administration   3.1 *Risks to the mother*   3.2 *Risks to the baby*   4 Antenatal corticosteroid administration in elective preterm delivery   4.1 *Hypertensive disease*   4.2 *Intrauterine growth retardation*   4.3 *Diabetes mellitus*   4.4 *Rhesus isoimmunization*   5 Other agents to promote pulmonary maturity   6 Conclusions

1 Introduction

Respiratory distress syndrome is the most common complication of preterm birth, affecting over 50 per cent of babies born before 34 weeks' gestation. It is a significant cause of death and severe morbidity in preterm infants.

A number of agents can promote fetal lung maturation and thereby reduce the risk of respiratory distress syndrome in the newborn. Only three of these have been evaluated in controlled trials: corticosteroids, ambroxol, and thyroid-releasing hormone administered in combination with corticosteroids. Of these, only corticosteroids have been evaluated thoroughly enough for the findings to be applicable to clinical practice.

2 Benefits of antenatal corticosteroid administration

2.1 *Respiratory distress syndrome*

Antenatal administration of corticosteroids that pass through the placenta to the fetus results in a clinically important and statistically significant decrease in the incidence of respiratory distress syndrome. Maximum benefit is achieved for babies delivered more than 24 hours, and less than seven days, after commencement of the medication, but a reduction in the incidence of respiratory distress also occurs among babies born before or after this optimum period.

Although there is a widespread view that corticosteroids are ineffective at very early gestational ages, the data show that among babies born at less than 31 weeks' gestation, corticosteroid administration is followed by a similar reduction in the risk of respiratory distress to that observed for preterm babies as a whole. Respiratory distress is uncommon among babies born after 34 weeks' gestation and so the beneficial effects will be less in absolute terms, but the relative reduction of risk is similar to that found at earlier gestational ages. Gender of the baby does not modify the effects of antenatal corticosteroid administration.

2.2 *Other neonatal morbidity and mortality*

An important secondary benefit of corticosteroids has been a reduction in the duration and cost of neonatal hospital stay. Corticosteroids reduce the risk not only of respiratory morbidity, but also of other serious forms of neonatal morbidity. The risk of periventricular haemorrhage is less than half that seen without the use of corticosteroids. This effect is probably related to the reduced risk of respiratory distress, although it might also reflect an effect of corticosteroids on the periventricular vasculature.

A similar effect has been observed on the incidence of necrotizing enterocolitis. As with periventricular haemorrhage, this benefit probably arises secondarily to the reduction of respiratory distress syndrome and the need for mechanical ventilation, but a more direct drug effect on the gastrointestinal tract or its vasculature may also be involved.

Not surprisingly, these marked reductions in serious forms of neonatal morbidity have been reflected in a substantial reduction in the risk of early neonatal mortality. This reduction in neonatal mortality has not been accompanied by any increase in the risk of fetal death and so it represents a decrease in overall perinatal mortality.

3 Potential risks of antenatal corticosteroid administration

3.1 Risks to the mother

Instances of pulmonary oedema have been reported in pregnant women receiving a combination of corticosteroids and labour-inhibiting drugs. It is difficult to estimate the magnitude of this risk, or to differentiate the separate effects of corticosteroids and the labour-inhibiting drugs.

Infection is another potential risk of antenatal corticosteroid administration. Trials reporting the incidence of maternal infection provide no strong evidence of either an increase or a decrease in the risk of infection.

Other pharmacological effects of corticosteroid administration in adults relate to long-term treatment, and they provide few grounds for concern when corticosteroids are used for a period of 24–48 hours to promote fetal maturation.

3.2 Risks to the baby

The immunosuppressive effects of corticosteroid therapy could, in theory, result in an increased susceptibility to infection or a delay in its recognition. This concern has received a great deal of attention, particularly in pregnancies complicated by prelabour rupture of the membranes. Data from the trials show no evidence that corticosteroid therapy increases the risk of perinatal infection. This holds for women both with and without prelabour rupture of the membranes. In the latter there may be a slight increase in the risk of neonatal infection, but it does not extend beyond chance observation (see Chapter 23).

Fetuses may be exposed to corticosteroids throughout pregnancy if their mothers are receiving long-term steroid therapy for ulcerative colitis, asthma, rheumatoid arthritis, or other conditions. A review of the literature shows no striking excess over expectation for any adverse outcomes.

Neonates have been exposed to corticosteroids in unsuccessful attempts to prevent retrolental fibroplasia and to treat respiratory distress syndrome. In one of the three relevant trials infection was more common in the infants exposed to steroids; in the other two trials infection was more common among those who had received placebo. None of the observed differences were statistically significant. As the doses of corticosteroid used in these neonatal trials were up to five times higher than the equivalent dose of betamethasone recommended for antenatal use, it is not possible to reach firm conclusions on the long term effects of antenatal therapy with data from neonatal treatments.

The most reliable evidence about the long term effects of antenatal corticosteroid therapy comes from follow-up of children whose mothers had been treated in the randomized trials. Because of the reduced neonatal mortality rate in corticosteroid-treated babies, survivors from the corticosteroid groups had a lower mean gestational age at delivery than survivors from the control group. Despite this, neurological and intellectual functions tend to be better in the corticosteroid-treated group than in the controls, although the improvement is not statistically significant. This is plausible in the light of the complications that sometimes accompany both respiratory distress and its treatment.

4 Antenatal corticosteroid administration in elective preterm delivery

Elective preterm delivery differs from spontaneous preterm birth in at least four main ways. First, the timing of elective preterm delivery can be controlled, thus securing the delay required to gain maximum benefit from corticosteroid administration. Second, caesarean section, which predisposes to respiratory distress, is the most common route of delivery in this group of babies. Third, elective preterm delivery usually takes place somewhat later in gestation than spontaneous preterm delivery, so that the absolute risk of respiratory distress is usually lower. Finally, elective preterm delivery is often undertaken for conditions such as severe hypertension, rhesus isoimmunization, or diabetes, in which corticosteroid administration may have unwanted effects.

4.1 *Hypertensive disease*

Hypertensive disorders in pregnancy constitute one of the major indications for elective preterm delivery. There was a statistically significantly increased risk of fetal death associated with corticosteroid use in the 90 women with pre-eclampsia studied in the first reported trial. All twelve deaths occurred in women with proteinuria of more than 2 grams/day for more than fourteen days, a severity of disease that was not found in any of the placebo-treated women. There were no fetal deaths of babies of a similar number of hypertensive women in the other trials from which data are available to address this issue. A consistent adverse effect of corticosteroids would have resulted in an increased incidence of stillbirth overall, but this did not occur.

Even in the absence of any adverse effect of corticosteroids in women with pre-eclampsia, the clinician may be faced with the possible risks of postponing delivery for the few hours required to achieve a useful effect of corticosteroid administration. In some cases this delay may constitute an unacceptably high risk of complications, such as eclampsia or cerebral haemorrhage in the mother.

4.2 Intrauterine growth retardation

Intrauterine growth retardation, like hypertensive disease in pregnancy, is a common indication for elective preterm delivery. Moreover, the two conditions often coexist. The lungs of fetuses with growth retardation in the absence of maternal hypertension may have accelerated maturation, but there might still be benefit from corticosteroid administration.

A potential disadvantage of antenatal corticosteroid therapy with intrauterine growth retardation is the risk of neonatal hypoglycaemia, which is an important complication in growth retarded infants. One trial reported eleven cases of neonatal hypoglycaemia among 75 corticosteroid-treated babies compared with five in 71 controls. Without information from other trials it is difficult to know whether this is anything more than a chance difference.

4.3 Diabetes mellitus

Maternal diabetes mellitus may predispose to the development of respiratory distress syndrome. The results of the randomized trials do not clarify whether or not the use of corticosteroids is of benefit in preterm delivery of diabetic women, as only 35 such women were included in the trials.

While the efficacy of antenatal corticosteroids in pregnancies complicated by diabetes mellitus is unknown, the potential side-effects should be a source of concern. Fetal hyperinsulinism may or may not cause cortisol resistance in the fetal lung. Administration of corticosteroids causes insulin resistance in the diabetic. Loss of diabetic control is to be expected with the doses of corticosteroids administered to promote fetal pulmonary maturation. Therefore antenatal corticosteroid therapy in the diabetic woman would require exceptionally close supervision, possibly with continuous intravenous insulin and frequent blood glucose estimation. Failure to maintain control can result in either ketoacidosis, which carries a high perinatal mortality rate, or a state of fetal hyperinsulinism, which would increase the likelihood of failure to respond to corticosteroid therapy. Corticosteroid administration, if used at all in diabetic women, should be used with great caution as it is not certain that it will do more good than harm.

4.4 Rhesus isoimmunization

Elective preterm delivery plays an important role in the management of rhesus isoimmunization. Unlike other conditions associated with chronic intrauterine stress, rhesus disease is not thought to provoke an acceleration of pulmonary maturation. While there is a trend towards a reduction in perinatal mortality and in the incidence of respiratory distress syndrome in steroid-treated infants compared with controls, the numbers reported in the trials are too small to provide any secure estimates of the likely effects. However, there are no specific contraindications to the administration of corticosteroids in women with rhesus isoimmunization.

5 Other agents to promote pulmonary maturity

Treatment with antenatal ambroxol shows a tendency towards a reduced incidence of respiratory distress syndrome, but the available data remain compatible with both a decrease and an increase in its incidence. Direct comparisons of ambroxol with corticosteroids show no clear differential effect, although there were methodological weaknesses in the trials that examined this. Because the evidence in favour of antenatal corticosteroids in anticipated preterm delivery is so strong, these drugs would seem to remain the prophylactic strategy of choice.

A review of currently available results provides some evidence that antenatal administration of thyrotropin-releasing hormone (in addition to corticosteroids) prior to very preterm delivery may further reduce the risks of respiratory distress syndrome and chronic lung disease, albeit at the expense of transient maternal side effects. Nevertheless, concerns remain about possible adverse effects, and thyrotropin-releasing hormone should be used only in the context of controlled trials until further evidence is available.

6 Conclusions

Antenatal treatment with 24 mg betamethasone, 24 mg dexamethasone, or 2 g hydrocortisone is associated with a significant reduction in the risks of neonatal respiratory distress. This reduction is of the order of 40–60 per cent, is independent of gender, and applies to babies born at all gestational ages at which respiratory distress syndrome may occur. While the most dramatic benefit is seen in babies delivered more than 24 hours and less than seven days after commencement of therapy, babies delivered before or after this optimum period also benefit. This reduction in the risk of respiratory distress is accompanied by reductions in periventricular haemorrhage and necrotizing enterocolitis, all of which are reflected in a lower mortality rate and in a reduced cost and duration of neonatal care.

These benefits are achieved without any detectable increase in the risk of maternal, fetal, or neonatal infection. Antenatal corticosteroid administration does not increase the risk of stillbirth.

Of other agents tested, ambroxol is of interest because it may be as effective as corticosteroids. The main disadvantage is the five-day period required to complete therapy. There is some evidence from comparisons of corticosteroids and thyroid-releasing hormone with corticosteroids used alone that the combined treatment may be superior to corticosteroids alone. The results of randomized comparisons that are currently being conducted should provide valuable information.

This chapter is derived from the chapter by Patricia Crowley (45) in EFFECTIVE CARE IN PREGNANCY AND CHILDBIRTH.

References to primary sources and more complete data for statements made in this chapter can be found in the source chapter and/or in the following reviews from the *Cochrane pregnancy and childbirth database*:

Crowley, P.
— Corticosteroids prior to preterm delivery. Review no. 02955.
— Ambroxol vs placebo prior to preterm delivery. Review no. 03276.
— Ambroxol vs betamethasone prior to preterm delivery. Review no. 03852.
— Ambroxol vs intralipid prior to preterm delivery. Review no. 03853.
— Corticosteroids after preterm prelabour rupture of membranes. Review no. 04395.
— Corticosteroids plus induction of labour after PROM preterm. Review no. 06871.

Crowther, C.A. and Grant, A.M.
— Antenatal thyrotropin-releasing hormone (TRH) prior to preterm delivery. Review no. 04749.

---

26 Post-term pregnancy

1 Introduction   2 Risks in post-term pregnancy
3 Effects of elective delivery   3.1 *Effects on the mother*
3.2 *Perinatal morbidity*   3.3 *Perinatal death*
4 Surveillance   5 Conclusions

1 Introduction

The reported frequency of post-term pregnancy (a pregnancy lasting 42 weeks or more) varies from 4 to 14 per cent, depending on the nature of the population surveyed, the criteria used for assessment of gestational age, and the proportion of women who undergo elective delivery. Apart from this, contradictory findings and conclusions on the risks associated with post-term pregnancy have led to opposing views on the most effective form of care. The difficulty of determining the incidence and significance of post-term pregnancy is compounded by variations in the way that the condition is defined, which range from 41 weeks to 43 weeks. Semantic problems also contribute to the confusion. The words 'post-term', 'prolonged', 'postdates', and 'postmature' are all used as synonyms, but are laden with different evaluative overtones.

The name 'post-maturity' has also been given to a clinical syndrome in the infant with a hierarchy of features ranging from loss of subcutaneous fat and dry cracked skin, through meconium staining and birth asphyxia, to respiratory distress, convulsions, and fetal death. Confusion is bound to arise when a clearly pathological syndrome is described by a word that is also used to make a simple statement about the chronological duration of a pregnancy.

2 Risks in post-term pregnancy

Perinatal mortality is increased in post-term pregnancy. Some of this increase is due to congenital malformations which are more frequent among post-term deliveries than among deliveries at term. The other main cause of death is asphyxia. The incidence of early neonatal seizures, a marker of perinatal asphyxia, is between two and five times higher in infants born after 41 weeks. The increased risk in post-term pregnancy relates predominantly to the intrapartum and neonatal period rather than the antepartum period. Thus the risk increases with the onset of labour. Meconium-stained amniotic fluid is a common feature among intrapartum and asphyxial neonatal deaths.

3 Effects of elective delivery

For many years, obstetricians have expressed irreconcilably different opinions on the role of induction of labour for post-term pregnancy. The results of even large studies using observational data shed little light on the question because of inherent selection biases and the influence of both duration of pregnancy and other aspects of care on outcome. The best evidence for the value of routine induction at or beyond term comes from randomized trials.

Trials have examined the effects of induction at or about 40 weeks, addressing the risks and benefits of pre-empting post-term pregnancy. Other trials have dealt with management during or after week 41. No trials have assessed the risks and benefits of alternative ways of managing post-term pregnancy at or beyond 42 weeks.

3.1 *Effects on the mother*

At and around term, women have been found to be much more likely to refuse a policy of routine induction to prevent post-term pregnancy than a policy of expectant management. As some studies suggest, this may differ once pregnancy has proceeded well beyond the expected date of delivery.

Policies of active induction of labour do not show any effect on the use of epidural analgesia. This is somewhat surprising, as women who have labour induced are in the labour ward throughout labour and during daylight hours, both factors which might be expected to increase their opportunity to avail themselves of an epidural service.

Active induction policies are not associated with an increased use of caesarean delivery; indeed, the trials show a small but statistically significant decrease in the use of caesarean section for women in whom labour is induced at or beyond term. This challenges a widely held belief that there is an inherent association between elective delivery and an increased risk of caesarean section. This unexpected finding may reflect the characteristics of the women who participated in these trials, for example the 'ripeness' of their cervices, and of the methods used for inducing labour.

A single trial of advice advocating breast stimulation plus no prohibition of sexual intercourse (compared with avoiding both breast stimulation and coitus) for women from 39 weeks to delivery showed a decrease in the incidence of postdate pregnancy. 'Sweeping the membranes' (digital separation of the fetal membranes from the lower pole of the uterus) in pregnancies at or beyond term may also reduce the likelihood of the pregnancy proceeding beyond 42 weeks (see Chapter 40).

3.2 *Perinatal morbidity*

Elective delivery, either at term or after 41+ weeks, reduces the risk of meconium-stained fluid. There is no evidence that policies of active induction increase the risk of fetal heart rate abnormalities during labour or the incidence of depressed Apgar score either at term or at 41+ weeks.

During the 1970s there were several reports of an association between elective induction of labour and unintended preterm delivery followed by respiratory distress and other neonatal morbidity. By the 1980s this had become less of a problem because of greater awareness of the dangers of elective induction of labour without firm grounds for being certain about the duration of gestation. No cases of iatrogenic respiratory distress syndrome are reported in the randomized trials of elective delivery, but it must be remembered that well-documented fetal maturity was one of the entry criteria for all of them.

No consistent effect of elective induction on the incidence of neonatal jaundice has been demonstrated in the available trials.

3.3 *Perinatal death*

A policy of elective induction of labour at or beyond term has been shown to result in a decrease in perinatal deaths not due to lethal anomalies. Although none of the trials individually was large enough to show a statistically significant difference, the combined results of the seventeen randomized trials that have assessed this outcome show a clear picture. There was only one such death among the almost 3500 women allocated to elective delivery, compared with nine among a similar number of women in the surveillance arm of the trials. This difference is both important and statistically significant.

The improved perinatal mortality appears to be confined to induction at and after 41 weeks gestation. There is no evidence of a beneficial effect of induction at 39–40 weeks gestation.

4 Surveillance

In all randomized trials of elective delivery at 41+ weeks, some form of fetal surveillance was used in the conservatively managed arm of the trial. This surveillance usually involved consultations at two- to three-day intervals after 41+ weeks, and varied from the mildly intrusive use of ultrasound or cardiotocography to the highly invasive procedures of amnioscopy or amniocentesis. There is some evidence that these tests can detect pregnancies in which there is 'something wrong', but less evidence that their use improves outcome or can eliminate the additional risk of post-term pregnancy.

5 Conclusions

In most cases, post-term pregnancy probably represents a variant of normal and is associated with a good outcome, regardless of the form of care given. In a minority of cases there is an increased risk of perinatal death and early neonatal convulsions.

Induction of labour at less than 41 weeks' gestation is not associated with any advantage apart from a small reduction in meconium staining of the amniotic fluid. The reduction in perinatal death appears to be confined to pregnancies of 41+ weeks' duration. A policy of routine induction at 40–41 weeks in normal pregnancies cannot be justified in the light of the evidence from controlled trials, and is unacceptable to many mothers.

Induction of labour after 41+ weeks' gestation is not associated with any major disadvantage and reduces the risk of perinatal death in women with post-term pregnancies. Provided that appropriate induction methods are used, it may also result in a small reduction in the high risk of caesarean section run by women with post-term pregnancies in institutions with high caesarean section rates. In the light of the available evidence, the best policy is to offer women a choice of induction of labour by the best method available once the duration of pregnancy has with certainty attained 41 weeks or more.

Obstetricians and parents should also be aware of the poor quality of the evidence available to support the use of all methods of fetal surveillance commonly offered to women with prolonged pregnancies.

This chapter is derived from the chapters by Leiv S. Bakketeig and Per Bergsjo (46), and by Patricia Crowley (47) in EFFECTIVE CARE IN PREGNANCY AND CHILDBIRTH.

References to primary sources and more complete data for the statements made in this chapter can be found in the source chapters and/or in the following reviews from the *Cochrane pregnancy and childbirth database*:

Crowley, P.
— Elective induction of labour at or beyond term. Review no. 04142.
— Elective induction of labour at < 41 weeks gestation. Review no. 04143.
— Elective induction of labour at 41+ weeks gestation. Review no. 04144.
— Breast stimulation for the management of post-term pregnancy. Review no. 06860.

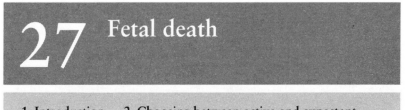

27 Fetal death

1 Introduction   2 Choosing between active and expectant care   3 Choice of methods for inducing labour
4 Conclusions

1 Introduction

The interval between a diagnosis of antepartum death and delivery is a time of great distress. When this diagnosis has been made and confirmed by ultrasound examination, women require the time and opportunity to adjust to it. Rushed decisions are unnecessary. Women should be made aware of the options available to them, and given time to consider these options and to decide what they want. They must be allowed the time to start to grieve, and to make decisions in an environment in which they feel secure. Many women will want to return home, even if only for a brief period. It is important to remember to ask the woman how she came to the hospital or clinic. She should not have to drive at such a time or go home unaccompanied.

2 Choosing between active and expectant care

From the physical standpoint, given appropriate means to induce labour after fetal death, there are no overwhelming benefits or hazards for induction of labour over expectant care. The advantages and disadvantages of both these approaches relate almost exclusively to their emotional and psychological effects. The woman herself is the best judge of these, and she is the one who should make the choice. Her caregivers should assist her by providing her with the information needed to make an informed choice. They should ensure that whatever option she chooses is implemented with as little psychological and physical discomfort as possible.

It is wrong to assume that all women desire the most rapid method of delivery when their babies have died *in utero*. For some women, the uncertainty and learning of the death are the worst moments; carrying the dead fetus still permits them a feeling of closeness to the baby that will be lost once it is born. A decision by the caregiver to induce labour, if made with little consultation or input by the woman and her partner, may be seen by the parents as a way of compensating for guilt or as a search for a quick solution.

In contrast, many women are anxious to end the pregnancy as quickly as possible. Some may even suggest that this should be done by caesarean section. Discussing the facts and alternatives with the woman and her partner conveys compassion and understanding. Often it will help to defuse initial feelings of anger, suspicion, inadequacy, and guilt which are typically felt by all, caregivers and women alike, after the sad diagnosis is made.

The main advantage of the expectant option is the absence of any need for intervention. The woman can stay at home, and she will avoid procedures that might turn out to be less effective and more risky than anticipated.

The disadvantages of expectant care are mainly psychological, and relate to the unpredictable and usually long time during which the woman may have to carry the dead baby. Sometimes she or her relatives may be under the impression that the baby will rot inside her and exude toxins that can poison her. It is important to dispel such fears, although this may not always be successful.

The only physical hazard of the expectant policy relates to a possible increase in the risk of disturbances in blood coagulation. These are most likely to occur when fetal death has been caused by placental abruption. Disorders of coagulation in association with other causes of fetal death are rare. The hypofibrinogenaemia that is held responsible for these disorders occurs very slowly, and is rarely clinically significant in the first four to five weeks after fetal death. By the time that clinically significant alterations in coagulation mechanisms have occurred, the chances are that delivery will have supervened.

The only advantages of an active policy to effect delivery in the care of women with a dead fetus are that it offers the option of ending a pregnancy that has lost its purpose and that a post-mortem diagnosis may be easier to achieve in the absence of maceration. The disadvantages of an active policy relate to the means through which it is effected. If labour is induced, the efficacy and safety of the method used will be the most influential factor in considering the relative merits of the policy.

3 Choice of methods for inducing labour

Before prostaglandins became available, attempts to induce labour after fetal death with oestrogen administration, intra-amniotic injection of hypertonic solutions, and high doses of oxytocin sometimes turned into iatrogenic nightmares. Instead of diminishing emotional anguish, the procedure aggravated it, leaving the woman exasperated and exhausted after every failed attempt to induce labour.

The availability of prostaglandins dramatically changed the prospects for inducing labour after intrauterine fetal death. Cumulative clinical experience with the administration of prostaglandins for this indication showed their overwhelming superiority compared with previously available methods. Successful induction (vaginal delivery following a single course of treatment) was achieved in 95 per cent or more, irrespective of the prostaglandin used and the dose in which it was administered.

Intravenous prostaglandin administration, although undoubtedly efficacious, has now largely been superseded because of a relatively high incidence of side-effects compared with local routes. Experience with intra-amniotic administration has been limited and was mostly gathered in the early days of clinical prostaglandin research. When the fetus has died it is often difficult to identify a clear amniotic pool in which to administer the drug, or to obtain a sample of clear fluid to guarantee that the right compartment has been found. This, and post-mortem changes in the fetal membranes that render the degree of absorption of the drug into the myometrium and into the maternal circulation unpredictable, mean that intra-amniotic administration is probably not suitable. In countries where 15-methyl-prostaglandin $F_{2\alpha}$ is the only prostaglandin analogue available, extra-amniotic infusion is preferable to intramuscular administration for inducing labour after fetal death because it is less likely to cause gastrointestinal side-effects. However, this recommendation needs qualifying if mobility during labour is an important consideration to the woman. In countries where they are available, the prostaglandin analogue sulprostone or the anti-progestational steroid mifepristone (if necessary combined with a prostaglandin preparation) are preferable alternatives.

Vaginal administration of prostaglandin E_2, usually in the form of 20 mg suppositories, is a widely used approach nowadays because of its convenience and ease of administration. It is not a completely satisfactory method because it results in a high rate of undesirable side-effects.

Extra-amniotic administration of natural prostaglandins, either by continuous infusion or by administration in a viscous gel, appears to produce the lowest incidence of systemic side-effects. The method requires insertion of a catheter into the extra-amniotic space, which in addition to being bothersome for the woman and limiting her freedom of movement carries a risk of introducing infection. It requires some experience to avoid rupture of the membranes and premature expulsion of the catheter. These complications occur in a sizable minority of cases.

When intrauterine fetal death occurs in late pregnancy, it is usually possible to induce labour with any of the prostaglandin regimens that are employed for other inductions. Methods with which one is thoroughly accustomed tend to perform better than those that are only rarely needed and require careful study before being applied. However, this is of little help at the earlier gestational ages when the sensitivity of the uterus to prostaglandins is much lower than it is at term. At these stages of pregnancy the choice between the various routes for administering natural prostaglandins is not easy. Despite their efficacy, they all have several drawbacks.

Structurally modified prostaglandin analogues have been synthesized in the hope that they might overcome some of the drawbacks of the natural compounds. These analogues are not suitable for induction of labour with a live fetus because of the unknown effects of their metabolites on infant development, and because of the long-lasting effect that makes them more prone to cause uterine hyperstimulation which may be hazardous to a live fetus. However, they offer a number of potential advantages for induction of labour after fetal death. Their longer duration of action results in less need for continuous administration, thus permitting intermittent doses and alternative routes of administration.

Results obtained with the prostaglandin E_2 analogue sulprostone appear to be superior to those obtained with prostaglandin $F_{2\alpha}$ analogues, although no formal comparisons between the drugs have been reported. Sulprostone is probably best administered intravenously in a dose of 1 μg /minute, as this appears to be as effective as larger doses but results in less vomiting and diarrhoea.

Mifepristone, a steroid compound that antagonizes progesterone action, has shown promise for induction of labour after fetal death, possibly in combination with prostaglandins to enhance the success rate further.

4 Conclusions

In the case of fetal death, the decision of whether or not to induce labour should be made on psychological or social grounds, and the woman herself is the best judge of these. Should induction be chosen, this should be implemented with the use of a prostaglandin, a prostaglandin analogue, or mifepristone. In the later weeks of pregnancy the most effective method is likely to be the method of which the caregiver has adequate experience from inducing labour in other pregnancies. Earlier in gestation, prostaglandin analogues (sulprostone in particular) or mifepristone, if available, appear to be the treatments of choice. Where such compounds are not available, a natural prostaglandin can be used by any one of a variety of routes, none of which will be ideal.

This chapter is derived from the chapter by Marc J.N.C. Keirse and Humphrey H.H. Kanhai (65) in EFFECTIVE CARE IN PREGNANCY AND CHILDBIRTH.

References to primary sources and more complete data for the statements made in this chapter can be found in the source chapter and/or in the following reviews from the *Cochrane pregnancy and childbirth database*:

Keirse, M.J.N.C.
— 15-methyl-prostaglandin F2alpha after fetal death. Review no. 06188.
— Low vs high dose sulprostone for induction after fetal death. Review no. 04474.
— Mifepristone for induction of labour after fetal death. Review no. 05533.

CHILDBIRTH

28 Social and professional support in childbirth

1 Introduction

Support during childbirth can be provided by the professionals responsible for the clinical care of the woman in labour, by other individuals specifically designated to provide support other than clinical care, or by the woman's partner, family, or friends. Controlled studies thus far have examined primarily the contribution of the second group — persons specifically designated to provide social support in labour. Insights into the nature and value of support by other professional and non-professional people have been gleaned from data from observational studies.

2 The nature of support in childbirth

A central feature of support in childbirth is the promise that the labouring woman will not, at any time, be left without available support. Mere physical presence is not enough. Supportive activities should encompass both physical comforting measures and emotional support. Physical comfort measures should be provided in response to the woman's own wishes. These will vary from culture to culture, and from individual to individual. The supportive companion may, for example, maintain eye contact, walk with her, and hold her hand. For some women, simply an ongoing supportive presence may be helpful. Emotional and cognitive support may include advocacy, explanation, praise, and encouragement. The supporting companion must ensure that the woman understands the purpose of every procedure and the result of every examination, that she is kept informed of the progress of her labour, and that she is praised for her efforts and encouraged to continue.

Support may or may not be seen as a normal part of care during childbirth, depending on the approach of the carers. For example, some professionals may see the technical tasks of caring for a woman in labour as having priority. Others may feel that care and support are intimately involved in helping the woman to progress successfully throughout labour, and so technical tasks and emotional/physical support cannot be separated. Technology may complicate the issue for those holding the latter view, as their time and attention may be distracted away from the woman towards the monitor, drip or other such devices.

Every woman should be able to choose her primary source of social support in labour. This may be her partner, another family member, or a friend. Midwives, doctors, and nurses should respect her choice, and provide appropriate additional support where it is needed in addition to physical care.

3 The birth environment: implications for support

Developments in obstetrical care over the past seventy-five years have increased the risk of isolation of the mother during labour. For perhaps a third of that time, this distressing feature of the birth environment was not apparent in many countries. For example, in North America the subjective experiences of labour and birth were submerged by twilight sleep and general anaesthesia as a result of doctors' efforts to assist women in labour. While women were unconscious, questions of psychological support were irrelevant. When the natural childbirth movement redefined the experience of giving birth as potentially positive, these aspects of the birth environment took on a new significance.

Many factors in the birth environment can induce stress. The setting and many of the people in it may be strange to the labouring woman. Common procedures such as insertion of intravenous lines, restriction of fluids and foods, shaving the perineum, enemas, vaginal examinations, restriction of movement, fetal monitoring, augmentation of labour, epidural anaesthesia, and the possibility of an operative outcome add further to the stress. Fear, pain, and anxiety may be increased by a mechanized clinical environment and unknown attendants, with potentially adverse effects on the progress of labour.

The perception of isolation and the reality of being left alone, even momentarily, can be compounded by the intermittent appearance and disappearance of unknown people, including obstetricians, midwives, nurses, and medical, nursing or midwifery students. A Canadian study found that women giving birth in hospital encountered an average of

6.4 unfamiliar professionals during labour (range 3–14). Another survey reported that one low-risk mother having her first child in a teaching hospital was attended by 16 people during 6 hours of labour, but was still left alone for most of the time. Women appreciate a constantly available supportive companion in labour, together with appropriate care from a small number of professionals.

A hospital environment where separation of family members and rigid protocols are enforced is one of the factors believed to cause the high intervention rates during labour that are seen in many industrialized societies. Yet the opening up of labour wards, first to fathers and then to other support people, has occurred at the same time as a staggering increase in caesarean birth. The assumption that rigid policies in the birth environment lead to an increased resort to major interventions is not necessarily correct. To date, only two trials have compared the effects of labour and birth in a home-like hospital birth room to that in a conventional labour ward. Allocating women to labour and to give birth in a home-like hospital birth room was associated with a decreased risk of perineal trauma (episiotomy, laceration, or both) and a decreased desire for a different setting for the next birth, but no significant effects on the use of epidural anaesthesia, forceps delivery, and caesarean delivery were found. Nor were there differences in the proportion of women 'rooming-in' on the postpartum ward.

In the years since the results of these trials were published, many hospitals have allocated scarce resources towards renovation of their labour wards, endeavouring to provide more attractive home-like settings. Undoubtedly such settings are more pleasant work environments for caregivers, and happier caregivers may provide better care. However, hospitals which are considering renovations of their labour wards should be aware that there is much stronger evidence to support the need for changes in caregivers' behaviour than there is to support the need for structural changes to labour wards. If renovations are desired, they should be targeted towards factors which would encourage changes in behaviour, such as removing lithotomy poles and replacing uncomfortable delivery beds with comfortable furniture and cushions. Efforts to change caregivers' behaviour, to help them to provide appropriate support to labouring women, should also be introduced.

4 Men during labour and at birth

The arrival of men as husbands and partners at labour and birth is a recent phenomenon in industrialized countries. As women in these countries have come to reclaim birth as a positive experience, the exclusion of a woman's sexual partner and the baby's father has come to be seen as incongruous.

In North America in particular, partners came to be expected to reinforce what had been taught in childbirth education classes, and if necessary to act as advocates for the childbearing woman. They were also expected to fill the gaps in care. The women planning a hospital birth in the Canadian study cited earlier rarely expected to have a nurse with them throughout labour. They felt that the nurses would be too busy, or they saw them as unwelcome strangers, or they viewed the nurse's role as purely technical in nature. They intended to rely on their partners for support, assistance with breathing techniques, and comfort measures.

Recognizing that labouring women require psychological support, and realizing that midwives and nurses often have little time to give it, hospitals have increasingly permitted and encouraged men to assume active roles in the care of their partners during labour. In many countries in the industrialized world the presence of women's partners during labour has, within twenty years, gone from being occasionally permitted to being normative and virtually universal.

Studies of the impact of the father's presence on labour and birth have been limited by small sample sizes and self-selection. Research in the 1970s, when the presence of fathers was unusual, studied groups where the presence of fathers in labour or at birth was associated with higher social class, attendance at childbirth education classes, and preference for non-medicated birth.

There has been almost no research on the support actually provided by husbands and partners. Also unresearched are the expectations that women bring to labour about the support that they will have and that they will need. Several surveys using retrospective interviews have asked women to rate the quality of support provided during labour and birth. The results are remarkably similar in very different settings, with partners almost uniformly rated very highly and mostly higher than midwives.

Some doubts have been expressed about handing over the supportive role to fathers. One concern relates to whether they are equipped for tasks that were formerly the responsibility of an experienced and professionally trained person. Another is the issue that the father should not be expected to provide the majority of the support when he is also emotionally involved. He is sharing the experience and may need support himself. Other questions relate to the possibility that the father's presence might have a negative influence on the labouring woman and interfere with the normal progress of labour. When there are major tensions in the couple's relationship, practical and emotional support in labour may be difficult for the partner to provide or for the woman to accept.

A recent study of pain in labour found that the presence of a labour partner who provided specific encouragement in pain control techniques, *not* just a reassuring presence and emotional support, was associated with lower epidural rates and with fewer women describing panic, exhaustion, or overwhelming pain. Self-selection presents a problem in the interpretation and generalizability of the results of such studies.

5 Other support people

Apart from institutionally employed support persons, midwives, and partners, two other categories of people are currently providing support in labour: other family members and friends, and paid professional companions usually called 'coaches' or 'monitrices'. Hospitals vary greatly in the extent to which they permit these other support people in labour wards.

At home, and in alternative birth settings such as birthing centres, it is customary for several people to be present for at least some of the time. The freedom to choose who will be present, and when, is often a factor in a woman's choice to give birth outside hospital. However, it would be unwise to assume that the presence of several people will necessarily provide additional support. Family and friends, like husbands and partners, may be there to share in the experience rather than to provide support.

To some extent, the move to have additional support arose from disillusionment with what the husband could provide, particularly when the labour was prolonged. After all, professional support persons and student midwives, usually work shifts rather than 24 hours a day. They also have their experience to guide their reaction to the woman's needs, and can more easily judge what is 'normal' and what is not. Support from family and friends is particularly appropriate where it is seen in terms of physical comfort, being there, maintaining eye contact, and praise and encouragement. When coping with labour is seen more in terms of utilizing learned skills, such as controlled breathing patterns and relaxation, then the chosen additional support person is likely to be a professional.

Just as the role of support persons arose from the splitting of care into management and support, so the role of the labour coach comes from the splitting of childbirth education from care. Once education for childbirth becomes a separate activity, those responsible for care and management in labour may not know what women or couples have been taught during antenatal classes, and may be unwilling or unable to support labouring women in the use of the skills that they have learned. They may belittle the usefulness of the education programme, at the same time complaining about the unreal expectations that it has created.

The potential for territorial rivalries over the provision of support is great. When a labour coach is recruited as an advocate for the labouring woman, rivalries with hospital staff are almost inevitable and the intended support may end up as a casualty of the conflict. One can legitimately ask if, given the constraints posed by institutional norms and policies, an employee of the hospital can provide the same quality of support and advocacy as a professional 'outsider'. However, the presence of an outsider can pose a threat to the institution, which may have a negative influence on the quality of care received by the labouring woman.

6 Controlled trials of support in labour

The role of a special support person in labour has now been assessed by controlled trials in several countries in a variety of settings. There was remarkable consistency in the descriptions of the experimental intervention in the various trials. In all instances the intervention, support, included continuous presence, if not for all of labour, then at least during active labour. In eight trials it also included as a minimum comforting touch and words of praise and encouragement.

The results of the trials were also remarkably consistent, despite the disparities in obstetrical routines, hospital conditions, the obstetrical risk status of the women, the differences in policies about the presence of significant others, and the differences in the professional qualifications of the persons who provided the support.

Regardless of whether or not a support person of the woman's own choosing could be present, the continuous presence of a trained support person who had no prior social bond with the labouring woman reduced the likelihood of medication for pain relief, of operative vaginal delivery, and of a 5 minute Apgar score below 7. In settings which did not permit the presence of significant others, the presence of a trained support person also reduced the likelihood of caesarean delivery. Another beneficial effect found in three trials was the decreased likelihood of negative ratings of the childbirth experience, of feeling very tense during labour, and of finding labour worse than expected. Individual trials have found many other benefits of intrapartum support, including less perineal trauma and a reduced likelihood of difficulty in mothering and of early cessation of breastfeeding.

7 Conclusions

Given the clear benefits and absence of known risks associated with intrapartum support, every effort should be made to ensure that all labouring women receive support, not only from those close to them but also from specially trained caregivers (midwives, nurses, or lay women). This support should include continuous presence (when wished by the mother), the provision of hands-on comfort, and praise and encouragement. Depending upon the circumstances, ensuring the provision of appropriate support may necessitate alterations in the current work activities of midwives and nurses so that they are able to spend less time on ineffective activities and more time providing support for women.

This chapter is derived from the chapters by Cornelis Naaktgeboren (48) and by Marc J.N.C. Keirse, Murray Enkin, and Judith Lumley (49) in EFFECTIVE CARE IN PREGNANCY AND CHILDBIRTH.

References to primary sources and more complete data for the statements made in this chapter can be found in the source chapters and/or in the following reviews from the *Cochrane pregnancy and childbirth database:*

Hodnett, E.D.
— Support from caregivers during childbirth. Review no. 03871.
— Birth room vs conventional delivery setting. Review no. 05735.

---

29 Hospital practices

1 Introduction   2 First impression   3 Clinical assessment
4 Preparation procedures   4.1 *Enemas*   4.2 *Pubic shaving*
5 Nutrition   5.1 *Risks of aspiration*   5.2 *Measures to reduce volume and acidity of stomach contents* 5.2.1 *Restriction of oral intake* 5.2.2 *Pharmacological approaches*   6 **Maternal position during first stage of labour** 6.1 *Effects on blood flow and uterine contractility* 6.2 *Effects on the mother and the baby*
7 **Conclusions**

1 Introduction

Most births now take place in hospitals. Like other large institutions, hospitals (and the professionals working in them) depend on rules and routines for efficient functioning — and it is probably essential that they continue to do so. Professionals need a structure within which to do their work. This structure necessarily involves working rules and at least some routines intended to serve the interests of other people working in and using the institution. Change can be slow because familiar rules and routines are comforting, and because it takes time to develop and agree on new policies — time that may be seen as better spent providing clinical care.

Therefore the marked variations in the type of care women receive tend to depend more on which maternity unit a woman happens to attend, and which professional she consults, than on her individual needs or preferences. These differences in practice are often so dramatic that they cannot possibly be explained by differences in medical indications or by the characteristics of women attending different hospitals.

2 First impression

A woman entering a hospital in labour may have experienced months or even years of anticipation, fear, and uncertainty about childbirth. All of it is focused on the moment when she walks past the 'point of no return' through the doors into the labour ward. This is the time when she feels and is most vulnerable. It is a woman in physical and emotional turmoil who needs to be welcomed into a strange environment and given comfort and care. This is particularly difficult if the woman has not met any of her caregivers before.

The midwife or labour room nurse may have an entirely different set of priorities. Her main concerns are probably to discover what stage of labour the woman is in, and to reassure herself that the mother and baby are well. She will also have record keeping tasks, and sometimes may be responsible for other women in labour. Providing appropriate care for each individual woman, with her own distinct needs, is a daunting task.

Various recommendations for changes in admission practices have been made to alleviate anxiety and fear when entering the hospital in labour. It is important that caregivers welcome and support mothers and their companions from the moment of arrival. Caregivers should introduce themselves and give information about others whom the mother might see during labour. This is common courtesy and should be universal. It would be helpful if midwives or nurses also asked women how they wish to be addressed.

Setting up systems of care from one midwife, or from a small team, throughout the childbearing period should improve the experience of arriving in hospital in labour. For example, the midwife is likely to have assessed the woman in her own home and advised her when to come in. Mother and midwife will have had the chance to get to know each other antenatally, and to discuss the sort of care that the woman would like in labour.

If the woman and the caregiver do not already know each other, admission in labour provides an opportunity to discuss a woman's requests and plans for (and worries and concerns about) labour and delivery. Midwives and nurses should inform and reassure women as they explain the various examinations and procedures that will be carried out.

The support of a partner or some other companion may be particularly important to a woman when she first comes in in labour, and during the initial examination, although some hospitals have policies which exclude companions at this time. Surveys show that only a small proportion of women prefer not to have anyone with them at this point. Most women interviewed expressed pleasure and relief when their partner could stay, and disappointment when he was unable to do so, whether because he was at work or looking after other children, or excluded as the result of a hospital regulation.

A woman is usually asked to undress when she first arrives in labour. If this is done insensitively it can be a humiliating experience for her. Many women prefer the option of bringing a comfortable nightdress from home rather than having to wear a hospital gown; this gives them a little more dignity and individuality. Provision for privacy is important; this privacy is sometimes lacking, with curtains and screens of various designs replacing a door that can be closed.

3 Clinical assessment

The 'diagnosis' of labour has received relatively little research attention, and some important practical questions remain unanswered. The advice given to a woman antenatally about the onset of labour, what she is told over the telephone when she calls in, and whether or not a caregiver can assess her at home first will influence her decision on when to come to hospital. What she experiences will depend on hospital policies and the decisions made by her caregivers. If she is judged not to be in labour, she may be sent home or to another hospital ward. A specific education programme can reduce the admission of women who are not in active labour.

The main clinical tasks when a woman comes into hospital are to assess her progress in labour, her condition and that of her baby, and to make decisions about care. To do so, caregivers have various means including a discussion with the woman about her history, symptoms, and obstetric records, observation of her temperature, blood pressure, and general condition, abdominal palpation, vaginal examination, and some form of monitoring of the fetal heart. An explanation of why these are necessary is not always given, and sometimes women are given no indication of the results. An explanation of *what* is being done is not sufficient; it is equally important to her *why* it is being done. Most women want to be involved in decisions about their care.

4 Preparation procedures

Admission to hospital in labour often included the routine use of bowel preparation with enemas or suppositories, and the shaving of the pubic and perineal area. Although mainly of historical interest in some countries, these practices have continued in others.

4.1 Enemas

The supposed benefits of bowel preparation were to allow the fetal head to descend, to stimulate contractions and thereby shorten labour, and to reduce contamination at delivery and so minimize infection in mother and baby. The practice is uncomfortable and not without risk. Cases of rectal irritation, colitis, gangrene, and anaphylactic shock have all been reported.

Two randomized controlled trials have evaluated the effects of routinely giving enemas on admission to hospital in labour. The available evidence suggests that the rate of faecal soiling is unaffected during the first stage of labour, but reduced during delivery. Without an enema the soiling was mainly slight, and it was easier to remove than the soiling after an enema. No effects on the duration of labour or on neonatal infection or perineal wound infection were detected.

Of the women who had enemas or suppositories, a small number were pleased or had requested this, half of the remainder either did not mind or were prepared to have whatever was necessary, while the other half expressed negative feelings such as embarrassment, discomfort, or reluctance. The majority of women who did not have an enema were pleased or relieved, but some did not have strong views.

As routinely administering enemas to women in labour confers no benefit, the results of trials of different types of enema are largely irrelevant. Nevertheless, for situations in which an enema is deemed necessary, or is requested by the mother, it is worth noting that no advantages have been shown for a medicated over a tap water enema, and that soapsuds enemas should not be used because of the frequency with which they result in cramps and griping.

4.2 Pubic shaving

The purpose of routine predelivery shaving has been to lessen the risk of infection and, presumably, to make suturing easier and safer. As early as 1922, these assumptions were challenged by a controlled trial. That trial, and the only other controlled trial examining this practice, were unable to detect any effect of perineal shaving on lowering puerperal morbidity; rather, there was a tendency towards increased morbidity in the shave groups. Shaving appears to decrease the proportion of women from whom Gram-negative organisms can be cultured, and to have no effect on the proportion from whom Gram-positive organisms are cultured.

The results of these trials are supported by the results of non-randomized cohort studies and of randomized trials of preoperative shaving in surgical patients. Other authors have drawn attention to the disadvantages in terms of discomfort as the hair grows back, as well as to the minor abrasions caused by shaving.

5 Nutrition

Withholding food and drink once labour has commenced is widely accepted in current hospital care. A small minority hold equally strongly that, except for women at high risk of needing general anaesthesia, the benefits of nourishment as women wish far outweigh the possible benefits of more restrictive policies.

Surveys of labour ward policies in England and the United States show that most units prohibit all solid foods. Almost 50 per cent allowed no oral intake except ice chips, most of the remainder allowed only sips of clear fluids, and only about one in ten units allowed women to drink as much fluid as they desired. None of the hospitals surveyed in the United States permitted women to eat and drink as they wished.

For many women these restrictions do not present a problem. Most do not want to eat during labour, particularly during its later phases. For those who do want to eat, however, enforced hunger during the first stage of labour can be a highly unpleasant experience. Why then are such restrictive policies employed when some women so obviously find them distressing? The explanation lies in the widespread concern that eating and drinking during labour will put women at risk of aspirating stomach contents during regurgitation.

5.1 *Risks of aspiration*

This concern is real and serious. However, the risk of aspiration is almost entirely associated with the use of general anaesthesia. Therefore the degree of risk relates directly to the frequency with which general anaesthesia accompanies childbirth, and to the care and skill with which the anaesthetic is administered. The level of risk has always been low, and is now very low. Aspiration plays a very small role in maternal mortality, but it remains a largely unquantified factor in maternal morbidity.

Policies that restrict oral intake during labour aim to reduce the risk of regurgitation and inhalation of gastric contents. Aspiration of food particles of sufficient size to obstruct a main stem or segmental bronchus may result in a collapse of lung tissue distal to the obstruction, possibly with severe hypoxaemia as a consequence. Even in the absence of food particles, if the stomach contents are sufficiently acidic, they will cause chemical burns in the airways, resulting in disruption and necrosis of the bronchial, bronchiolar, and alveolar lining. It is this syndrome of acid aspiration in particular, described by Mendelson over forty years ago, that constitutes the greatest risk in pregnant women who undergo general anaesthesia.

Over the years a number of specific measures have been introduced to avoid aspiration. It has been pointed out repeatedly that failure to apply proper anaesthetic technique is the major reason that deaths from aspiration of gastric contents still occur. Most cases of aspiration could be prevented by a combination of decreasing the frequency of procedures that require anaesthesia (particularly caesarean section), the use of regional anaesthesia whenever feasible, and meticulous attention to safe anaesthetic technique.

5.2 Measures to reduce volume and acidity of stomach contents

Measures to reduce the volume and acidity of gastric contents cannot compensate for inadequate anaesthetic technique. However, such measures are widely used.

5.2.1 Restriction of oral intake

Fasting is the most commonly used measure to reduce the stomach contents. Fasting during labour does not have the desired effect of ensuring an empty stomach and, to quote the conclusions of one study, 'the myth of considering the time interval between the last meal and either delivery or the onset of labour as a guide to gastric content volume should now be laid firmly to rest'. There is no guarantee that withholding food and drink during labour will result in an empty stomach when general anaesthesia should become necessary. No time interval between the last meal and the onset of labour guarantees a stomach volume of less than 100 ml.

The use of a low-residue low-fat diet with the aim of providing palatable and attractive small meals at frequent intervals is a reasonable alternative to fasting. Such a diet could consist of tea, fruit juice, lightly cooked eggs, crisp toast and butter, plain biscuits, clear broth, and cooked fruits. Some women prefer high calorie snacks and drinks.

Nor can fasting during labour be relied on to lower the acidity of the gastric contents. One author commented provocatively: 'Is it not intriguing that, in England and Wales, the number of maternal deaths from acid-aspiration apparently rose only after the institution of severe dietary restriction in labour, amounting in most units almost to starvation?'

Restricting food and drink during labour may result in dehydration and ketosis. Whether the degree of ketosis that occurs in some women during labour is a harmless physiological state or a pathological condition that interferes with uterine action is uncertain. There are no published data about the nutritional needs of labouring women. For some women, these are likely to be similar to those of an individual engaged in strenuous athletic activity.

The most common response to the problem of ketosis in maternity units where eating during labour is prohibited is the use of intravenous infusion of glucose and fluid. The effects of this practice should be carefully weighed against those of the alternative option of allowing women to eat and drink as they desire.

The maternal effects of intravenous infusion of glucose solutions during labour have been evaluated in a number of controlled trials. The rise in mean serum glucose levels appears to be accompanied by a rise in maternal insulin levels and a reduction in mean levels of 3-hydroxybutyrate. The available data show no consistent effect on either maternal pH or lactate levels.

Infusion of glucose solutions to the mother not only results in an increase in plasma glucose levels in the baby but may also result in a decrease in umbilical arterial blood pH. Hyperinsulinism in the fetus can occur when women receive more than 25 grams of glucose intravenously during labour. This can result in neonatal hypoglycaemia and raised levels of blood lactate. Furthermore, the excessive use of salt-free intravenous solutions can result in serious hyponatraemia in both the mother and the fetus.

Thus the use of intravenous infusion of glucose and fluids to combat ketosis and dehydration in the mother may have potentially serious unwanted effects on the baby. These potential hazards might be obviated by the more natural approach of allowing women to eat and drink during labour.

5.2.2 *Pharmacological approaches* The frequency of unpredictably large volumes and equally unpredictable acidity of the stomach contents, whether or not women fast during labour, has led to the deployment of a number of agents in attempts to decrease both the content and the acidity of the stomach in labouring women.

An increased rate of emptying can be achieved with both cimetidine and ranitidine, and this results in quite striking decreases in stomach content. The stomach contents can be emptied mechanically with a stomach tube, or vomiting can be induced with preoperative apomorphine. A comparison of these two methods for women in labour who required a general anaesthetic showed no statistically significant difference in the mean gastric aspirate during operation. The majority of the women having the stomach tube passed found it 'very unpleasant', whereas the majority of those receiving apomorphine found the procedure only 'slightly unpleasant'. It should be noted that neither method guarantees that the stomach will be empty.

The stomach contents can be made less acid by the use of aluminium hydroxide, magnesium trisilicate, sodium citrate, and hydrogen ion antagonists such as cimetidine or ranitidine. Randomized comparisons of different agents have provided no evidence that any particular agent or class of agents influences gastric pH more effectively than others.

However, the effectiveness of these agents in reducing acidity does not necessarily mean that they will have an effect on the incidence or severity of Mendelson's syndrome. Although from 1966 onwards there has been a movement towards the routine administration of antacids to all women in labour, cases of Mendelson's syndrome still occur in women who have had a full regimen of antacid treatment.

6 Maternal position during first stage of labour

Interest in maternal position during the first stage of labour has existed throughout the twentieth century, but until recently there has been relatively little well-controlled research to assess the validity of various strongly held opinions. At the present time lying down in labour continues to be a policy in many maternity units, particularly in some countries, and is required by many of the professionals who provide care during labour. The available data cast doubt on the wisdom of this policy.

6.1 *Effects on blood flow and uterine contractility*

The supine position (lying flat on the back) causes a significant reduction in cardiac output. It is associated with a greater decline in femoral than in brachial arterial pressure, which does not occur in the lateral position or when the uterus is tilted to the left. This observation suggests that the supine position can compromise uterine blood flow during labour.

Contraction intensity is consistently reduced and contraction frequency is often increased when the labouring woman sits or lies supine after being upright. Standing and lying on the side are associated with greater contraction intensity. The efficiency of the contractions (their ability to accomplish cervical dilatation) is also increased by standing and by the lateral position.

The results of several studies suggest that the supine position can adversely affect both the condition of the fetus and the progression of labour by interference with the uterine blood supply and by compromising the efficiency of uterine contractions. Frequent changes of maternal position may be a way of avoiding the adverse effects of supine recumbency. No evidence from controlled studies suggests that the supine position should be encouraged.

6.2 *Effects on the mother and the baby*

The results of controlled trials show that women who were asked to stand, walk, or sit upright during labour had, on average, shorter labours than women asked to remain lying flat. Trials in which an upright position was compared with lying on the side showed no striking differences in the length of labour.

In the only trial in which labour was found to be longer in the ambulant than in the non-ambulant group, women in the recumbent group were permitted to get up if they desired and women in the ambulant group were allowed to rest in bed 'whenever they wanted'. Women in the upright group preferred to recline in bed as labour progressed, often at about 5–6 centimetres dilatation. This suggests that free choice of position may be the most important consideration.

Women allocated to an upright posture used less narcotic analgesics or epidural anaesthesia, and received fewer oxytocics to augment labour. In part this may be because it is easier to administer such drugs to women who are lying in bed. The available data provide no evidence of a consistent effect of position during the first stage of labour on the likelihood of instrumental delivery.

Similarly, there is no consistency in the findings with respect to the condition of the baby. Only one trial reported significantly lower incidences of fetal heart rate abnormalities and depressed Apgar scores associated with an upright position. Other investigators, some of whom used telemetry (a means of monitoring the fetal heart while the woman is mobile) in conjunction with ambulation, did not detect differences in fetal heart rate patterns or Apgar scores. No information is available about the effect of position during the first stage of labour on more substantive indicators of the babies' well-being.

7 Conclusions

Hospital routines are necessary for efficient functioning. The challenge faced by professionals working in maternity units is firstly to maintain and introduce only those routines and rules that have been shown, on balance, to do more good than harm, and secondly to apply those routines flexibly in a way that takes the needs of each individual childbearing woman into account.

The presence of a companion when women first come into hospital and during the initial examination is important to many women. The presence of companions that women choose should be encouraged and facilitated.

Caregivers should pay attention to ways of maintaining the woman's dignity, of providing more privacy, and of treating women more as adults, for example in styles of address and introductions. Abandoning the traditional hospital gown is a step towards this goal.

Women appreciate the efforts of caregivers to inform and consult them about their progress in labour and the care that they are to receive. When choices about care are offered, they should be presented in a manner that allows women to ask for what they want and to discuss their uncertainties.

There is no evidence to support the continuation of outmoded practices such as administering enemas routinely or perineal shaving, which cause discomfort and embarrassment for women.

No presently known measures can ensure that a labouring woman's stomach is empty or that her gastric juices will have a pH greater than 2.5. Enforced fasting in labour, the use of antacids, or preanaesthetic mechanical or chemical emptying of the stomach are only partially effective. All of these have unpleasant consequences, and are potentially hazardous to the mother and possibly her baby.

The syndrome of aspiration of stomach contents under general anaesthesia is rare but serious. It is wise to avoid general anaesthesia for delivery whenever possible, and to use a proper anaesthetic technique with meticulous attention to the known safeguards when general anaesthesia must be used.

Professional requirements that women lie flat during the first stage of labour are less widespread than they used to be, but they still exist. The available evidence suggests that this policy compromises effective uterine activity, prolongs labour, and leads to an increased use of oxytocics to augment contractions.

Controlled trials are required, not only to evaluate the effects of routine hospital policies and practices, but also to evaluate methods of implementing changes when these practices are ineffective, inefficient, or counterproductive.

This chapter is derived from the chapters by Iain Chalmers, Jo Garcia, and Shirley Post (50), Sally Garforth and Jo Garcia (51), Claire Johnson, Marc J.N.C. Keirse, Murray Enkin, and Iain Chalmers (52), and Joyce Roberts (55), in EFFECTIVE CARE IN PREGNANCY AND CHILDBIRTH.

References to primary sources and more complete data for the statements made in this chapter can be found in the source chapters and/or in the following reviews from the *Cochrane pregnancy and childbirth database*:

Hay-Smith, J. and Renfrew, M.J.
— Medicated vs tapwater enema on admission in labour. Review no. 07653.
— Medicated vs soapsuds enema on admission in labour. Review no. 07654.
— Soapsuds vs tapwater enema on admission in labour. Review no. 07655.

Nikodem, C.
— Upright vs recumbent position during first stage of labour. Review no. 03334.

Renfrew, M.J.
— Routine perineal shaving on admission in labour. Review no. 03876.

Renfrew, M.J. and Hay-Smith, J.
— Routine enema on admission in labour. Review no. 03877.

---

30 Care of the fetus during labour

1 Introduction

The aim of monitoring the fetus during labour is to identify fetal problems which, if uncorrected, might cause death or short or long-term morbidity. In theory, at least, it should then be possible to avert the adverse outcomes by appropriate and timely measures.

A variety of methods of monitoring fetal well-being have been evaluated in randomized trials. These trials have generated consistent evidence about the effects of alternative methods of monitoring on fetal and maternal outcome, but controversy still remains about how monitoring should be performed and about the appropriate response to an 'abnormal' result.

2 Clinical methods of fetal monitoring during labour

2.1 *Intermittent auscultation of the fetal heart*

Intermittent auscultation of the fetal heart had become the predominant method of monitoring the fetus during labour by the start of the twentieth century. Although some authorities felt that changes in fetal heart rate during contractions might give an earlier warning sign, auscultation was usually performed between contractions. The criteria for 'fetal distress' were a fetal heart rate above 160 or below 100–120, an irregular heart beat, or the passage of fresh meconium. Despite a trend during the 1970s and early 1980s to replace intermittent auscultation by continuous electronic monitoring, intermittent auscultation continues to be widely used.

2.2 *Assessment of the amniotic fluid*

Passage of meconium is associated with an increased risk of intrapartum stillbirth, neonatal death, and various measures of neonatal morbidity, such as low Apgar score or lowered acid–base status. Part, but by no means all, of this association is explained by respiratory problems due to meconium aspiration.

Thick meconium recognized at the onset of labour carries the worst prognosis, and is associated with a five- to sevenfold increase in the risk of perinatal death. Thick undiluted meconium also reflects reduced amniotic fluid volume at the onset of labour, which in itself is a significant risk factor. Slight staining of the liquor at the onset of labour probably carries a small increase in risk, but this has been disputed.

Meconium staining of the fluid at the onset of labour reflects events that have occurred prior to the onset of labour. It may be a sign of impaired placental function that exposes the fetus to the risk of hypoxia during labour. Passage of meconium for the first time after the onset of labour is less common, and it seems to carry an associated risk intermediate between heavy and light early passage of meconium. Whatever the degree or time of passage of meconium, the risks associated are increased if fetal heart rate abnormalities are also present.

Because of these associations between liquor status and adverse outcomes, routine assessment of the amniotic fluid in early labour, if necessary by amnioscopy or artificial rupture of the membranes, has been recommended as a screening test for identifying fetuses at increased risk. Unfortunately, no controlled evaluation of such a policy has been reported.

3 Continuous assessment of the fetal heart

The development in 1960 of an electrode that could be attached to the fetal scalp led to a great deal of research into the relationship between fetal heart rate changes and events during labour. Various fetal heart rate changes were deemed to indicate 'fetal distress'. These changes were of three types: changes in baseline rate, periodic changes related to uterine contractions, and changes in the variability of the baseline rate. Although tachycardia (abnormally rapid fetal heart rate) and bradycardia (abnormally slow fetal heart rate) were classical signs of 'fetal distress', a distinction could be made between constant or 'baseline' bradycardias, which were almost invariably associated with good fetal outcome, and bradycardias which represented a change in rate from a previously higher level. Periodic changes included late fetal heart rate decelerations that followed, repetitively, the peaks of contractions and which were thought to be due to uteroplacental insufficiency, and variable decelerations with a varying shape and relationship to contractions which were ascribed to umbilical cord compression. A reduction in the normal 'beat-to-beat' variation of the fetal heart rate was attributed to fetal 'depression', reflecting a reduced influence of the central nervous system.

Continuous electronic fetal heart monitoring during labour is now most commonly achieved either 'externally' by Doppler ultrasound or 'internally' by electrocardiography. Doppler ultrasound provides the most reliable method for monitoring the fetal heart rate during late pregnancy and for external monitoring of the fetal heart rate during labour. Ultrasound fetal heart rate monitors are satisfactory for determining the fetal heart rate, but give a poorer impression of fetal heart rate variability than those which use electrocardiographic and phonocardiographic recording. Additional problems are that a maternal heart rate may occasionally be counted in error (which may lead to an inappropriate diagnosis of fetal distress) and that, when the fetus or mother moves, the signal may be lost or artefactual, necessitating frequent repositioning of the transducer.

External monitoring is usually employed during early labour, particularly before the membranes have ruptured. In many centres internal monitoring is preferred later in labour because it provides a more reliable trace and allows the mother greater freedom of movement.

4 Fetal scalp blood acid–base assessment

The technique of sampling blood from the fetal scalp for assessment of the acid–base status during labour was first described in the early 1960s. A scalpel or stylet is passed through the cervical os to make a small incision in the fetal scalp. A sample of blood is then collected in a capillary tube and analysed to determine its acid–base status. The technique has changed little since it was first described, and remains somewhat cumbersome, time consuming, and difficult and uncomfortable for the woman.

5 Comparison of auscultation and electronic fetal monitoring

The two broad approaches to fetal monitoring during labour currently practised are firstly, the use of electronic monitoring in as high a proportion of women as possible, and secondly its use only in women whose pregnancies are deemed to be at high risk.

Continuous electronic fetal heart rate monitoring provides more information than intermittent auscultation with a fetal stethoscope. Listening for a minute every fifteen minutes between contractions, as is commonly employed with intermittent auscultation during the first stage of labour, samples the fetal heart rate for only about 7 per cent of the time and provides relatively little information about the relationship between changes in the fetal heart rate and uterine contractions, or about fetal heart rate variability. The important question is whether or not the increased information provided by continuous electronic monitoring during labour leads to any improvement in outcome for the baby.

Although continuous electronic fetal heart monitoring gives a substantially more accurate measurement of the fetal heart rate, the interpretation of fetal heart rate traces is open to great variation. Tracings are often interpreted differently, not only by different obstetricians, but also by the same obstetrician at different times. The problem with electronic fetal heart monitoring is not with its ability to measure, but in its interpretation.

5.1 Effects on labour and delivery

At least ten randomized controlled trials comparing electronic fetal monitoring with intermittent auscultation of the fetal heart rate, involving over 17 000 women in six countries, have been reported. Caesarean section and operative vaginal delivery rates were both higher in all the electronically monitored groups. The increase in caesarean section rate is much greater when scalp pH estimations are not available. The higher operative delivery rates in the electronically monitored groups are reflected in somewhat higher rates of maternal infection postpartum.

One concern about the use of continuous electronic monitoring (without telemetry, which allows the woman to be mobile) has been the possibility that it prolongs labour by restricting women's movement, but data from the trials, when considered together, do not support this. There is no clear effect of the monitoring method on either the use or the method of analgesia.

5.2 Effects on the fetus and neonate

There is little evidence that the extra caesarean sections associated with electronic fetal monitoring lead to any substantive benefits for the baby. The 61 perinatal deaths that occurred among the 17 000 births in the trials of electronic fetal heart monitoring with fetal pH estimation were evenly distributed between the electronically monitored and control groups. There is no evidence that intensive fetal heart rate monitoring, with or without fetal pH estimation, reduces the risk of low Apgar score or the rates of admission to special care nurseries.

The one measure of neonatal outcome that does seem to be improved by more intensive intrapartum monitoring is neonatal seizures. This effect seems to be restricted to electronic fetal monitoring backed by fetal pH estimation. For babies monitored in this way the odds of neonatal seizure appear to be reduced by about a half. No effect was seen in preterm babies, and so the estimated protective effect may be even greater in term babies. A secondary analysis of the largest trial suggested that the reduced risk of neonatal seizures was limited to labours that were induced or augmented with oxytocin, or that were prolonged.

The finding of a 50 per cent reduction in the risk of neonatal seizures associated with continuous monitoring of the fetal heart and fetal acid–base estimation is potentially very important. Indeed, between a quarter and a third of babies who suffer neonatal seizures die, and a further quarter to a third are seriously impaired in childhood. Nevertheless, the follow-up data available suggest that the neonatal seizures prevented by intensive monitoring are not those associated with long-term impairment. In fact, more babies developed cerebral palsy in the electronically monitored groups than in the intermittently auscultated groups, although the increase was not statistically significant.

Neonatal infection was uncommon in all the trials, but there is no evidence to suggest that intensive monitoring increased this risk. The data are insufficient to explore differences between types of electrode in this respect.

5.3 Mothers' views

Most studies of women's opinions of intrapartum fetal monitoring have been uncontrolled surveys of their views. These surveys suggest that continuous monitoring is acceptable to most women, but that it can also have important adverse consequences for some.

Many women reported that continuous monitoring and recording of the fetal heart rate was reassuring because it demonstrated that the baby was alive, and provided the information that caregivers need during labour. These feelings are enhanced if women are given a clear view of the monitor during labour. Women at relatively high risk of problems during labour and those most knowledgeable about continuous monitoring seem most likely to be reassured. However, detailed information given just before the start of labour appears to have little positive effect on women's perceptions of intrapartum monitoring.

Continuous electronic monitoring of the fetal heart rate can generate anxiety in a number of ways that cannot be predicted in advance for individual women. Some women interviewed in the surveys reported discomfort and restriction of movement, or worries that an electrode would damage the baby's scalp. Others found the monitor a distraction that interfered with their relationships with caregivers and their companion in labour. The trace may become worrying, and this may be particularly disquieting if there is uncertainty about the significance of the 'abnormality' among caregivers. It may be of poor quality or even artefactual, or the monitor itself may malfunction, sometimes repeatedly. An external abdominal transducer may become displaced or a scalp electrode detached. These problems are not uncommon.

How much weight should be given to the various maternal views of electronic fetal heart rate monitoring revealed in these uncontrolled surveys? Three of the randomized controlled trials included an assessment of women's views of the alternative approaches. There were no clear differences between the groups, but there was a tendency for women allocated auscultation to have a more positive experience of labour. Women in the electronically monitored group tended to be left alone more often, but nearly all women interviewed reported that they were able to contact a nurse or doctor at any time. No difference was detected in the proportions of women who reported 'worries or anxieties' during labour or that labour had been 'unpleasant'. In line with the observational studies, more women in the continuously monitored group felt 'too restricted' during labour. Overall, the method of monitoring was less important to women than was the support that they received from staff and companions during labour.

5.4 Technique of electronic fetal monitoring

Direct comparisons of electronic fetal monitoring with and without fetal blood sampling as an adjunct show that access to scalp sampling reduces the number of caesareans for fetal distress, with no clear differences in neonatal outcome. Data from indirect comparisons point both to better maternal and better neonatal outcome if fetal blood sampling is used.

Electronic fetal heart rate monitoring is now often performed intermittently (e.g. for 15 minutes each hour) rather than continuously. Intermittent cardiotocography has not been assessed in randomized trials.

Comparisons of various types of scalp clips for internal cardiotocography show that the Surgicraft clip is less likely than spiral electrodes to require replacement because of either poor quality tracings or detachment of the clip. There were no clear differences in maternal discomfort during monitoring. The only difference in neonatal outcome was more scalp trauma associated with the clip.

One trial compared cardiotocography plus simultaneous analysis of the fetal electrocardiogram (ECG) waveform with cardiotocography alone. Operative delivery for 'fetal distress' was markedly reduced in the cardiotocography plus ECG waveform group, but this effect was to some extent attenuated by operative deliveries for other reasons. Overall there is little reason, as yet, to recommend this approach and further evaluations are necessary.

5.5 Comment

Evidence from the randomized comparisons of alternative methods of fetal heart rate monitoring, particularly the large trial conducted in Dublin, suggests that intrapartum death is prevented equally effectively by intermittent auscultation and by continuous electronic fetal heart rate monitoring, provided that importance is attached to the prompt recognition of intrapartum fetal heart rate abnormalities whatever monitoring policy is adopted. (During the two years of the Dublin trial, for example, the intrapartum death rate was lower than in the preceding and following years.)

The reliability of intermittent auscultation may be increased by the use of hand-held ultrasound monitors when heart sounds are difficult to hear with a conventional stethoscope. There are arguments for always using these devices, partly because they cause less maternal discomfort than a fetal stethoscope. Compliance with intermittent auscultation should be straightforward if a caregiver has responsibility for only one woman during labour. Such individualized attention is likely to have other benefits for a woman. It is to be deplored that staffing and other policies for intrapartum care in many delivery wards make this ideal impossible to meet. The implication is that auscultation may not be performed as frequently or regularly as it should be to provide safe fetal monitoring.

The complexity of continuous electronic monitoring makes it susceptible to technical and mechanical failures. Therefore machine maintenance and replacement and in-service training of personnel are important. Electronic fetal monitoring may also provide suboptimal surveillance if it reduces the frequency with which the caregiver formally checks the fetal heart rate. A fetal heart monitor should be an adjunct to, and not a substitute for, personal care.

The wide variation in the interpretation of continuous fetal heart rate records, even amongst 'experts', demonstrates that this is a major problem with current methods of continuous monitoring.

The policy implications of this review will depend on the importance attached to the observed reduction in the risk of neonatal seizures. Limited follow-up of the children in the Dublin trial who suffered neonatal seizures suggests that the neonatal seizures which are potentially preventable by more intensive monitoring are *not* associated with long-term problems. Nevertheless, some people will consider that neonatal seizures are sufficiently important in their own right for their prevention to form the basis for current policy. On this basis, there is a good case for using more intensive monitoring when labour ceases to be 'physiological', for example during induction or augmentation of labour, if labour is prolonged, if there is meconium-staining of the liquor, or with multiple pregnancy.

For the majority of labours for which no such indications apply, the current evidence suggests that more intensive monitoring increases obstetric intervention with no clear benefit for the fetus. Therefore regular auscultation by a personal attendant, as used in the randomized trials, seems to be the policy of choice in these labours. Such a policy will be difficult to reimplement in the many hospitals whose current policy is universal electronic monitoring. Firstly individualized care during labour is often perceived as not possible, and secondly midwives and others may have lost the ability and confidence to monitor labour by intermittent auscultation.

The choice of technique for fetal heart monitoring has much wider implications than the direct effects on surveillance and physical health of the fetus. Depending on the prevailing system of care for women during childbirth, it may influence the roles and relationships of those involved. With intermittent auscultation the midwife is at the centre of caregiving, with the obstetrician playing a consultative role if the midwife is worried that there may be problems. In contrast, use of continuous electronic monitoring changes the delivery room into an intensive care unit. The midwife takes on a more technical role with obstetricians becoming more centrally involved in routine care. The presence of a monitor may also change the relationships between the woman and her partner on the one hand, and the woman, midwife, and doctor on the other. These wider implications must be recognized.

6 Other methods of fetal monitoring and diagnosis in labour

6.1 Admission test

Intrapartum fetal distress commonly reflects problems which predate the onset of labour. For this reason, there is a strong case for careful risk assessment at the beginning of labour.

A short (15–20 minute) period of external electronic fetal heart monitoring upon admission in labour has been recommended as a screening test for women who are deemed to be at low risk. The rationale for this practice is that it would identify a subgroup of fetuses who would benefit from more intensive monitoring, and might identify major fetal problems that would be missed by intermittent auscultation.

In principle, a screening test on admission in labour as a basis for deciding on selective intensive monitoring is attractive because it should identify fetuses which embark on labour in an already compromised state. However, there are two other components of monitoring that must be fulfilled if the policy is to be effective: whether the test is interpreted accurately, and whether it is acted on appropriately when it is used in clinical practice. These questions can only be addressed satisfactorily in trials to compare the outcomes for women randomly allocated to a policy of screening test plus appropriate response, or to a control group who do not receive the test.

6.2 Intrapartum fetal stimulation tests

Fetal heart rate acceleration is commonly accepted as an indicator of fetal well-being in antepartum non-stress testing. These accelerations in a non-stress test are commonly associated with fetal movements or uterine contractions, but may be evoked by other stimuli such as sound. The observations that fetal heart rate acceleration sometimes coincided with fetal scalp blood sampling, and that scalp blood pH tended to be normal if an acceleration occurred, prompted a prospective study using 'firm digital pressure on the head followed by a gentle pinch of the scalp with an atraumatic clamp'. Response (by fetal heart rate acceleration) to either of these stimuli was associated with a scalp blood pH of greater than or equal to 7.19. Of the fetuses which showed no response, about 40 per cent had pH estimations below 7.19. A study using sound stimulation by an 'electronic artificial larynx' placed over the fetal head had broadly similar results. Fetal heart rate acceleration in response to sound stimulation was associated with fetal pH values greater than 7.25.

These studies require replication. If their results are confirmed, fetal stimulation tests could be useful. They could reduce the need for scalp blood sampling or could be used as an alternative when scalp sampling is either not available or technically impossible. On the basis of currently available evidence a non-reactive stimulation test should be followed by fetal scalp blood acid–base estimation.

7 Conservative management of fetal distress

The most common treatment for intrapartum fetal distress, diagnosed by persistent fetal heart rate abnormalities or depressed fetal scalp blood pH, is prompt delivery. However, many fetal heart rate abnormalities will resolve with simple conservative measures, such as a change in maternal position (to relieve aortocaval compression and pressure on the umbilical cord), interruption of oxytocin administration to increase uteroplacental blood flow, and short-term maternal oxygen administration (to improve oxygen transport to the placenta).

Maternal hypotension often follows the induction of epidural anaesthesia, with consequent fetal heart rate abnormalities. Preloading with intravenous fluids has been shown in a well-conducted trial to counteract the relative hypovolaemia that follows epidural block, and to reduce substantially the frequency of fetal heart rate abnormalities.

Intravenous betamimetics are a useful treatment for 'buying time' when persistent fetal heart rate abnormalities indicate elective delivery. In a randomized controlled trial involving twenty labours characterized by both ominous fetal heart rate changes and a fetal scalp blood pH of less than 7.25, ten of the eleven treated with intravenous terbutaline showed improvement in the heart rate pattern compared with none in the control group. At birth, the babies were less likely to be acidotic and to have low Apgar scores. The results of this trial are supported by other studies which were less well controlled. This short-term improvement could be very useful in situations where facilities for emergency caesarean section are not immediately available, or to allow time to set up regional anaesthesia. The improvement in the trace pattern is sometimes sustained. In these circumstances, labour can be allowed to continue without further intervention.

Another temporizing manoeuvre, amnioinfusion to correct oligohydramnios, may be useful as a method of preventing or relieving umbilical cord compression during labour. Saline or Ringer's lactate is infused through a catheter into the uterine cavity. The technique has been used prophylactically in various conditions which are commonly associated with oligohydramnios, and therapeutically for repetitive variable fetal heart rate decelerations during labour which are attributed to umbilical cord compression.

The use of amnioinfusion for intrapartum umbilical cord compression, either potential or diagnosed by cardiotocography, or for meconium-stained liquor in labour has been evaluated in several controlled trials. This procedure significantly decreases the rate of persistent variable decelerations of the fetal heart. It also improves more substantive outcomes, such as the rate of caesarean section (both overall and for fetal distress) and postpartum endometritis for the mother, and results in fewer babies with birth asphyxia, low Apgar score, or low umbilical cord pH. In the presence of meconium-stained liquor, amnio-infusion also reduces the incidence of meconium aspiration syndrome. No clear adverse effects of the procedure have been noted, but there was a non-significant trend towards an increase in neonatal sepsis.

The prophylactic use of amnio-infusion for women with intrapartum oligohydramnios but no cardiotocographic abnormalities has not shown any advantages over its therapeutic use with the appearance of variable decelerations, and it is associated with a significant increased risk of chorio-amnionitis.

A third approach to the conservative treatment of persistent fetal distress has been to 'treat' the fetus to prevent any adverse effects. Piracetam, a derivative of gamma-aminobenzoic acid given intravenously, is thought to promote the metabolism of the brain cells when they are hypoxic. It has been evaluated in a single placebo-controlled trial. The results suggest that piracetam treatment reduces the need for caesarean section and improves neonatal outcome as judged by the Apgar score and neonatal 'respiratory problems, and signs of hypoxia', but these results must be confirmed by other studies before the approach can be applied clinically. Pyridoxine administration during labour appears to decrease oxygen affinity in cord blood. This may have therapeutic implications, but no information is available as to its clinical importance.

An interesting trial reported in 1959, which compared operative with conservative policies of management for fetal distress, may still be of more than historical interest. In this study, operative delivery rates for fetal distress were 61 per cent in the operative group and 20 per cent with the conservative policy. The rate of perinatal mortality was similar in the two groups.

The changes in obstetric practice and methods of fetal evaluation since this trial was carried out make it difficult to relate the results to contemporary obstetric practice. For those working in situations without modern obstetric facilities, however, it is useful to note that the ready use of operative delivery in the event of meconium-stained liquor or slowing of the fetal heart causes a considerable increase in operative deliveries and has not been shown to reduce perinatal mortality. While these conclusions are not directly applicable today, they give cause to question the interventionist policies for the management of suspected fetal distress which have become accepted practice without being subjected to randomized evaluation.

8 Conclusions

Amniotic fluid which is sparse or contains meconium is associated with an increased risk of perinatal mortality and morbidity. The status of the liquor when the membranes have ruptured spontaneously should be assessed early in labour, and the presence of meconium or low liquor volume should prompt more intensive fetal surveillance. Whether or not routine amnioscopy or artificial rupture of the membranes to assess the liquor is justified is not clear from the available evidence.

Intrapartum death is prevented equally effectively by intermittent auscultation and by continuous electronic fetal heart rate monitoring provided that intrapartum fetal heart rate abnormalities are promptly recognized and followed by an appropriate clinical response, whatever the monitoring policy. The use of electronic fetal monitoring with fetal scalp sampling is associated with a lower rate of neonatal seizures, but not with a lower rate of serious long-term neurological disability.

Electronic monitoring results in an increase in caesarean section rates and postpartum morbidity for the mother, with no compensating benefits to the baby other than the decrease in neonatal seizures. Whether or not it should be used will depend on the importance attached to the prevention of seizures. Selective use of electronic fetal monitoring could be based on assessment of risk by clinical history, and possibly by early intrapartum assessment.

Despite its practical problems, on the basis of current evidence fetal acid–base assessment is an essential adjunct to fetal heart rate monitoring and should be much more widely used during both the second stage and the first stage of labour. When electronic monitoring is used, both false positives (false alarms) and false negatives (a misplaced sense of confidence in the baby's welfare) are reduced by the use of fetal blood sampling as an adjunct.

Intrapartum amnioinfusion is an effective means of treatment for cord compression in the presence of oligohydraminios. Prophylactic amnioinfusion for women with oligohydramnios but without signs of cord compression should not be used. The use of betamimetics for fetal distress in labour is a useful means of 'buying time' to permit definitive management of the situation.

A number of lessons can be learned from the trials of intrapartum electronic fetal heart rate monitoring. First, 'more information' is not necessarily beneficial and can have harmful effects. Second, if a test result is predictive of an adverse outcome, it should not be taken as self-evident that intervention based on the results of that test will prevent or ameliorate that outcome. Third, the relationship between measures in the neonatal period and long-term outcome is not straightforward and measures in the neonatal period should not be used as surrogates of long-term outcome. Fourth, death and serious childhood morbidity are (thankfully) rare and very large numbers of labours must be studied if the evaluation is to be reliable. Fifth, there should be a healthy scepticism about new methods of intrapartum surveillance such that, when their development has reached the point that they are considered ready for use in clinical practice, they are introduced into clinical practice only in the context of large-scale randomized controlled trials.

This chapter is derived from the chapters by Adrian Grant (54) and by Robert Bryce, Fiona Stanley, and Eve Blair (76) in EFFECTIVE CARE IN PREGNANCY AND CHILDBIRTH.

References to primary sources and more complete data for statements made in this chapter can be found in the source chapters and/or in the following reviews from the *Cochrane pregnancy and childbirth database:*

Grant, A.M.
— EFM vs intermittent auscultation in labour. Review no. 03884.
— EFM + scalp sampling vs intermittent auscultation in labour. Review no. 03297.
— EFM alone vs intermittent auscultation in labour. Review no. 03298.
— Fetal blood sampling as adjunct to heart rate monitoring. Review no. 07018.
— Liberal vs restrictive use of EFM in labour (low-risk labours). Review no. 03886.
— Liberal vs restrictive use of EFM in labour (all labours). Review no. 03885.
— Surgicraft clip vs all types spiral fetal electrode. Review no. 06014.
— Berkeley helix scalp electrode vs Corometrics spiral. Review no. 06009.
— Surgicraft clip vs Corometrics spiral fetal electrode. Review no. 06010.
— Surgicraft clip vs Rocket helix fetal electrode. Review no. 06011.
— Surgicraft clip vs Hewlett-P helix fetal electrode. Review no. 06012.
— Hewlett-Packard vs Rocket helix fetal electrode. Review no. 06013.

Hofmeyr, G.J.
— Intravenous betamimetics for intrapartum fetal distress. Review no. 04138.
— Amnioinfusion in intrapartum umbilical cord compression. Review no. 04137.
— Amnioinfusion for meconium-stained liquor in labour. Review no. 05379.

— Prophylactic vs therapeutic amnioinfusion for intrapartum oligohydramnios. Review no. 07642.
— Piracetam for fetal distress in labour. Review no. 04141.
— Operative vs conservative care for 'fetal distress' in labour. Review no. 04139.
— Maternal hydration in oligohydramnios. Review no. 06646.

Mahomed, K.
— Pyridoxine administration during labour. Review no. 06506.

Neilson, J.P.
— Intrapartum fetal ECG plus heart rate recording. Review no. 07137.

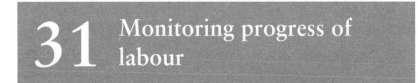

31 Monitoring progress of labour

1 Introduction   2 Recognition of the onset of labour
3 Condition of the mother   4 Uterine contractions
5 Cervical dilatation   6 Descent of the presenting part
7 Normal labour   8 Recording the progress of labour
9 Conclusions

1 Introduction

Labour is a special time, both emotionally and physically, for each woman. It is a time of intense physical activity, stress, and pain, and it may prove to be a time of overt or hidden danger. The care that a woman receives during labour should not only help her to cope with the effort, stress, and pain, but should also minimize or remove the danger.

The purpose of monitoring progress in labour is to recognize incipient problems, so that their progression to serious problems may be prevented. Prolonged labour is strongly associated with several adverse outcomes. It can lead to maternal exhaustion, perinatal asphyxia, and even death. Thus the anticipation of prolonged labour cannot be considered a trivial issue, as inefficient uterine action can be corrected and some adverse outcomes can be prevented. This monitoring must be carried out with thought and consideration, rather than as an unthinking routine or a procrustean attempt to make all women fit predetermined criteria of so-called normality.

2 Recognition of the onset of labour

Women usually make a diagnosis of labour themselves on the basis of painful regular contractions. Sometimes they make the diagnosis after a show of mucus or blood, or after rupture of the membranes. On admission to hospital the women's self-diagnosis of labour may, or may not, be confirmed by the professional staff.

One of the most important decisions in care in labour is to recognize whether or not labour has started. True labour must be differentiated from false labour. The clinical diagnosis of active labour is easy when 'pains' plus progress in cervical dilatation are present. To confirm or deny the diagnosis of labour in a woman self-admitted as 'in labour' is much more difficult when the cervix is uneffaced and closed.

Labour is, by definition, the presence of regular uterine contractions, leading to progressive effacement and dilatation of the cervix, and ultimately to delivery of the baby. While there is no difficulty in confirming the presence of labour when it is strong and well established, the diagnosis is seldom as clear cut as this definition would suggest. Often, the time of the onset of labour is not precisely known, and at least two assessments are required to establish whether or not progressive effacement and dilatation of the cervix is occurring.

Despite the difficulties in establishing reliably when labour has started, the most convenient and most frequently used marker of the onset of labour for women delivering in hospital (although an arbitrary rather than a biologically correct starting point) is the time when the woman is admitted in labour. For women planning to give birth at home, the time at which the midwife arrives (having been called by the mother) may similarly be used. This serves as a semi-objective surrogate index of the onset of labour, and is a practical starting point from which subsequent progress can be monitored.

The point in labour at which a woman presents herself for admission to hospital, or asks her caregiver to attend, will vary from woman to woman. Several factors may influence this decision, including the way she feels, her expectations of labour, her anxiety about arriving too early or too late, and any complications that may have arisen. It will also depend on the advice that she has been given as to how and when she should recognize herself to be in labour and when to come to the hospital, which in turn will depend on the admission policy of each maternity unit. All these factors will affect when a woman is first seen in labour and hence the apparent length of her labour.

The timing of hospital admission may have important consequences for the progress of labour. Studies show that although the women who come to hospital early have a shorter total length of labour than those admitted in more advanced labour, they have more diagnoses of 'difficult labour' recorded, receive more intrapartum interventions, and more caesarean sections.

It is unlikely that any universal 'best' time for asking the midwife to come or for hospital admission in labour will be determined. For most women, the 'best' time is when they feel that they would be happier or more comfortable in hospital or with the midwife in attendance.

3 Condition of the mother

The physical and mental state, the comfort, and the well-being of the woman must be just as carefully monitored during labour as the progress of contractions or the state of the cervix. The possible causes of symptoms such as nausea, dyspnoea, or dizziness should be fully assessed and treatment provided if necessary. Fear can be allayed and stress alleviated by the presence of companions and competent caring staff. The intensity of pain that she experiences will determine her need for, and the timing of, pain relief.

Adequate attention must be paid to her physical condition. In most circumstances this will include, at least, assessment of her blood pressure, pulse, and temperature. Although such assessments have become traditional, there is little agreement as to how frequently they should be performed. The value, if any, of frequent assessments of pulse and blood pressure in normal labour to screen for problems such as intrapartum pre-eclampsia is unknown. It is likely to be small. In the presence of known or suspected abnormality (such as antepartum or intrapartum haemorrhage, or pre-eclampsia) such assessments should be made as frequently as necessary, or even continuously, rather than being dictated by a rigid schedule that is applied to all women. It is questionable whether any useful purpose is served by routinely repeated observations of these parameters in healthy women in apparently normal labour.

4 Uterine contractions

Labour is initiated, and progress maintained, by the contractions of the uterus. Almost always the woman herself is aware of the contractions, their frequency, their duration, and their strength. These parameters can be confirmed by abdominal palpation. Self-report by the woman, supplemented by abdominal examination when required, is quite sufficient to monitor the contractions in most situations.

However, abdominal palpation cannot accurately quantitate the changes in uterine pressure resulting from the contraction, and this constraint also applies to the record of uterine contractions made by an external tocodynamometer. It may provide an accurate record of the frequency and, to a lesser extent, of the duration of contractions, but not of their intensity.

The limited information available from controlled trials shows no advantage to be gained from the use of intrauterine pressure catheter monitoring in the treatment of delays in labour progress.

5 Cervical dilatation

The rate of dilatation of the cervix is the most exact measure of the progress of labour. Cervical dilatation is usually estimated in centimetres, from 0 cm when closed to 10 cm at full dilatation. However, assessment of cervical dilatation is not as precise as one would like to believe. To our knowledge, no studies of either inter- or intra-observer variation have been reported, but personal experience has shown substantial variations in estimates by different observers in the same situation, and even by the same observer on repeat examination. There is no clear guidance from the literature as to the most accurate time to assess the dilatation in relation to a contraction, but consistent timing of observation is probably important when assessing progress.

Cervical dilatation and effacement can be assessed directly by vaginal examination or indirectly by rectal examination. Rectal examinations were advocated toward the end of the nineteenth century in the belief that, unlike vaginal examinations, they did not cause contamination of the genital tract. Several studies comparing vaginal and rectal examinations were made in the United States from the mid-1950s to the 1960s. All showed a similar incidence of puerperal infection whether rectal or vaginal examinations were employed during labour. Women's preference for vaginal rather than rectal examinations was clearly demonstrated in a randomized clinical trial.

On the basis of these studies, vaginal examinations have become standard practice for the assessment of cervical dilatation during labour, although a few units have continued the use of rectal examination for assessment. Overall, the available evidence suggests that rectal examinations have no place in monitoring the progress of labour.

In many units masks are worn when vaginal examinations are performed, but there is no evidence that they are of any benefit. In view of the fact that masks have not been shown to be of value during vaginal surgery or in the delivery room it is highly unlikely that any infections are prevented by this practice.

The recommended frequency of vaginal examinations to assess the progress of cervical dilatation varies greatly among units and in the literature. This variation illustrates the lack of consensus for the optimal timing of vaginal examinations in labour. Like all assessments in labour, it would seem most sensible that care should be individualized. The number and timing of vaginal examinations should be frequent enough to permit adequate assessment of progress and to detect any problems promptly, but not more frequently than necessary to accomplish this end.

6 Descent of the presenting part

If the head is presenting, its relationship to the brim of the pelvis can be determined by abdominal or vaginal examination. Descent can be estimated abdominally by determining the amount of the baby's head that is still above the pelvic brim. Abdominal assessment avoids the need for vaginal examination, and is not influenced by the presence of a caput succedaneum or moulding. On vaginal examination the level of the presenting part can be related to the ischial spines. Moulding of the fetal head, an important observation in following the progress of labour if cephalopelvic disproportion is suspected, can also be determined by vaginal examination. Given the complementary information that can be obtained, both methods of examination should be carried out before operative delivery is undertaken.

7 Normal labour

Normal labour can be defined either in terms of the total length of labour or as a rate of progress of cervical dilatation (usually expressed in centimetres per hour). The latter measure is clinically more useful, as the total length of labour can only be known in retrospect.

A rate of 1 cm/hour in the active phase of labour is often accepted as the cut-off between normal and abnormal labour. The validity of this can certainly be challenged. Many women who show slower rates of cervical dilatation proceed to normal delivery. A rate of 0.5 cm/hour may be more appropriate as a lower limit for defining normal progress, but this should also be interpreted with discretion in the context of the woman's total well-being.

8 Recording the progress of labour

When monitoring the progress of labour, recording the findings is almost as important as making the assessments. The primary reasons for doing this are to make the degree of progress readily apparent, so that problems will be recognized early, and to facilitate transfer of information to other caregivers. Several methods of recording measures of progress are in current use.

A time-based diary of events permits a detailed documentation of all important maternal and fetal assessments, but the recording and inspection of such a record can be tedious. It is often difficult to follow, particularly when labour is prolonged or when there is a change of staff. A more structured representation of events and progress can facilitate early recognition of potentially correctable problems.

The partogram, a structured graphical representation of the progress of labour, has been adopted in many units throughout the world. In addition to the graph depicting cervical dilatation in relation to time, space can be provided for notes on the frequency of contractions, medications, the fetal heart rate, and other important events. With the use of a partogram the progress of labour can be seen at a glance on one sheet of paper, failure to progress can be readily recognized, and the writing of lengthy descriptions can be avoided. It is simple to use, a practical teaching aid, and an efficient means of exchange of technical information about labour progress between teams of caregivers. However, too much reliance on partograms, and particularly on strict protocols of action related to partogram patterns, can be an agent for regimenting labour rather than for caring for the woman in labour.

9 Conclusions

The well-being of the mother as well as that of the fetus must be carefully monitored during labour. This monitoring does not necessarily require the use of special equipment, but it always requires careful and individualized observation.

Monitoring the progress of labour requires more than the assessment of uterine contractions and cervical dilatation. The rate of progress must be considered in the context of the woman's total well-being, rather than as a mere physical phenomenon. A dilatation rate of 1 cm/hour in a woman who is having strong contractions and is in severe distress is far more worrying than a rate of 0.3 cm/hour in a woman who is comfortable, walking around, drinking cups of tea, and chatting with her companions (see Chapter 35).

Vaginal rather than rectal examination should be used to assess the progress of labour, but no more often than deemed necessary. Slow progress should alert one to the possibility of abnormal labour, but should not automatically result in intervention.

This chapter is derived from the chapter by Caroline Crowther, Murray Enkin, Marc J.N.C. Keirse, and Ian Brown (53) in EFFECTIVE CARE IN PREGNANCY AND CHILDBIRTH.

References to primary sources and more complete data for statements made in this chapter can be found in the source chapter and/or in reviews from the *Cochrane pregnancy and childbirth database*.

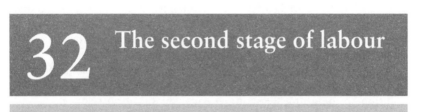

32 The second stage of labour

1 Introduction   2 Diagnosis of the onset of the second stage of labour   3 Pushing during the second stage of labour   4 Position during the second stage of labour   5 Duration of the second stage of labour   6 Care of the perineum
6.1 *Guarding and massaging the perineum*   6.2 *Episiotomy*
6.3 *Liberal use of episiotomy*   7 **Delivery**   8 **Conclusions**

1 Introduction

In some respects it is undesirable to separate consideration of care during the second stage of labour from care during the first stage. Nevertheless, the second stage is a period during which the whole tempo and nature of activities surrounding labour tend to change. It is a time when women may become vulnerable and dependent on the influence of those who assist them. Discussion about aspects of care is not easy at this time, and this leaves the caregiver with even more than usual responsibility to safeguard the interests of the mother and baby.

2 Diagnosis of the onset of the second stage of labour

By definition, the second stage of labour, which ends with the birth of the baby, begins when the cervix is fully dilated. This 'anatomical' onset may or may not coincide with the onset of the expulsion phase, when the mother begins to feel the urge to bear down. In some women, the urge to bear down occurs before the cervix is fully dilated; for others, this urge may not come until well after full cervical dilatation is achieved.

   The mother herself may signal the transition into the expulsive phase in words, by action, by a change in the expression on her face, or in the way she may squeeze her companion's hand. If the presenting part is visible at the introitus, full dilatation is easily confirmed. If the mother feels that she wishes to start pushing when the progress of labour gives reason to believe that the cervix may not be fully dilated, cervical dilatation should be checked by vaginal examination. If the cervix is only 6 or 7 cm dilated, the woman should be asked to find the position in which she feels most comfortable and try to resist the urge to push by trying alternatives such as breathing techniques; an epidural may be given if necessary. If there is only a rim of cervix left and the woman has an irresistible urge to push, she may feel better doing so; it is unlikely that any harm will come from this spontaneous pushing before full dilatation, as long as she does not exhaust herself.

When an epidural anaesthetic has been administered for pain relief in labour, maternal bearing-down efforts are reduced, delayed, or abolished. Because the woman is usually in a supine or semi-recumbent position, abdominal palpation is a satisfactory way of gauging descent of the presenting part. Full dilatation can be tentatively diagnosed in this way and confirmed either by the appearance of the presenting part at the vulva or by vaginal examination.

3 Pushing during the second stage of labour

In a study of healthy nulliparous women who had received no formal childbirth education and were allowed to push spontaneously without any directions from those caring for them, three to five relatively brief (4–6 seconds) bearing-down efforts were made with each contraction. The number of bearing down efforts per contraction increased as the second stage progressed, and most were accompanied by the release of air. The minority of bearing down efforts that were not accompanied by the release of air were accompanied by very brief periods of breath holding (lasting less than 6 seconds). Despite this pattern of breathing, the average length of the second stage of labour was 45 minutes, and did not exceed 95 minutes for any of the 31 women studied.

The duration of breath holding (less than 6 seconds) in the women who spontaneously used this technique contrasts with the 10–30 second duration which is widely advocated for sustained directed bearing-down efforts. Although sustained bearing-down efforts accompanied by breath holding result in shorter second stages of labour, the wisdom of the commonly given advice to make these efforts can be questioned. In addition to respiratory-induced alterations in heart rate and stroke volume, maternal bearing down efforts, particularly when the mother is lying flat on her back, are associated with compression of the distal aorta and reduced blood flow to the uterus and lower extremities. In combination with sustained maternal breath holding, these effects may compromise fetal oxygenation.

In all the published controlled trials comparing different approaches to bearing down in which cord umbilical arterial pH assessments were available, mean cord umbilical arterial pH was lower in the group in which sustained or early bearing down had been encouraged. Sustained bearing down efforts also appear to predispose to abnormalities of the fetal heart rate and depressed Apgar score.

In women with epidural anaesthesia, rotational forceps deliveries are more common among women who have been encouraged to bear down relatively early. There is no evidence that a policy of early bearing down has any compensating advantages for either the mother or the baby.

Despite the limitations of the available evidence, a consistent pattern emerges. The widespread policy of directing women to use sustained and early bearing-down efforts may well result in a modest decrease in the duration of the second stage, but this does not appear to confer any benefit; indeed it seems to compromise maternal–fetal gas exchange.

4 Position during the second stage of labour

The use of upright positions such as standing, kneeling, sitting on a specially designed chair, or squatting for delivery is common in many cultures. Yet, in institutions women have been expected to adopt recumbent positions for childbirth. Constraining women to adopt positions that they find awkward or uncomfortable can only be justified if there is good evidence that the policy has important advantages for the health of either the mother or her baby.

Upright posture has been compared with the recumbent position during childbirth in several trials. In most of these, either specially designed obstetric chairs or a back-rest or wedge were used to support the upright position.

The mean duration of the second stage of labour was found to be shorter with the upright position in most of the trials. There is no evidence that posture during the second stage of labour affects the incidence of either operative delivery or perineal trauma.

More mothers in the groups that used upright positions during the second stage of labour expressed a positive response about the position. The perceived advantages included less pain and less backache. A majority of women wished to use a birth chair or upright position for a subsequent birth.

Women using birth chairs during the second stage of labour are at increased risk of postpartum haemorrhage. This tendency to postpartum haemorrhage is likely to arise from perineal trauma exacerbated by obstructed venous return. Excessive perineal oedema and haemorrhoids have been observed in women who are upright in birth chairs for extended periods of time.

Abnormal fetal heart rate patterns seem to be less frequently observed, and mean umbilical arterial pH tends to be higher in babies born to women who use the upright position for delivery, although the differences observed do not achieve statistical significance. These apparent effects may be due to the avoidance of the aortocaval compression associated with lying down. This alteration in fetal acid–base status was also observed in two trials comparing the effects of the supine position with a 15° left lateral tilt. Babies whose mothers were lying flat on their backs had lower umbilical cord arterial pH values than those whose mothers were lying on their left sides.

Adoption of the squatting position for excretion, resting, and other reasons is not common in industrialized societies, and many people find it uncomfortable for long periods of time. The relative merits and possible disadvantages of the squatting position for birth have not yet been systematically explored.

5 Duration of the second stage of labour

The second stage of labour has been considered to be a time of particular risk to the fetus for well over a century. Echoes of this view exist today in the widespread policies of imposing arbitrary limits on the length of the second stage.

Statistical associations have been demonstrated between prolonged second stage of labour and obviously undesirable outcomes such as perinatal mortality, postpartum haemorrhage, puerperal febrile morbidity, and neonatal seizures, as well as with outcomes of less certain significance relating to the acid–base status of the baby at birth. On their own, these associations are not sufficient justification for concluding that the length of the second stage of labour *per se* is the crucial variable.

Curtailing the length of the second stage of labour by active pushing or operative delivery can modify the decline in fetal pH that tends to occur over the course of labour. However, without some evidence that this policy has a beneficial effect on important outcomes, the maternal trauma and occasional fetal trauma resulting from the increased surgical interference can hardly be justified. One trial, available in abstract form only, found a shortened duration of second stage and decreased operative delivery rate with the use of an inflatable abdominal girdle during the second stage.

Decisions about curtailing the second stage of labour should be based on the same principles of monitoring the well-being of mother and baby that apply during the first stage of labour. If the mother's condition is satisfactory, the baby's condition is satisfactory, and there is evidence that progress is occurring with descent of the presenting part, there are no grounds for intervention. A single trial has assessed the use of prophylactic betamimetics with the aim of reducing fetal distress during the second stage; no positive effects were found.

Maternal exhaustion can occur at any time during labour, but is more likely to occur during the second stage when the extra effort of pushing is added to the stress of the contractions. If the mother is not unduly distressed and is not actively pushing (particularly when she has epidural analgesia), there is no reason to think that the second stage is any more likely to cause exhaustion than the first stage.

Monitoring the fetal heart using intermittent auscultation may on occasions pose difficulties, as it is sometimes hard to find the fetal heart when the baby moves down into the pelvis. It can be frustrating and uncomfortable for a woman to have people continually trying to listen to her baby's heart, or to have to change her position in order to facilitate fetal auscultation. In these circumstances, electronic fetal monitoring is often more comfortable and less disruptive for the woman.

Failure of the presenting part to descend may be due to inadequate or incoordinate uterine contractions, to malposition or malpresentation of the baby, or to cephalopelvic disproportion. The cause of this failure to progress must be diagnosed and appropriately treated. Malpresentation or minor degrees of cephalopelvic disproportion may sometimes be overcome by encouraging the mother to vary her position. Intravenous oxytocin can be used if contractions are inadequate. Instrumental or manual manipulation, or sometimes caesarean section, may be necessary.

6 Care of the perineum

Reducing the risk of perineal trauma is important because the consequent discomfort can dominate the experience of early motherhood and result in significant disability during the months and years that follow. This risk can be minimized by intervening to expedite delivery only on the basis of clear maternal or fetal indications rather than 'because of the clock', and by the use of the vacuum extractor rather than forceps when instrumental delivery is required.

Perineal damage may occur either from spontaneous lacerations or from episiotomy. Although some individual accoucheurs appear to be particularly skilful in assisting birth in a way that minimizes perineal trauma, in most hospitals at least two-thirds of all women giving birth for the first time sustain trauma sufficient to require suturing.

6.1 *Guarding and massaging the perineum*

The practice of guarding the perineum, with the birth attendant's fingers held against the perineum during contractions, is widespread. Many believe that this practice supports the tissues sufficiently to reduce the risk of spontaneous trauma. The practice might seem logical if combined with gentle pressure applied to the fetal head to control the speed of crowning, as this is the time that the perineal tissues are most at risk of spontaneous damage. Unfortunately there have been no formal evaluations of this or alternative strategies, such as not touching either the perineum or the baby's head unless necessary to prevent rapid expulsion.

'Ironing out' or massaging the perineum as the second stage of labour advances, sometimes with an emollient such as olive oil or the application of a hot pad, is also designed to stretch the tissues and reduce the risk of trauma. These techniques have enthusiastic advocates, as well as detractors. The latter suggest that touch may be a disruptive distraction, and that the increase in vascularity and oedema in tissues that are already at risk of trauma is counterproductive. Neither of these diametrically opposed opinions can be supported with evidence from controlled comparisons.

6.2 *Episiotomy*

If monitoring during the second stage of labour suggests that either the fetus or the mother has become distressed during the second stage of labour, or that progress has ceased, it may be necessary to hasten delivery either by instrumental delivery or by episiotomy.

Like any surgical procedure, episiotomy carries a number of risks: excessive blood loss, haematoma formation, and infection. Trauma involving the anal sphincter and rectal mucosa may lead to rectovaginal fistulae, loss of rectal tone, and perineal abscess formation.

Midline and mediolateral episiotomies have been compared in only one controlled trial. The perineum was significantly less bruised in the women who had the midline operation. However, third-degree lacerations occurred in almost a quarter of the women in the midline episiotomy group and in less than 10 per cent of those in the mediolateral group. The amount of pain experienced was similar in the two groups, as was the proportion of women requiring analgesics.

Women in the midline episiotomy group began intercourse significantly earlier than those who had a mediolateral incision. At follow-up, the investigators judged the cosmetic appearance and texture of the scar to be somewhat better following the midline operation. Well-controlled research to assess the short- and long-term advantages and disadvantages of midline and mediolateral episiotomies is long overdue.

Episiotomies are sometimes performed using scissors, sometimes with a scalpel. Those who favour scissors maintain that they are less likely to damage the presenting part of the baby and more likely to promote haemostasis in the wound edges because of their crushing as well as cutting action. Those who favour the scalpel say that it minimizes trauma and thus is followed by better healing of the perineal wound. There are no data on which to base any judgements about the validity of these claims.

6.3 *Liberal use of episiotomy*

Episiotomy is carried out on between 50 and 90 per cent of women giving birth to their first child. Although this extensive use of episiotomy has been questioned, the prevailing view appears to be that the policy is justified.

Liberal use of an operation with the risks described above could only be justified by evidence that such use confers worthwhile benefits. There are four postulated benefits of a liberal use of episiotomy: prevention of damage to the anal sphincter and rectal mucosa (third- and fourth-degree lacerations), easier repair and better healing than with a spontaneous tear, prevention of trauma to the fetal head, and prevention of serious damage to the muscles of the pelvic floor. There is no evidence to support these postulated benefits of liberal use of episiotomy.

Liberal use of episiotomy is associated with higher overall rates of perineal trauma. There is no evidence that this policy reduces the risk of serious perineal or vaginal trauma. Where data from randomized trials are available, women allocated to either a liberal or restricted use of episiotomy experienced a comparable amount of perineal pain at ten days and three months postpartum. Women allocated to liberal use of episiotomy resumed intercourse somewhat later than those who had experienced restricted use of the operation, but the proportions of women experiencing pain on intercourse three months and three years after delivery was almost identical.

There is no evidence to support the suggestion that liberal use of episiotomy minimizes trauma to the fetal head. Data from the randomized trials show similar distributions of Apgar scores and rates of admission to the special care nursery.

A frequent claim in support of liberal use of episiotomy is that it prevents pelvic relaxation during delivery and thereby prevents urinary incontinence and genital prolapse. If the intention really were to protect the pelvic floor, a much more extensive incision than is usual would almost certainly be required. The question of whether stress incontinence is caused or aggravated by vaginal delivery remains controversial.

Liberal and restricted use of episiotomy are associated with contrasting patterns of trauma: liberal use is associated with a lower frequency of anterior vaginal and labial tears. This raises the possibility that episiotomy may have a more specific protective effect on the tissues around the bladder neck. However, there is no good evidence that more liberal use of episiotomy is protective against this distressing condition. In a three-year follow-up of a comparison of liberal with restricted use of episiotomy, rates and severity of incontinence were almost identical in the two trial groups.

7 Delivery

Women may choose a variety of positions for giving birth if they are encouraged to discover for themselves which is most comfortable for them. Accoucheurs should be sensitive to these differences among women. There is no justification for requiring, or actively encouraging, a supine or lithotomy position during childbirth; these positions are often particularly painful and disruptive at this point. Women who choose to lie down for delivery often appear to find a lateral position comfortable.

The woman will depend on the midwife's guidance to moderate her pushing effort to allow an unhurried gentle delivery of the head. This can be achieved by interspersing short pushing efforts with periods of panting, thus giving the tissues time to relax and stretch under pressure. Using this approach, several contractions may occur before the head crowns and is delivered.

After delivery of the head, the shoulders rotate internally. If the umbilical cord is tightly wound around the baby's neck, it may be possible to loosen it and then loop it over the baby's head. If necessary, it can be clamped and severed. Once rotation is complete, the shoulders are delivered one at a time to reduce the risk of perineal trauma. When the mother is in the semi-recumbent position, the anterior shoulder may deliver first; in the squatting or kneeling position, the posterior shoulder may be released first. The mother may then wish to grasp her baby and complete the rest of the delivery herself.

Difficulty with delivery of the shoulders is rare following spontaneous birth of the head. Delivery of the shoulders should not be attempted until they have rotated into the anteroposterior axis. Posterior traction on the head, combined with the mother's expulsive efforts, is usually sufficient to effect delivery of the anterior shoulder. The accoucheur should be aware of techniques to overcome the problem of shoulder dystocia on the rare occasions in which it does occur. These include wide abduction of the mother's thighs and complete flexion of her hips, manual rotation of the posterior shoulder anteriorly; and, if necessary, sustained pressure exerted by an assistant directly above the pubic bone.

8 Conclusions

There are no data to support a policy of directed pushing during the second stage of labour, and some evidence to suggest that it may be harmful. The practice should be abandoned.

Similarly, there is no evidence to justify forcing women to lie flat during the second stage of labour. With some reservations, the data tend to support the use of upright positions. There is a tendency for recumbency to lengthen the second stage of labour, to reduce the incidence of spontaneous births, to increase the incidence of abnormal fetal heart rate patterns, and to reduce umbilical cord blood pH. Although some birth attendants report that upright positions sometimes caused them inconvenience, there has been a consistently positive response from the women who have used an upright position for birth. However, at least some of the birthing chairs that have been introduced during recent years appear to predispose to perineal oedema and venous engorgement which, in conjunction with perineal trauma, can result in the loss of substantial amounts of blood. Use of a birthing chair is not the only way of adopting an upright position during labour. The mother should be encouraged to use the position that she prefers.

There is no evidence to suggest that, when the second stage of labour is progressing and the condition of both mother and fetus is satisfactory, the imposition of any upper arbitrary limit on its duration is justified. Such limits should be discarded.

There is no evidence to support the practices of guarding and 'ironing out' or massaging the perineum, or to support claims that liberal use of episiotomy reduces the risk of severe perineal trauma, improves perineal healing, prevents fetal trauma, or reduces the risk of urinary stress incontinence after delivery. Episiotomy should be used only to relieve fetal or maternal distress, or to achieve adequate progress when it is the perineum that is responsible for lack of progress.

This chapter is derived from the chapter by Jennifer Sleep, Joyce Roberts, and Iain Chalmers (66) in EFFECTIVE CARE IN PREGNANCY AND CHILDBIRTH.

References to primary sources and more complete data for the statements made in this chapter can be found in the source chapter and/or in the following reviews from the *Cochrane pregnancy and childbirth database:*

Hay-Smith, J. and Renfrew, M.J.
— Insufflatable abdominal girdle in second stage labour. Review no. 07656.

Hofmeyr, G.J.
— Prophylactic iv betamimetics in 2nd stage of labour. Review no. 06644.

Nikodem, C.
— Early vs late pushing with epidural anaesthesia in 2nd stage of labour. Review no. 03403.
— Sustained (valsalva) vs exhalatory bearing down in 2nd stage of labour. Review no. 03336.
— Upright vs recumbent position during second stage of labour. Review no. 03335.
— Birth chair vs recumbent position for second stage. Review no. 04735.
— Lateral tilt vs dorsal position for second stage. Review no. 03402.

Johanson, R.
— Forceps vs spontaneous vaginal delivery. Review no. 07087.

Renfrew, M.J.
— Vacuum extraction compared to normal delivery. Review no. 06517.
— Liberal use of episiotomy for spontaneous vaginal delivery. Review no. 03695.

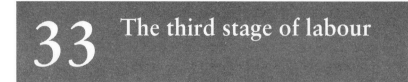

33 The third stage of labour

1 Introduction

After the climactic experience of giving birth to a baby, the delivery of the placenta (the third stage of labour) may seem tame and rather dull. However, this period is a time of great potential hazard. Postpartum haemorrhage remains an important cause of maternal morbidity and mortality; retained placenta can necessitate manual removal, and inversion of the uterus is a rare but frightening and life-threatening complication. The effects of care during this period can have important consequences. The best approach to third-stage management would be one which minimizes serious adverse effects, such as blood loss and retained placenta, and which interferes as little as possible with physiological processes, and with interaction between mother and baby.

In practice, caregivers have a number of choices to make in relation to the third stage. The first is whether or not to take an active or an expectant (physiological) approach. The second choice is to decide which of the components of that approach to use.

Components of active management include prophylactic use of oxytocic drugs (oxytocin and/or ergometrine), early clamping and division of the umbilical cord, and controlled cord traction for delivery of the placenta. Some caregivers may also apply fundal pressure. Components of expectant management include watchful waiting, no use of prophylactic drugs, cord traction, or fundal pressure, maternal effort aided by gravity for delivery of the placenta, and clamping and division of the umbilical cord after delivery of the placenta.

Those who adopt an active approach may differ among themselves in the components they use, as may those who adopt an expectant approach. Therefore in practice the two approaches are not as separate as they appear at first; some commentators have suggested a third category, the 'piecemeal' approach where a combination of both approaches is used by some caregivers. As a result of this overlap, evaluation of the 'packages' of care is difficult and its results are not always easy to interpret.

2 Components of active management of the third stage

2.1 *Routine prophylactic use of oxytocics*
While few would dispute the valuable contribution of oxytocic drugs in the *treatment* of postpartum haemorrhage, the routine prophylactic administration of these drugs to reduce the risk of postpartum haemorrhage has not been so universally accepted. During the 1940s several uncontrolled studies claimed a beneficial effect for routine administration of ergometrine in the management of the third stage of labour. Later investigators reported advantages from combining ergometrine with oxytocin (Syntometrine). The value of the combined preparation was claimed to lie in the rapid effect of oxytocin and the sustained effect of ergometrine.

Ten trials, including a total of well over 4000 women, have compared women who did or did not receive prophylactic oxytocic preparations (in combination with other, variable, components of active management). The available data suggest that routine administration of oxytocics results in an important reduction in the risk of postpartum haemorrhage. The odds of this risk were reduced by about 60 per cent, and the odds of needing therapeutic oxytocics by about 70 per cent.

The effect of prophylactic oxytocics on retention of the placenta is still not clear. There is some suggestion from limited data that routine administration of oxytocics increases the risk of retained placenta, but this finding might easily reflect selective presentation of outcome data or be a result of chance.

The studies that provided information about hypertension as a potential side-effect showed a statistically significant hypertensive effect of oxytocics. More general data on blood pressure from other studies also suggest that the prophylactic use of oxytocics leads to a rise in blood pressure.

The advantage of prophylactic oxytocics in terms of reduced risk of postpartum haemorrhage must be weighed against the rare but serious morbidity that has sometimes been associated with their administration. Maternal deaths from cardiac arrest and intracerebral haemorrhage have been attributed to ergometrine, as have non-fatal instances of cardiac arrest, myocardial infarction, postpartum eclampsia, and pulmonary oedema. Because these events are so rare, the available randomized trials cannot provide useful estimates of the extent to which they may be attributed to oxytocic administration.

Other rare, but definite, adverse consequences of routine oxytocic administration include intrauterine asphyxia of an undiagnosed second twin and neonatal convulsions in a baby mistakenly injected with an oxytocic instead of prophylactic vitamin K.

In principle, randomized trials should be able to provide useful information about adverse effects which are less serious but more commonly encountered, such as nausea and vomiting, and headache. In fact, little usable information is available from the trials. Apart from the investigation of hypertensive effects, there were few systematic attempts to quantify side-effects. In view of the fact that ergometrine is known to lower serum prolactin levels, it is unfortunate that few of the trials have investigated whether or not it interferes with breastfeeding. The two large trials in which this was examined found no difference in breastfeeding at discharge from hospital.

2.2 Comparisons of different oxytocics

From the trials in which oxytocin has been compared with ergot alkaloids, there is no evidence that these two kinds of oxytocics differ greatly in their effects on the incidence of postpartum haemorrhage. The use of oxytocin was associated with a trend towards less postpartum haemorrhage, and was less likely than ergot alkaloids to lead to a delay in placental delivery or to a rise in blood pressure. None of these differences achieved statistical significance.

Ergot and oxytocin together (Syntometrine) and ergot alkaloids alone have similar effects on the rate of postpartum haemorrhage. Ergot + oxytocin is somewhat less likely than ergot alkaloids alone to be associated with a prolonged third stage. In the only trial that considered effects on blood pressure, ergot + oxytocin was less likely than ergometrine to be associated with an elevation of diastolic blood pressure of 20 mmHg or more.

Ergot + oxytocin has been compared with oxytocin alone in a number of trials, and data on postpartum haemorrhage are available from four. These suggest that ergot + oxytocin reduces the risk of postpartum haemorrhage more effectively than oxytocin used alone, although the authors of one trial have suggested that further studies are needed to clarify the most effective doses to use. Other studies have shown a lower mean blood loss with ergot + oxytocin than with oxytocin, although one trial, which presented no data on postpartum haemorrhage as such, reported that oxytocin was superior in this respect.

As yet there is little evidence comparing the effects of prostaglandins with any of the above oxytocics.

2.3 *Early clamping and division of the umbilical cord*

Active management of the third stage of labour usually entails clamping and dividing the umbilical cord relatively early, before beginning controlled cord traction. Pre-empting physiological equilibration of the blood volume within the fetoplacental unit in this way may predispose to retained placenta, postpartum haemorrhage, fetomaternal transfusion, and a variety of unwanted effects in the neonate, respiratory distress in particular. Delayed cord clamping results in a placental transfusion to the baby varying between 20 and 50 per cent of neonatal blood volume, depending on when the cord is clamped, at what level the baby is held before clamping, and whether oxytocics have been administered.

Early cord clamping leads to higher residual placental blood volumes and heavier placentas, but these observations have no clinical relevance. The duration of the third stage is reduced by the use of early cord clamping. The time of cord clamping does not appear to influence the frequency of postpartum haemorrhage, although numbers are small.

Allowing free bleeding from the placental end of the cord reduces the risk of fetomaternal transfusion, which may be important with regard to blood group isoimmunization.

Early cord clamping results in lower haemoglobin values and haematocrits in the newborn, but these effects are minimal at six weeks of age and undetectable at six months after birth. Neonatal bilirubin levels are lower in babies born after early cord clamping. It is difficult to draw relevant information from the trials about the effect on clinical jaundice. No detectable differences were noted in the trials that reported on this.

This issue is of particular interest in the care of preterm babies, where early clamping is often carried out to facilitate resuscitation. Theoretical considerations suggest that a delay of as little as 30 seconds may have important clinical benefits for these babies. Further information is needed.

2.4 *Controlled cord traction*

Use of controlled cord traction involves traction on the cord while maintaining counter-pressure upwards on the lower segment of the uterus using a hand placed on the lower abdomen. There have been two controlled trials in which controlled cord traction was compared with less active approaches, one of which sometimes entailed use of fundal pressure. Controlled cord traction was associated with a lower mean blood loss and shorter third stage, but the trials provide insufficient data to warrant firm conclusions about its effects on either postpartum haemorrhage or manual removal of the placenta. One of the investigators noted that the umbilical cord had ruptured in 3 per cent of the women managed with controlled cord traction, but that women were more likely to find fundal pressure uncomfortable.

3 Active versus expectant management of the third stage

The effects of prophylactic oxytocics, early clamping of the cord, and controlled cord traction were considered separately in the controlled trials from which the above conclusions were drawn. If these three components are interdependent, as has been suggested, these conclusions need to be viewed with caution.

Active management of the third stage (including all three of these elements) has also been directly compared with a policy of expectant or physiological management (which includes no prophylactic oxytocics, cord clamping after placental delivery, no cord traction, and maternal effort aided by gravity). To date, all trials of active versus expectant management have been carried out in centres where active management was the normal practice. In addition, there has been considerable variation in the components of expectant management used in these trials. This 'piecemeal' approach has been criticized. To address this issue, a large trial is needed in a setting where both expectant and active managements are normal. Such a trial is in progress.

In the available trials, active management of the third stage of labour was found to be associated with important reductions in clinically estimated postpartum blood loss, low haemoglobin levels postpartum, and use of blood transfusion. It was also associated with a reduction in the use of therapeutic oxytocics.

Some adverse effects have been observed. Active management results in an increase in nausea and vomiting, headache, and hypertension postpartum. Overall, manual removal of the placenta was more frequent with active management, but this effect was not significant and was not observed in all trials. Secondary postpartum haemorrhage was more common with active management in one large trial, but not in the combined results of all trials. Active management was associated with an increased incidence of low neonatal haematocrit (below 50 per cent) and a decreased incidence of high haematocrit (above 65 per cent). There was a trend towards less jaundice and fewer admissions to the special care nursery after active management, but the differences were not statistically significant. No effects on Apgar score or neonatal respiratory problems were detected. In the two trials which assessed breastfeeding on discharge from hospital, no differences were found.

The views of mothers and midwives did not strongly favour either form of management.

4 Complications of the third stage

4.1 *Postpartum haemorrhage*

The care of a woman with postpartum haemorrhage depends on a rapid but careful assessment of the cause, and prompt arrest of the bleeding before the situation becomes critical. If the source of the bleeding is traumatic, this will require surgical repair; if it is due to uterine atony, contraction of the uterus must be achieved by ensuring that the uterus is empty and well contracted.

Oxytocin and ergometrine have been the traditional first line approaches for achieving contraction of the uterus when the haemorrhage is due to uterine atony. Since numerous reports of the superior haemostatic effect of prostaglandins have appeared, these agents should be the first line of approach in severe postpartum haemorrhage due to uterine atony. Although the superiority of prostaglandins for arresting postpartum haemorrhage has not been demonstrated in controlled trials, their dramatic effect when all other measures have failed suggest that these drugs are worthwhile. Injection of prostaglandins into the myometrium may obviate the need for uterine packing, internal iliac artery ligation, or even hysterectomy.

4.2 *Retained placenta*

The conventional treatment for retained placenta is manual removal following digital separation of the placenta from the uterine wall, usually under either general anaesthesia or epidural block. Other methods have been proposed.

The available evidence suggests that injection of oxytocin into the umbilical vein reduces the use of manual removal of the placenta. No evidence of any adverse effects of the oxytocin were reported, although manual removal following oxytocin may be more difficult than usual because of a firmly contracted uterus. Trials have also evaluated the use of saline injection into the cord; no effect was found on the incidence of manual removal of the placenta.

Waiting for 60 minutes before resorting to manual removal may greatly reduce the number of women who will require manual removal with its attendant anaesthetic risks. In the absence of bleeding, this may be a more effective approach than embarking on manual removal too early. A large trial to compare expectant management with the use of umbilical vein oxytocin or saline for situations in which the placenta has been retained for 30 minutes is in progress.

4.3 *Inversion of the uterus*

Inversion of the uterus is now very rare. It may occur as a result of excessive cord traction in the presence of a relaxed uterus, vigorous fundal pressure, or exceptionally high intra-abdominal pressure as a result of coughing or vomiting. Inappropriate cord traction without counter-pressure to prevent fundal descent is said to result in the occasional case of uterine inversion. Treatment involves replacement of the inversion.

5 Conclusions

The routine use of oxytocic drugs in the third stage of labour will result in a reduced risk of postpartum haemorrhage, when compared with other techniques (including some components of active management) without a prophylactic oxytocic drug.

The evidence available provides no support for the continued prophylactic use of ergometrine alone. This drug used alone offers no advantage over oxytocin or ergot + oxytocin in reducing blood loss, and is associated with a greater risk of hypertension and vomiting.

The available evidence does not reveal any effect of early cord clamping upon blood loss or postpartum haemorrhage although numbers are small. Early cord clamping should be avoided in rhesus-negative women, unless the placental end of the cord is allowed to bleed freely, because it increases the risk of fetomaternal transfusion. More information is needed about the effects of the timing of cord clamping for preterm babies.

As a package, the active management of the third stage has been shown to have a significant protective effect against postpartum haemorrhage when compared with components of expectant management. The implications for practice depend on the relative weight placed on the different outcomes considered. In terms of postpartum blood loss, active management is clearly better in the situations studied so far. However, in terms of nausea, vomiting, and hypertension, expectant management may be preferable.

Further research is needed in settings where both expectant and active management are normal, and where the components of both active and expectant management are clearly differentiated.

Prostaglandins should be used for the treatment of severe intractable postpartum haemorrhage, but it has not been established which preparation, dose, or route of administration is most effective.

This chapter is derived from the chapter by Walter Prendiville and Diana Elbourne (67) in EFFECTIVE CARE IN PREGNANCY AND CHILDBIRTH.

References to primary sources and more complete data for statements made in this chapter can be found in the source chapter and/or in the following reviews from the *Cochrane pregnancy and childbirth database*:

Elbourne, D.R.
— Prophylactic oxytocics in third stage of labour. Review no. 02974.
— Prophylactic syntometrine vs oxytocin in third stage of labour. Review no. 02999.
— Prophylactic oxytocin vs ergot derivatives in third stage of labour. Review no. 03000.
— Prophylactic syntometrine vs ergot derivatives in third stage of labour. Review no. 03001.
— Early umbilical cord clamping in third stage of labour. Review no. 03818.
— Cord traction vs fundal pressure in third stage of labour. Review no. 03004.
— Active vs conservative 3rd stage management. Review no. 05352.
— Active vs conservative 3rd stage management - low risk women. Review no. 05353.
— Umbilical vein injection (oxytocin or saline) for retained placenta. Review no. 03826.
— Umbilical vein oxytocin for retained placenta. Review no. 05831.
— Free bleeding from placental end of umbilical cord. Review no. 04004.

PROBLEMS DURING CHILDBIRTH

34 Control of pain in labour

1 Introduction

Women experience a wide range of pain in labour, and an equally wide range of responses to it. A woman's reactions to labour pain may be modified by a variety of circumstances, including the environment and the support that she receives from her caregivers and companions (see Chapter 28). Support may include helping the woman in her wish to avoid pharmacological pain relief, or helping her choose among pharmacological and non-pharmacological methods of pain relief. She will require accurate information to choose what is best for her and continuing support.

2 Non-pharmacological methods

The study of pain transmission and its modulation have produced many exciting findings in the past twenty-five years. Many of these are applied in a variety of non-pharmacological approaches to relieving the pain of childbirth. Some of these approaches are a revival of traditional methods, while others have been newly developed. They can usefully be classified as techniques that reduce painful stimuli, techniques that activate peripheral sensory receptors; and techniques that enhance descending inhibitory neural pathways. Many of these techniques are taught in antenatal childbirth preparation classes (see Chapter 4).

2.1 *Techniques that reduce painful stimuli*

The most obvious solution to the problem of pain is to avoid or reduce the stimuli that cause it. In labour the painful stimuli arise from the uterine contractions or from pressure exerted on the cervix, vagina, and pelvic joints by the presenting part of the fetus. They cannot be avoided, but techniques to reduce these painful stimuli are at least theoretically possible. This is the intended purpose of various maternal positions, movement, counter-pressure, and abdominal decompression.

2.1.1 *Maternal movement and position changes*   Labouring women find that they experience less pain in some positions than in others, and if left to their own devices will usually select the positions that they find most comfortable. Many labouring women today are restricted to bed, either because of cultural expectations or because of obstetrical practices such as electronic fetal monitoring, intravenous hydration, and medications that render movement out of bed difficult or unsafe.

Despite these restraints, women seem to prefer freedom of movement when it is allowed. Given the opportunity to assume any position in or out of bed during the course of their labour without interference or instruction by caregivers, labouring women spontaneously adopt upright postures such as sitting, standing, and walking, often returning to a recumbent position in advanced labour.

When the mother changes position, she alters relationships among gravity, uterine contractions, the fetus, and her pelvis, which may enhance the progress of labour and reduce pain. For example, pressure of the fetal head against the sacroiliac joint may be relieved if the mother moves from a semi-recumbent to a 'hands and knees' posture. The effects of maternal position on perceived pain are influenced by a number of factors including fetal size, position, size and shape of the fetal head and the maternal pelvis, and the strength of the uterine contractions. Knowing this, experienced caregivers place trust in the mother's ability to find pain-reducing positions. They try not to restrict women, but encourage them to seek comfort, suggest possible positions, and trust their judgment.

Few data from controlled trials are available about the effects of ambulation on pain relief. No effects have been demonstrated on the use of pharmacological analgesia or epidural block. In the second stage of labour, an upright position is associated with fewer episodes of severe pain (see Chapter 29).

2.1.2 *Counter-pressure* Counter-pressure consists of steady strong force applied to a spot on the low back during contractions using the fist, the 'heel' of the hand, or a firm object, or of pressure applied to the side of each hip, using both hands, by the labouring woman's support person or caregiver. While there are no controlled trials of its effectiveness, counter-pressure appears to alleviate back pain in some labouring women. It seems to be most effective when a woman suffers extreme back pain, possibly related to an occiput posterior position.

2.1.3 *Abdominal decompression* Abdominal decompression (see Chapter 16) was introduced in the mid-1950s as a non-pharmacological method of shortening labour and reducing labour pain. Although anecdotal reports were positive, it has now largely disappeared from use, partly because of the lack of good evidence that it is beneficial but also because some women found the decompression apparatus to be cumbersome, constrictive, noisy, and uncomfortable.

2.2 *Techniques that activate peripheral sensory receptors*

2.2.1 *Superficial heat and cold* Superficial heat is generated from hot or warm objects, such as hot-water bottles, hot moist towels, electric heating pads, heated silica gel packs, warm blankets, baths, and showers. Superficial cold can come from ice bags, blocks of ice, frozen silica gel packs, and towels soaked in cool or ice water.

In addition to possible direct effects on pain perception, several physiological responses elicited by heat and cold may indirectly result in pain relief. The therapeutic uses of heat and cold have not been evaluated in randomized controlled trials. Observational evidence suggests that they may both be effective. The use of hot compresses applied to the low abdomen, groin, or perineum, a warm blanket over the entire body, or ice packs on the low back, anus, or perineum relieves pain in labour for some women.

Heat and cold are widely accepted as comfort measures. Because they provide only partial relief from labour pain, they must be considered only as adjuncts to other measures.

2.2.2 *Immersion in water during labour and birth*   The healing and pain-relieving properties of water — hot or cold, flowing or still, sprayed or poured — have been hailed for centuries. In recent years, immersion in water during labour and birth has aroused interest in many countries in response to women's requests for this form of comfort. Practice varies widely and includes the use of showers, baths, whirlpools, and specially designed 'birth pools'. Women at home may use either their own household tub or a birth pool. Some hospitals allow women to bring in a birth pool, and an increasing number of hospitals are installing large permanent pools. This requires the use of a dedicated delivery room.

There is no consistency in the criteria developed to guide practice. Guidelines may exclude some women from using a pool because of conditions such as raised blood pressure or the need for electronic fetal monitoring. They may specify a minimum cervical dilatation before entry to the water, the temperature of the water, and whether or not women are allowed to stay in the water for second stage, delivery, or third stage. These guidelines have been derived from experience and theoretical considerations as there is currently little or no evidence on which to base them.

Advocates stress the relaxing effect of water, which may reduce the use of pharmacological methods of pain relief. Some suggest that immersion in water may accelerate labour, decrease blood pressure, increase the mother's control over the birth environment, result in less perineal trauma and intervention in general, and introduce the baby into the world gently. Critics maintain that there may be an increased risk of infection for both mother and baby, possible inhibitions of effective contractions, increased risk of perineal trauma, postpartum haemorrhage, water embolus, and trauma to the baby from a labour and/or birth in water. Caregivers may also be at risk from infection and back injury.

Recent case reports raise two other possible problems. The use of hot water over several hours of labour has been linked to hyperthermia, possibly causing brain damage or death of the baby. A few baby deaths have been reported when the baby was held under the water after birth, presumably in the belief that a slow transition from the water into air would result in a gentle introduction to life. In the light of these reports, there is ample reason to assess the safety of the procedure. In the meantime, it would seem sensible to restrict the temperature of the water to body temperature or lower, and to bring the baby to the surface as soon as it is born.

Immersion in water during labour but not during birth has been assessed in two small randomized trials. One found that women who used a birth pool received less augmentation of labour, although the other found that the duration of labour was increased for women using a whirlpool bath. No other benefits or complications were identified. Further large trials are needed to determine the safety and efficacy of immersion in water during labour and birth.

2.2.3 *Touch and massage* The use of touch in various forms conveys pain-reducing messages, depending on the quality and circumstances of the touch. A hand placed on a painful spot, a pat of reassurance, stroking the hair or a cheek in an affectionate gesture, a tight embrace, or more formal purposeful massage techniques all communicate to the receiver a message of caring, of wanting to be with her and to help her. On the other hand some women may find this unpleasant or intrusive. The object of massage is to make people feel better, or to relieve pain and facilitate relaxation.

Massage takes the form of light or firm stroking, vibration, kneading, deep circular pressure, continual steady pressure, and joint manipulation. It can be carried out using fingertips, entire hands, or various devices which roll, vibrate, or apply pressure. In theory, the various forms of massage stimulate different sensory receptors. When they are discontinued, the woman's awareness of her pain increases. In addition, the phenomenon of adaptation may diminish the pain-relieving effects of massage over a period of time. Therefore use of intermittent massage, or variation in the type of stroke and location of the touch, may prolong the pain-reducing effects.

None of the touch or massage techniques has been subjected to careful scientific evaluation, but the intervention seems to be harmless and is well received by labouring women. It can easily be discontinued if she wishes.

2.2.4 *Acupuncture and acupressure* Acupuncture consists of the insertion of strategically placed needles in any of more than 365 points along the twelve 'meridians' of the body. It is often combined with an electrical current, which is believed to augment the pain relieving effect. Practitioners vary in their choice of points, size of needle, and method of insertion. Acupuncture appears to block both sensory and emotional components of pain, but the mechanism is poorly understood.

No controlled trials of acupuncture during labour have been published, despite suggestions that it might provide good analgesia. The techniques involved are complex and time-consuming, and the use of multiple needles attached to electrical stimulators is inconvenient and may immobilize the woman.

Acupressure has been called 'acupuncture without needles'. The technique involves the application of pressure or deep massage to the traditional acupuncture points, using thumb, fingertip, fingernail, or palm of hand. There are no published reports of its effects, either anecdotal or scientific. Acupressure can be learned and applied by a non-professional companion of the labouring woman. Formal evaluation of its effectiveness would be worthwhile to establish its place (if any) as a comfort measure for labour.

2.2.5 *Transcutaneous electrical nerve stimulation*

Transcutaneous electrical nerve stimulation (TENS) is a non-invasive method which is easy to use and can be discontinued quickly if necessary. Used originally for the relief of chronic pain, trauma, and postsurgical pain, TENS has also been introduced for pain relief in labour.

The TENS unit consists of a portable hand-held box containing a battery-powered generator of electrical impulses. A low-voltage electric current is transmitted to the skin using surface electrodes, and this results in a 'buzzing' or tingling sensation. The labouring woman may vary the intensity, pulse frequency, and patterns of stimulation, so that she can increase, decrease, or pulse the sensations as she wishes.

Safety concerns have focused on theoretically possible effects of high intensity TENS on fetal heart function, particularly when electrodes are placed on the low abdomen close to the fetus. A further concern is that it may interfere with signals of electronic fetal monitors. Though no untoward effects on the fetus have been reported, there has been only one investigation of the fetal safety aspects of TENS. In fifteen births, the investigators established a limit or maximum current density level at 0.5 $\mu A/mm^2$ when TENS was used in the suprapubic region. No adverse fetal effects were detected. Less concern has been raised when electrodes are placed only on the back, which has been the procedure in most studies.

TENS for childbirth pain has been subjected to more controlled trials than any of the other modalities of non-pharmacological pain relief. Unfortunately, the results of the trials still remain inconclusive. TENS has been reported to increase rather than decrease the incidence of intense pain, yet it is favourably assessed by the women who use it. Direct comparison of TENS and pethidine failed to demonstrate any differential effect on pain experienced, and the use of TENS has no obvious effect on the use of other forms of analgesia.

2.2.6 *Intradermal injection of sterile water*

Controlled trials have demonstrated a dramatic analgesic effect on low back pain in labour from intradermal injection of small amounts (0.1 ml) of sterile water at four spots in the low back area, approximately corresponding to the borders of the sacrum. This simple measure deserves further evaluation, but even on the basis of presently available data it would seem worth trying.

2.2.7 *Aromatherapy* Aromatherapy has been used increasingly in recent years. The term refers to the use of essential oils such as lavender, rose, camomile, and clary sage. These can be administered in a variety of ways, including in oil during a massage, in hot water as a bath or footbath, directly on a taper or a drop on the palm or forehead of the labouring woman, or applied with a hot face cloth. Oils are considered to have a number of specific properties. For example, camomile is said to be calming, and clary sage is said to strengthen contractions by dissipating stress and tension. Aromatherapy may also reduce stress and tension among caregivers and labour companions.

There are no trials of the use of these oils in labour, although there are reports that women find them comforting and effective. Randomized trials are needed to clarify the possible benefits and risks. For example, assessment of the effects on uterine action, effects on the use of other forms of pain relief, and possible adverse side-effects including allergic reactions is needed.

2.3 Techniques that enhance descending inhibitory pathways

2.3.1 *Attention focusing and distraction* Many methods for coping with pain involve the conscious participation of the individual in attention focusing or mind-diverting activities, designed to 'take one's mind off the pain'.

Attention focusing may be accomplished by deliberate intentional activities on the part of the labouring woman. Examples include patterned breathing, attention to verbal coaching, visualization and self-hypnosis, performing familiar tasks (such as grooming and eating), and concentration on a visual, auditory, tactile, or other stimulus.

Distraction may be a more passive form of attention focusing, with stimuli from the environment (television or a walk out of doors) or from other people drawing a woman's attention away from her pain. It does not require as much mental concentration as deliberate attention focusing measures, and is probably ineffective when pain is severe. Attention focusing and distraction are usually used in combination with other strategies.

2.3.2 *Hypnosis* Hypnosis was introduced into obstetrics in the early nineteenth century and has been used in various ways ever since. It is defined as 'a temporarily altered state of consciousness, in which the individual has increased suggestibility'. Under hypnosis a person demonstrates physical and mental relaxation, increased focus of concentration, ability to modify perception, and ability to control normally uncontrollable physiological responses such as blood pressure, blood flow, and heart rate.

Hypnosis is used in two ways to control pain perception in childbirth: self-hypnosis and post-hypnotic suggestion. Most hypnotherapists teach self-hypnosis, so that women may enter a trance during labour and reduce awareness of painful sensations. Among the techniques used are relaxation, visualization (helping the woman imagine a pleasant safe scene and placing herself there, symbolizing her pain as an object that can be discarded, or picturing herself as in control or free of pain), distraction (focusing on something other than the pain), and glove anaesthesia (creating a feeling of numbness in one of her hands through suggestion, and then spreading that numbness wherever she wishes by placing her numb hand on the desired places of her body). The woman is taught to induce these techniques herself; only rarely do hypnotherapists accompany their clients in labour.

Other therapists rely almost completely on post-hypnotic suggestion. These hypnotherapists do not teach their clients to enter a hypnotic state routinely during labour, because they will not need to. Most women, they claim, will be comfortable as a result of the effectiveness of the post-hypnotic suggestions. Exceptions to this are circumstances such as forceps delivery or episiotomy and repair, for which it would be necessary to go into a trance.

To date, only one randomized trial of hypnosis in labour has been reported. There was no difference in analgesia use between the experimental and control groups. The mean duration of pregnancy and the mean duration of labour were both statistically significantly longer in the hypnosis group.

Hypnosis had lost its popularity among obstetricians by the early 1970s, probably because of the development of better methods of anaesthesia and the amount of time required for adequate hypnosis preparation.

2.3.3 *Music and audioanalgesia* Music and audioanalgesia are used to control pain in numerous situations, including dental work, post-operative pain, treatment of burns, and occasionally in childbirth. Many childbirth educators use music in antenatal classes to create a peaceful and relaxing environment, and also advocate it for use during labour as an aid to relaxation.

Audioanalgesia for pain relief in obstetrics consists of the use of soothing music between contractions combined with 'white sound', the volume of which is controlled by the labouring woman, during contractions. The only published placebo-controlled trial of audioanalgesia noted a trend towards better pain relief in the audioanalgesia group, but the effect was confined to primigravidae and was not statistically significant. As the 'placebo' consisted of a lower intensity of white sound, which might have had pain-relieving qualities in itself, the true benefit of the audioanalgesia might have been masked. A number of other investigators have reported a decreased use of analgesic medication and less pain with the use of audioanalgesia in non-randomized cohort studies. These results, although equivocal, merit further better controlled trials.

The pleasing qualities of music may offer an added dimension beyond the distraction brought about by white sound. Music from a tape recorder or phonograph creates a pleasant and relaxing ambience, and music transmitted through earphones can block out disturbing, distracting, or unpleasant sounds. When carefully chosen, music can be used to reinforce rhythmic breathing patterns and massage strokes, or to facilitate visualizations and induction of hypnosis. Thus music may have the potential to reduce stress and to enhance other pain relieving measures. Music may also elicit more relaxed and positive behaviour from the staff and from the woman's chosen companions.

The few small studies of the pain relieving effects of music in labour have found positive effects. These small studies of childbirth pain, when combined with findings of the effects of music or auditory stimulation on other types of pain (for example, post-operative pain, dental pain, pain associated with burn therapy) suggest that music has the capacity to reduce pain, at least in some women. However, its efficacy seems to depend on the degree of prior education, preparation, and accommodation to the personal musical tastes of each woman.

2.3.4 *Biofeedback in prenatal class attenders* Trials of electro-myographic biofeedback taught during prenatal classes failed to demonstrate any significant effect on the use of pharmacological analgesia or other interventions during childbirth.

3 Pharmacological control of pain in labour

Pharmacological control of pain in childbirth has a long history. The use of opiates was mentioned in early Chinese writings, the drinking of wine was noted in Persian literature; and wine, beer, and brandy were commonly self-administered in Europe during the Middle Ages. Various concoctions and potions have been used over time, some inhaled, some swallowed, and some applied to the labouring woman's skin.

There have been more clinical trials of pharmacological pain relief during labour and childbirth than of any other intervention in the perinatal field. The benefits of pain relief are obvious, but the possible adverse effects on the mother or infant have received little attention in this research. The clinically important question is: 'What method will achieve an acceptable degree of pain relief while least compromising the health of the mother and child?'

3.1 *Systemic agents*

3.1.1 *Narcotics*   Systemic narcotics can provide reasonable pain relief but can have unwanted side-effects. As these effects are dose related, the amount of analgesia achievable is limited by the side-effects of the drug. Maternal side-effects include orthostatic hypotension, nausea, vomiting, dizziness, and delayed stomach emptying. Narcotics cross the placenta, and this may cause respiratory depression in the baby. Trial data have shown lower Apgar scores and more neonatal behavioural abnormalities in babies of mothers who received narcotic analgesia in labour than in babies of mothers who received a placebo.

Self-administered intravenous pethidine results in better pain relief and a lower total dose of narcotic than intramuscular pethidine administered by a caregiver. This suggests that self-administration may be preferable, at least for women who are already receiving an intravenous infusion.

Narcotic antagonists (naloxone, nalorphine, or levallorphan) have been administered to counteract the depressant effects of narcotics, either with each dose of narcotic or 10–15 minutes before delivery. The rationale was to provide analgesia with minimal respiratory depression, but narcotic antagonists also reverse the analgesic effects of the narcotics. After birth, there is a role for administration of narcotic antagonists (naloxone in particular) to neonates depressed by narcotics.

3.1.2 *Sedatives and tranquilizers*   Many clinicians feel that the use of a tranquillizer, particularly in early labour, is helpful in reducing the woman's anxiety and in promoting sleep.

The barbiturates (secobarbital, pentobarbital, and amobarbital) are no longer popular for use in obstetrics because they have no analgesic properties and because they may have a profound depressant effect on the newborn. They are still used in some countries in the early latent phase of labour for their sedative effect.

The phenothiazine derivatives (promethazine, propiomazine, chlorpromazine, promazine, and prochlorperazine) are, with the benzodiazepines (diazepam and droperidol), the major tranquilizers currently used in obstetrics. They are often administered intramuscularly or intravenously in combination with narcotics. They have antiemetic as well as sedative properties, and do not appear to cause severe neonatal depression.

Diazepam (a benzodiazepine) can cause neonatal respiratory depression, hypotonia and lethargy, and hypothermia, although these effects have not been noted in data from the available randomized trials.

3.2 Inhalation analgesia

The use of inhalation analgesia has been decreasing in recent years, primarily because it does not offer reliable or complete pain relief. In addition, side-effects of nausea and vomiting, as well as the possibility of aspiration of gastric contents in cases of accidental overdose, decrease the usefulness of the method. Concerns have been raised about possible long-term effects of exposure to inhalation agents on the medical and nursing staff.

Advantages of inhalation analgesia are that the mother remains awake and in control of the analgesia, that neither uterine activity nor 'bearing down' during the second stage is affected, that the duration of the effect is short and that better control is thus possible, and that no clinically obvious side-effects on the mother or fetus have been noted.

The most commonly used agent is nitrous oxide, usually in a 50 per cent concentration with 50 per cent oxygen. Trials comparing 50 per cent with 70 per cent nitrous oxide administration showed no significant differences in pain relief in normal labours. Other inhalation analgesic agents (methoxyflurane, enflurane, isoflurane, and trichloro-ethylene) have also been employed, but their use has been curtailed or eliminated because of potential toxicity.

3.3 Regional analgesia

Over the past twenty years techniques of regional analgesia have emerged as the major approach to pain relief in obstetrics, not only for labour but also for operative vaginal deliveries and for caesarean section. There are many reasons for the popularity of regional anaesthesia. The most important are that it is more effective in relieving pain than other agents and that the mother remains conscious.

3.3.1 *Epidural analgesia*   Epidural block provides better and more lasting pain relief than narcotic analgesia in the first stage of labour, although a few women fail to obtain adequate relief. There is a tendency for the first stage of labour to be somewhat longer, and for oxytocin to be used more frequently with epidural analgesia. When maintained beyond the first stage of labour, it tends to result in a longer duration of the second stage, predisposes to malrotation of the presenting part, and results in a substantially increased use of operative vaginal delivery.

The data also show an increase in the use of caesarean section when epidural analgesia is used. A recent study suggests that much of this increase is due to failure to progress in labour (dystocia). The timing of the epidural may be important, as the increase was found only in women who had early epidurals. Further study is required to confirm or refute this. There are as yet no randomized studies on the timing of initial administration of the epidural.

Epidural analgesia can be administered either by continuous administration with an infusion device or by intermittent 'top-up' doses by way of an indwelling catheter. Care must be taken with regular top-ups, as with all other infusion techniques, that the height of the block is carefully monitored. Regularly scheduled top-ups provide better pain relief than top-ups on maternal demand. A trial comparing regular top-ups of epidural analgesia at 90 minute intervals with top-up on maternal demand showed that episodes of both moderate and severe pain were reported by fewer of the women who received regular top-ups.

The dose of local anaesthetic used in infusions may vary significantly among hospitals. This may influence the degree of motor block and therefore the likelihood of a instrumental delivery or other adverse effects. The increased risk of instrumental delivery associated with epidural block may well be reduced by careful timing of top-up doses and a liberal attitude to length of the second stage of labour.

Preloading with intravenous fluids is an effective means of reducing the incidence and extent of the hypotension that so commonly occurs with epidural analgesia in labour. There is some evidence of its value in healthy women, although this is currently a controversial issue. The relative risks and benefits of preloading prior to epidural anaesthesia in the presence of pregnancy complications such as pre-eclampsia and cardiac disease require further evaluation.

No differential effects on fetal heart rate abnormalities or passage of meconium during labour were detected in the few trials in which these outcomes were reported. Epidural is less likely than narcotic analgesia to result in a low umbilical arterial pH or low 5 minute Apgar score.

Epidural administration of opiates appears to potentiate the analgesic effect of the local anaesthetic. This may allow sufficient reduction in the dose of the anaesthetic agent to decrease the extent of motor block and thus allow the woman to become more mobile. These effects are not adequately evaluated as yet. Unwanted effects of epidural narcotics include pruritis, urinary retention, and delayed respiratory depression in the mother.

Virtually no data from randomized trials are available to explore the possible effects of epidural analgesia on either mother or baby in the long term. A single report of neurobehavioural assessments of 18-month-old children failed to detect differences between the epidural and non-epidural groups at this age.

A number of complications of epidural analgesia have been reported. Well-established complications include dural puncture, hypotension with associated nausea and vomiting, localized short-term backache, shivering, prolonged labour, and increased use of operative delivery and caesarean section. Rare complications include neurological sequelae, toxic drug reactions, respiratory insufficiency, and maternal death. Possible, but as yet unproven, complications are bladder dysfunction, chronic headache, long-term backache, tingling and numbness, and 'sensory confusion'. Randomized comparisons will be needed to confirm or refute these observations. The fetus may suffer complications as a result of maternal effects (e.g. hypotension) or direct drug toxicity, although it may benefit from increased placental blood flow.

Inadvertent dural puncture during the placement of an epidural block may result in severe, often incapacitating, headache. The use of a prophylactic blood patch is highly effective in reducing the incidence of postdural puncture headache. Unfortunately, no data about possible long-term effects are available.

Numerous trials have compared dose regimens of various anaesthetic agents. The effort spent on these trials is disproportionate to their value in view of the importance of the safety and efficacy of the method itself compared with the effects of specific drugs or dose regimens.

Epidural block is a more effective form of pain relief than alternatives, increases the likelihood of caesarean section for dystocia, and, continued into the second stage of labour, increases the use of instrumental delivery. Little more can be deduced with confidence about its effects. It is particularly worrying that there are so few experimentally derived data available to assess its effects on infants or its long-term effects on the mother.

3.3.2 *Other routes of regional analgesia* **Caudal block** is rarely used today for labour analgesia. It requires a larger dose of the anaesthetic agent, results in more blocked neural segments, and the spread of the anaesthetic is less easily controlled than with the usual form of epidural analgesia. Failures occur in 5–10 per cent of women owing to variation in sacral anatomy. The only advantage of the caudal approach over the lumbar epidural technique is the decreased likelihood of dural puncture. There may be some useful applications for caudal block for perineal analgesia, but these have yet to be evaluated.

Paracervical block provides adequate analgesia and has the advantage that it can be administered by the obstetrician, thus avoiding the need for anaesthetic personnel. This form of analgesia was popular in the 1950s and 1960s, but has fallen out of favour because of reports of fetal bradycardia, acidosis, and fetal death associated with its use.

Spinal anaesthesia is mainly used for caesarean section and occasionally for an instrumental vaginal delivery, although there has been recent interest in intrathecal opiates which appear to give excellent analgesia for the first stage of labour. As with epidural opiates, mothers may develop side effects of pruritus, nausea and vomiting, and urinary retention, as well as respiratory depression in the early postpartum period.

There has been a recent upswing in interest in the use of spinal anaesthesia with the introduction of new needles which are reported to result in fewer dural puncture headaches.

4 Conclusions

Satisfaction in childbirth is not necessarily contingent upon the absence of pain. Many women are willing to experience some pain in childbirth, but they do not want it to overwhelm them. For women whose goals for childbirth include the use of self-help measures to manage pain with minimal drug use, and for those who have little or no access to pharmacological methods of pain relief, many of the non-pharmacological methods are useful alternatives. They cannot match epidural analgesia for analgesic effectiveness, but they seem to help some women and are not likely to have harmful side-effects.

Systemic narcotics can reduce pain in labour, although they do not provide as effective analgesia as epidural block. Their use in effective doses is limited by their side-effects of maternal drowsiness, nausea and vomiting, and neonatal respiratory depression. These effects, plus their effect in delaying stomach emptying, must be kept in mind, particularly if general anaesthesia might be required for delivery. Self-administered intravenous narcotics appear to give better pain relief with lower doses than does intermittent use of narcotics, although the lack of availability of suitable infusion devices may limit widespread use.

Barbiturates have no analgesic effect. Diazepam can result in neonatal respiratory depression, hypotonia, and hypothermia. Inhalation agents, such as 50 per cent nitrous oxide in oxygen, are only moderately effective analgesics, but are simple to use, have a short duration of action, and are under the control of the mother. No major side-effects have been noted.

Remarkably little is known about the short and long-term effects on women and their babies of epidural block during labour. All that can be concluded with confidence is that epidural block is likely to provide more effective pain relief during labour than alternative methods, but may result in a substantial increase in operative delivery. In view of the many important unanswered questions about the effects of epidural block during labour, more randomized comparisons with alternative methods of pain relief are required.

Consideration of the needs of each individual labouring woman, along with knowledge of the analgesic effectiveness and the adverse side-effects of each form of analgesia, will help the woman to make an informed choice among the alternatives available to her.

This chapter is derived from the chapters by Penny Simkin (56) and Kay Dickersin (57) in EFFECTIVE CARE IN PREGNANCY AND CHILDBIRTH.

References to primary sources and more complete data for the statements made in this chapter can be found in the source chapters and/or in the following reviews from the *Cochrane pregnancy and childbirth database:*

Hofmeyr, G.J.
— Prophylactic intravenous preloading before epidural anaesthesia in labour. Review no. 04140.

Howell, C.J.
— Transcutaneous nerve stimulation (TENS) in labour. Review no. 02553.
— Transcutaneous nerve stimulation (TENS) vs pethidine in labour. Review no. 05254.
— Biofeedback in prenatal class attenders. Review no. 05620.
— Systemic narcotics for analgesia in labour. Review no. 03398.
— Diazepam in labour. Review no. 03401.
— Methoxyflurane vs nitrous oxide/oxygen for analgesia in labour. Review no. 03400.
— Epidural vs non-epidural analgesia in labour. Review o. 03399.
— Epidural top-ups on maternal request vs scheduled top-ups. Review o. 04010
— Discontinuing epidural block at 8 cm dilatation. Review no. 04009.
— Prophylactic blood patch for dural puncture. Review no. 07093.

Nikodem, C.
— Upright vs recumbent position during first stage of labour. Review no. 03334.
— Upright vs recumbent position during second stage of labour. Review no. 03335.
— Birth chair vs recumbent position for second stage. Review no. 04735.

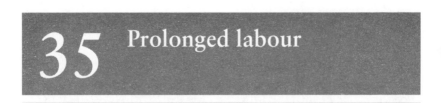

35 Prolonged labour

1 Introduction

Slow progress in the first stage of labour can occur in either the latent or active phase. It does not necessarily mean abnormal labour or the presence of a problem. However, it should draw attention to the possibility of a problem.

Attention devoted to the prevention of prolonged labour is at least as worthwhile as the attention which is usually devoted to its cure. Friendly support, and allowing women to move about as they please, have been demonstrated to be effective. Both these measures may be seen as characteristics of a welcoming environment, rather than as specific interventions.

2 Prolonged labour

2.1 *Prolonged latent phase*

The latent phase of labour, from the start of uterine contractions until progressive dilatation of the cervix commences, is poorly understood. As this phase usually starts before the woman is admitted to hospital, its precise time of onset is often difficult to determine. The duration of the latent phase varies so widely from woman to woman that a normal range is difficult to define.

According to some studies, a prolonged latent phase is not associated with increased perinatal morbidity, mortality, or other adverse outcome. Other studies have shown a significantly higher incidence of caesarean section and lower 5 minute Apgar scores in both primigravidae and multigravidae with a prolonged latent phase. Whether these adverse effects were due to the underlying condition or the result of injudicious treatment is uncertain.

Differentiating between a prolonged latent phase and false labour is often difficult. This distinction can only be made in retrospect. No trials of alternative forms of care for prolonged latent phase have been reported, although one such trial is in progress. There is an urgent need for controlled studies about this common, distressing, and poorly understood problem in labour including its aetiology, its significance, and the best policy of care.

2.2 Prolonged active phase

Slow progress or failure to progress in the active phase of labour can be defined either as an overall measurement (e.g. longer than x hours), or as a rate related measurement (e.g. a rate of cervical dilatation of less than y cm/hour). The commonly cited 12 hour duration of the active phase is roughly equivalent to a rate of 0.5 cm/hour, half as fast as the 1 cm per/hour that is also commonly used.

Deviation from this arbitrarily defined 'normal' rate of dilatation should be an indication for evaluation rather than for intervention. While there is certainly an association between prolonged labour and adverse outcome, the extent to which that the relationship is causal is by no means certain.

Cephalopelvic disproportion must be considered when progress in labour is slow. Intrapartum X-ray pelvimetry has not proved to be useful for women with the fetus in a cephalic presentation. The available controlled trials show a significant increase in caesarean section or symphisiotomy rates with the use of intrapartum X-ray pelvimetry, but no significant benefits in terms of reduced neonatal morbidity. The diagnosis of cephalopelvic disproportion is made by excluding functional causes (uterine hypotonia) for slow progress in labour. Uterine contractions of adequate intensity must be assured, by stimulation if necessary, before this diagnosis can be made. If gross cephalopelvic disproportion or marked moulding of the fetal skull is present, a caesarean section is necessary.

3 Prevention and treatment of prolonged labour

Protracted labour has been recognized as a problem for centuries, and a bewildering variety of treatments has been proposed to correct the condition. The assumption underlying all these treatments is that 'inadequate' progress is bad in some way and that 'something should be done about it'. In the past proposed remedies have included homeopathic medications, various spasmolytic drugs, sparteine sulphate, oestrogens, relaxin, caulophyllum, dimenhydrate, nipple stimulation, intracervical injections of hyaluronidase, vibration of the cervix, and acupuncture. The most commonly used measures today are amniotomy and intravenous oxytocin infusion.

Many factors can influence myometrial contractility and the progress of labour. Consideration of these factors suggests a number of measures that can help to prevent prolonged labour and obviate much of the need for augmentation. The presence of a supportive companion and ambulation during labour have been shown to result in shorter labours and a lesser use of oxytocics. Less well documented, but generally known to those who provide care in labour, is the observation that uterine contractility tends to subside when a women is taken from home to the hospital. More often than not, allowing her and her companion(s) the necessary time to settle in and feel at home in the new environment will do more good than a cascade of interventions aimed at procuring so-called 'adequate progress'.

When more active intervention is required, this may consist of measures to increase the power of the uterine contractions or to reduce the degree of resistance to cervical dilatation and descent of the presenting part.

3.1 *Increasing uterine contractility*

There is a close relationship between low levels of uterine contractility and slow progress in labour. Treatments to increase uterine contractility may be based on either increasing the endogenous production of prostaglandins by amniotomy or the administration of uterine stimulants such as exogenous oxytocin or prostaglandins.

3.1.1 *Amniotomy* A policy of early amniotomy leads to a reduction, on average, of between 60 and 120 minutes in the duration of labour, and also to a reduction in the incidence of dystocia (defined as a mean rate of dilatation of less than 0.5/hour). Oxytocin augmentation is used less frequently after early amniotomy. No effects of early amniotomy have been noted on the use of analgesia, rates of instrumental vaginal delivery, abnormal fetal heart tracings in labour, or the use of caesarean section.

The effects of early amniotomy on the baby are also open to interpretation. Meta-analysis of the controlled trials shows that early amniotomy is associated with fewer low 5 minute Apgar scores, but has no effect on the incidence of low arterial cord pH. There is a non-significant trend toward an increase in the frequency of cephal-haematoma and meconium aspiration. A variety of adverse effects have been postulated, but they have not been confirmed by controlled studies. The studies that purported to show these harmful effects of amniotomy on the fetus and neonate were all subject to considerable selection bias, precluding an adequate assessment of the effects of amniotomy.

In the studies that assessed mothers' views of the policy of membrane management, there was no evidence that the differences in policy affected their opinions. However, a policy of early amniotomy is associated with a statistically significant reduction in the risk of experiencing 'horrible or excruciating' pain at some time in labour.

None of the reported studies specifically addressed the question of whether or not amniotomy is effective in augmenting slow or prolonged labour, and to the best of our knowledge this issue has never been addressed in a randomized controlled trial. Given the evidence that is available from the controlled trials in spontaneous labour and from the data on induction of labour, it is highly likely that amniotomy would enhance progress in prolonged labour as well.

3.1.2 *Oxytocin* Intravenous infusion of synthetic oxytocin, usually after either spontaneous or artificial rupture of the membranes, is the most widely used treatment to expedite labour when progress is deemed to be inadequate. Despite this, there is remarkably little evidence about the effects of oxytocin from controlled trials.

The effect of oxytocin stimulation on duration of labour is still not clear. Of the three trials that provide data on the length of labour when intravenous oxytocin infusion was used in cases of poor progress, only one showed a shorter mean duration in women allocated to early oxytocin augmentation compared with controls. In one trial in which women in the control group were encouraged to get up and to move around, stand, or sit as they wished, the mean duration of labour was slightly shorter in the control group than in the augmented group.

The available data on the rate of cervical dilatation show a similar pattern. Early oxytocin use results in a slightly faster rate of dilatation when compared with recumbent controls, but a slower rate when compared with ambulatory controls. There is no evidence from these trials that a policy of early oxytocin use influences the method of delivery.

Neither Apgar scores nor the incidence of admission to a special care nursery were detectably different between oxytocin augmentation and control groups in the trials that reported on these outcomes. No other categorical data on infant outcomes are available from controlled trials.

Only one of the studies sought women's views on the augmentation procedures. Over half of the women asked about their opinion on the oxytocin treatment said that it was unpleasant and indicated that they would like to try without the drug when next giving birth. Over 80 per cent felt that it had increased the amount of pain that they had experienced, whereas less than 20 per cent of the women in the ambulant group felt that walking about had increased their pain.

From the data available thus far, it does not appear that liberal use of oxytocin augmentation in labour is of benefit to the women and babies so treated. This does not imply that there is no place for oxytocin augmentation in slow progress of labour. In each of the controlled trials a high proportion of the women assigned to be controls ultimately received oxytocin for subsequent failure to show adequate progress in labour (as defined by the authors). Thus these studies indicate that many women who are not treated early with oxytocin for inadequate progress will still receive an oxytocin infusion before delivery. However, they do suggest that other simple measures, such as allowing the woman freedom to move around and to eat and drink as she pleases, may be at least as effective and certainly more pleasant for a sizeable proportion of women considered to be in need of augmentation of labour.

Situations will undoubtedly remain in which pharmacological augmentation will be necessary to correct inadequate uterine activity in order to prevent maternal exhaustion and risks of fetal and maternal infection. Logic would dictate that, in such circumstances, the smallest effective drug dose be given in the most effective manner. As individual sensitivity to oxytocin varies greatly from woman to woman, oxytocin titration by means of an intravenous infusion is the treatment of choice. It is less clear, however, what the initial dose should be, how large the increments should be, and at what interval they should be implemented. Little data are available to answer these questions. Two principles should guide the clinician when using oxytocin to augment labour: first, hyperstimulation should be avoided; second, the therapeutic trial should be sufficient to minimize the risk of a false positive diagnosis of cephalopelvic disroportion. Insufficient doses will take an unacceptably long time to achieve an adequate response, while excessive doses will result in hyperstimulation.

Slow progress in cervical dilatation is not necessarily due to subnormal levels of uterine activity. The level of activity that is needed both to ensure adequate progress and to avoid hyperstimulation has not yet been established. A major additional factor in the rate of progress of labour is the amount of resistance that must be overcome.

3.1.3 *Active management of labour* Uncontrolled studies have suggested that active management of labour, with liberal use of both amniotomy and oxytocin augmentation, may be instrumental in achieving low caesarean section rates. This may well prove to be the case, but it has not been demonstrated by the results of the published controlled trials. Only one of the published trials showed a small and statistically insignificant reduction in the caesarean section rate; the others showed no effect. The results for other outcomes (use of epidural, neonatal morbidity) also showed no effects of the policy. Unfortunately, the total number of women included in these trials is too small to give a clear indication of the effect of active management of labour on the rates of caesarean section or operative vaginal delivery.

3.2 Influencing resistance

Between a third and half of the women with slow progress in labour have levels of uterine activity usually judged to be adequate. One care alternative for these women would be the use of high doses of oxytocin to augment uterine activity to levels well in excess of those encountered during normal spontaneous labour. A more logical approach would be to reduce cervical resistance.

Three methods of influencing resistance in order to accelerate cervical dilatation have been tested in controlled trials: the use of intravenous or intramuscular porcine relaxin, local cervical injections of hyalouronidase, and cervical vibration. The trials did not demonstrate any advantage from the use of these modalities. However, observational studies have suggested that vibration may be useful when cervical dilatation is not achieved despite satisfactory uterine activity in the absence of cephalopelvic disproportion, and that the procedure appears to be safe. Evaluation awaits a controlled trial of adequate size.

Despite the fact that their use would seem reasonable, measures to reduce soft tissue resistance have not been adequately explored.

4 Conclusions

Slow progress or lack of progress in the first stage of labour usually, but not necessarily, results from a lower level of uterine contractility than that seen in normally progressing labours. It may be due to a higher resistance in the soft parts of the birth canal or to cephalopelvic disproportion. It will often be necessary to ensure that adequate uterine contractility exists, if necessary by oxytocic stimulation, in order to differentiate these dissimilar mechanisms. Every effort should be made to correct uterine hypotonia before resorting to caesarean section for dystocia.

It is too early to recommend active management of labour on the basis of currently available evidence. Approximately half the women judged to have slow labour or poor progress in cervical dilatation will progress equally well whether or not oxytocic drugs are administered.

When augmentation becomes necessary, the first approach should be to rupture the membranes. The data that are available suggest that amniotomy will shorten the length of spontaneous labour, and that it may forestall the need for oxytocin infusion in some of these women. Moreover, the combination of amniotomy with oxytocin may provide a better stimulation of labour than oxytocin alone.

There is no evidence that high and rapidly escalating doses of oxytocin confer any advantage over a more moderate approach in which small doses are increased at 30 minute intervals in response to uterine contractility. The risks of hyperstimulation and increased pain are greater with larger doses of oxytocin.

Other methods for augmenting uterine activity, including the use of prostaglandins, are certainly worth considering, but they have been inadequately explored up to the present time.

In view of the importance of labour progress and the amount of discomfort that slow progress can provoke in the woman, the fetus, and the caregivers, it is important that the many suggested guidelines for care be substantiated by solid research evidence.

This chapter is derived from the chapters by Caroline Crowther, Murray Enkin, Marc J.N.C. Keirse, and Ian Brown (53), and by Marc J.N.C. Keirse (58) in EFFECTIVE CARE IN PREGNANCY AND CHILDBIRTH.

References to primary sources and more complete data for the statements made in this chapter can be found in the source chapters and/or in the following reviews from the *Cochrane pregnancy and childbirth database*:

Fraser, W.D.
— X-ray pelvimetry in cephalic presentations. Review no. 03229.
— Amniotomy to shorten spontaneous labour. Review no. 04134.
— Early oxytocin to shorten spontaneous labour. Review no. 04136.
— Early amniotomy and early oxytocin for delay in labour compared to routine care. Review no. 06949.
— Relaxin to shorten spontaneous labour. Review no. 04132.
— Cervical vibration to shorten spontaneous labour. Review no. 04133.

Repair of perineal trauma

1 Introduction

A third to nearly all women in some countries in the developed world
are likely to require repair of perineal trauma after vaginal birth. The
majority of these women experience perineal pain or discomfort in the
immediate postpartum period. Even three months later, as many as
20 per cent still have problems, such as pain during intercourse, which
can be related to perineal trauma and its repair.

2 Technique of perineal repair

Perineal trauma is most commonly repaired in layers. The vagina may
be repaired with a continuous suture or (less commonly) with
interrupted sutures. In theory a continuous stitch might 'concertina' the
vagina, and for this reason a locking stitch is usually recommended.

The deeper perineal tissues are usually closed with interrupted
sutures, but sometimes continuous 'running' sutures are used. The skin
may be closed with interrupted transcutaneous sutures, or a continuous
subcuticular suture using an absorbable material may be used. Another
approach, simply to appose the deeper tissues with no separate suturing
of the skin, has been claimed to result in satisfactory healing with
minimal discomfort.

In addition to the nature and extent of the trauma, the technique of
repair and the choice of suture material will have a bearing on the
extent of morbidity associated with perineal trauma. A variety of
techniques and suture materials are in current use, and there is no
consensus as to which are best.

Continuous subcuticular suture for closure of the perineal skin results
in fewer short-term problems than interrupted transcutaneous suturing
techniques. Women randomized to perineal repair with continuous
subcuticular sutures experienced less pain and used less analgesia in the

immediate postpartum period than those randomized to interrupted sutures. No substantial differences between the two techniques were found with respect to long-term pain or pain during intercourse.

3 Choice of suture material

Trials comparing the use of absorbable sutures (Dexon) with non-absorbable skin sutures (silk, nylon, or Supramid) show that the groups repaired with absorbable sutures generally had less pain and used less analgesia in the first few days after delivery. They were also less likely to require resuturing. No clear differences were noted on other longer-term morbidity, although women commonly reported the need for removal of some absorbable material in the three months after delivery.

The absorbable materials most commonly used for perineal closure are polyglycolic acid (Dexon, Vicryl) and chromic catgut. The use of polyglycolic acid sutures results in less short-term pain and less use of analgesia than chromic catgut. In the one trial that included an adequate follow-up after discharge from hospital, both perineal pain and pain during intercourse were equally common in the two groups three months after delivery.

Removal of some suture material was reported more frequently after the use of polyglycolic acid sutures than with chromic catgut sutures. This was particularly marked in the first ten days, but persisted up to three months postpartum. The commonest reasons given were 'irritation' and 'tightness'. The number of women requiring resuturing was small, but this occurred more frequently after suturing with chromic catgut than with polyglycolic acid.

In summary, the evidence suggests that polyglycolic acid sutures cause less pain than chromic catgut in the immediate postpartum period, but may cause irritation sufficient to lead to the removal of some suture material in an important minority of women. Polyglycolic acid sutures cause less tissue reaction. This may explain why their use is associated with less pain in the immediate postpartum period. Another suggested explanation for the trial findings is that they reflect differences in the tightness of the stitches rather than differences in the materials *per se*.

The only other published trials of two absorbable materials compared glycerol-impregnated catgut with chromic catgut, with both materials being used for all layers. In the better conducted trial, the use of glycerol-impregnated catgut was associated with more pain ten days after delivery and with a higher frequency of pain during intercourse three months postpartum. The increased prevalence of pain during intercourse persisted, and was reported nearly twice as commonly three years after delivery by women sutured with glycerol-impregnated catgut. On the basis of these findings, glycerol-impregnated catgut sutures should not be used for repair of perineal trauma.

Of the non-absorbable materials, polyamide sutures, such as nylon or Supramid, would be expected to cause less pain than silk because they cause less tissue reaction and pass more easily through the tissues. However, handling and knotting polyamides is less easy. They tend to be stiff and have a 'memory', and they thus require three or four throws in a knot. In contrast, the handling properties of silk are probably the best of all suture materials; it knots easily and securely. These latter characteristics almost certainly explain silk's continuing popularity for perineal repair, despite the fact that it results in increased discomfort.

A tissue adhesive, Histoacryl, has been compared with chromic catgut sutures in one trial. It suggested excellent results in terms of reduced pain and analgesic use in the 48 hours after delivery. This approach is a promising development, but there is no information about long-term outcome and the single trial does not provide adequate evidence for introducing Histoacryl into practice.

4 Who should perform the repair?

It is likely that the skills of the operator are as important, if not more important, than the materials and techniques used. However, there is little research evidence on the effects of skill on symptoms associated with perineal repair. Experience does not necessarily result in a better outcome — the same mistakes may be made with increasing confidence. There is an urgent need for this to be clarified in respect of the repair of perineal trauma.

Perineal repairs are often delegated to a junior or trainee obstetrician or midwife. Training is likely to have an important effect on the outcome of perineal repair, but has been a neglected area of research. The usual approach often is 'see three, do three, and now you are on your own!' Video recordings have been introduced in some places to supplement this, and apparatus on which to practice suturing is becoming available. Ideally, the usefulness of these developments should be carefully assessed before they are introduced widely.

5 Episiotomy breakdown

Episiotomy breakdown is a rare but unpleasant complication. One trial has compared a policy of primary resuturing plus antibiotic cover with wound cleansing and expectant treatment. Women managed with primary resuturing spent less time in hospital and made fewer visits to the hospital as outpatients. They resumed sexual intercourse sooner, and intercourse was more likely to be pain free. Four women out of twenty had a 'superficial rupture' in the resutured group, but none suffered serious wound breakdown.

The results of this single trial are not definitive, but they suggest that serious consideration should be given to primary resuturing (with antibiotic cover) following rupture of perineal trauma during the puerperium.

6 Conclusions

A continuous subcuticular stitch is preferable to interrupted trans-cutaneous sutures for episiotomy skin closure because it results in less short-term pain without any clear difference in the long term. On balance, absorbable sutures are preferable to non-absorbable material for this purpose.

On the basis of currently available evidence, polyglycolic acid sutures (Dexon or Vicryl) should be chosen for both the deep layers and the skin. Questions still remain about the long-term effects of polyglycolic sutures, but the available evidence is reassuring. The relatively frequent need to remove polyglycolic acid material in the puerperium because of irritation indicates either that this material is not ideal or that the stitches were tied too tightly. Whichever material is chosen for the skin, polyglycolic acid currently appears to be the material of choice for the deeper tissues.

Further research is required to evaluate the differential long-term effects of various suture materials and techniques, and to confirm or refute the suggested benefits of primary resuturing as opposed to conservative management when perineal trauma ruptures during the puerperium.

This chapter is derived from the chapter by Adrian Grant (68) in EFFECTIVE CARE IN PREGNANCY AND CHILDBIRTH .

References to primary sources and more complete data for statements made in this chapter can be found in the source chapter and/or in the following reviews from the *Cochrane pregnancy and childbirth database*:

Grant, A.M.
— Continuous vs interrupted sutures for perineal repair. Review no. 03252.
— Polyglycolic acid vs catgut for perineal repair. Review no. 01845.
— Glycerol-impregnated catgut vs chromic catgut for perineal repair. Review no. 03694.
— Absorbable vs non-absorbable sutures for perineal repair. Review no. 03253.
— Chromic catgut vs Supramid for perineal repair. Review no. 03692.

— Polyglycolic acid vs nylon for perineal repair. Review no. 03693.
— Polyglycolic acid vs silk for perineal repair. Review no. 03794.
— Subcuticular prolene vs interrupted catgut for skin closure during perineal repair. Review no. 07670.
— Povidone iodine prior to perineal suturing. Review no. 05574.
— Primary resuturing vs expectancy for ruptured episiotomy. Review no. 07017.
— Histoacryl vs chromic catgut for perineal skin closure. Review no. 07056.

37 Preterm birth

1 Introduction   2 Nature and range of preterm birth
2.1 *Types of preterm birth* 2.1.1 *Antepartum death and lethal malformations* 2.1.2 *Multiple pregnancy* 2.1.3 *Elective delivery* 2.1.4 *Maternal and fetal pathology*   2.2 *Gestational age* 2.2.1 *Range of gestation* 2.2.2 *Estimated weight or gestational age as a basis for care options*   3 Place of and preparations for birth   3.1 *Place of birth*   3.2 *Preparations for birth*   3.3 *Prevention of intraventricular haemorrhage*
4 Route of delivery   4.1 *Abdominal delivery*   4.2 *Vaginal delivery* 4.2.1 *Epidural and other analgesia* 4.2.2 *Elective forceps delivery* 4.2.3 *Routine use of early episiotomy*
5 Immediate care at birth   6 Conclusions

1 Introduction

Preterm birth is the most important single determinant of adverse infant outcome, in terms of both the likelihood of survival and the quality of life.

Preterm delivery, defined as the birth of an infant with a gestational age of less than 37 completed weeks, is not a single simple entity. It may occur as a result of spontaneous preterm labour or a deliberate intervention. In comparison with birth at term, it is more frequently associated with other conditions such as inadequate fetal growth, prelabour rupture of the membranes, multiple pregnancy, placenta praevia, placental abruption, fetal congenital malformations, abnormal fetal lie, and severe disease of the mother, all of which add their own hazards to the baby.

Only a few decades ago the prognosis for survival of very preterm infants was considered to be too poor to warrant special care for their birth. Any measures taken to increase the survival rate, it was believed, would result only in increasing numbers of handicapped children who would be a burden to their families and society. The pendulum has swung from this therapeutic nihilism to possibly overzealous intervention. Recommendations for major interventions without any clear evidence of benefit, such as 'caesarean section may be preferable to vaginal delivery for all singleton infants of very low birthweight', imply a wanton disregard for the well-being of the mother.

2 Nature and range of preterm birth

2.1 *Types of preterm birth*

Reported outcomes of preterm birth vary widely, depending largely on the vantage point of those who accumulate the data. Neonatal data often do not include stillbirths or grossly malformed infants, and they rarely refer to the significant pathology that may be present in mother or fetus before birth. Such data pertain to the outcomes for liveborn infants who have received modern intensive neonatal care, rather than for the fetuses for whom the obstetrician must make a decision.

2.1.1 *Antepartum death and lethal malformations* For between 10 and 15 per cent of all preterm births the form of care used cannot have any influence on the outcome for the baby because the infant had either died before the onset of labour or before admission, or had malformations that were incompatible with life. These 10–15 per cent of preterm births account for over 50 per cent of the total perinatal mortality associated with preterm birth.

For these births the prognosis for the baby is already determined, but the form of care chosen can have a profound effect on the mother's well-being. The objectives of care should be directed at maternal rather than fetal or neonatal interests.

The frequency with which preterm birth is associated with a dead or malformed fetus mandates a careful ultrasound examination before undertaking any form of care that carries a substantial risk of morbidity to the mother.

2.1.2 *Multiple pregnancy* Nearly half of all multiple births occur preterm, and multiple births are 15 times more frequent among preterm births than among births at term. Infants from multiple pregnancies, when born alive, have gestational age-specific mortality rates comparable with singleton infants who are at least 1–2 weeks less mature. These babies, who constitute 20 per cent of all liveborn preterm infants, have a higher incidence of respiratory distress syndrome and a higher mortality rate than singleton infants, even after correcting for gestational age (see Chapter 17).

2.1.3 *Elective delivery* As shown above, infants with antepartum fetal death, lethal malformations or multiple pregnancy make up about 25 per cent of preterm births. About a third of the remainder result from a deliberate obstetric decision to end the pregnancy, either by induction of labour or by elective caesarean section. This category of preterm birth constitutes an entirely different obstetric problem from that of delivery following the spontaneous onset of labour. The issue of how best to achieve delivery is secondary to that of whether or not one should attempt to achieve delivery at that time.

The prognosis for the baby, if elective delivery is undertaken, will not necessarily be similar to that after preterm birth following the spontaneous onset of labour. Data on survival, morbidity, and follow-up of infants of low birthweight, very low birthweight (less than 1500 gram), and extremely low birthweight (less than 1000 gram) may not be relevant to the outcome for electively delivered infants of comparable weight or maturity.

Thus, for more than half of the infants born preterm, care decisions for delivery are dominated by considerations such as the likelihood of infant survival (including disability-free survival) at the particular gestational age, the complexities of multiple pregnancy, and the question of whether and how to put an end to a pregnancy that is at no inherent risk of spontaneous preterm labour and delivery.

2.1.4 *Maternal and fetal pathology* For about half of the remaining preterm infants, preterm birth ensues (causally or incidentally) from pathological processes in the mother, such as hypertension or antepartum haemorrhage, or in the baby, such as inadequate fetal growth. The occurrence of preterm labour in these circumstances is often interpreted as demonstrating that nature is trying to remove the fetus from a 'hostile intrauterine environment'. Whether that interpretation is or is not correct in any individual situation may be difficult to ascertain, but it can profoundly influence the type of care provided.

2.2 Gestational age

2.2.1 Range of gestation

The likelihood of preterm birth increases with increasing gestational age up to the internationally defined cut-off point of 37 weeks. Less than a quarter of preterm births occur below 32 weeks. The delivery of a very preterm infant (gestational age of less than 32 completed weeks) presents the greatest challenge, but this arbitrary cut-off point has little relevance to clinical care. There is no specific gestational age or estimated fetal weight above and below which a 'hands-off approach' suddenly changes from being appropriate to being negligent.

Measures that have been clearly demonstrated to be beneficial for the infant and safe for the mother, such as corticosteroid administration, can be applied with far greater confidence at very low gestational ages than other measures, such as caesarean section, which confer far more dubious, if any, benefits to the infant and significant hazards to the mother.

2.2.2 Estimated weight or gestational age as a basis for care options

Although a wealth of information is available on the short- and long-term outcomes of infants weighing less than 1000 grams, less than 1500 grams, less than 2000 grams, or less than 2500 grams, experienced clinicians are aware of the pitfalls inherent in the birthweight-specific data on which they sometimes are expected to base their care options. These data often refer to infants delivered to neonatal units after all the selection processes that occur between the decision for delivery and arrival in a neonatal unit. They include data on infants whose intrauterine growth was restrained and therefore who were born at a more mature gestational age than would be expected from birthweight alone.

Moreover, estimates of what the actual weight of the baby will be at birth are notoriously inaccurate. Clinical estimates are often far off the mark. While advances have been made in the accuracy of fetal weight estimation by ultrasound, in most hands its accuracy is still far from satisfactory. This applies in particular to the low weight ranges, and, as is often the case for preterm birth, when the measurements must be made with some urgency and by whoever happens to be available.

Even if birthweight could be estimated with 100 per cent accuracy before birth, this information would still be inferior to gestational age as a determinant of infant outcome and care options. Although birthweight is a better predictor of mortality than gestational age when the whole range of gestation from 20 to over 40 weeks is considered, organ maturity is more important than organ weight for the very preterm infant. Estimated gestational age is a better predictor than birthweight of both mortality and morbidity.

A carefully taken menstrual history and ascertainment of any prior assessment of gestational age during pregnancy (such as an early ultrasound examination) will be more useful for the woman in preterm labour than attempts to determine fetal weight. In the absence of any evidence that estimation of fetal weight improves outcome, it is not justified to substitute weight estimation for a careful assessment of the menstrual history and gestational age as the basis for care decisions.

3 Place of and preparations for birth

3.1 *Place of birth*

Whenever possible, delivery of the very young and very small fetus should be in a centre with adequate facilities and equipment, persons capable of managing and handling the equipment, sufficient manpower to ensure round the clock utilization of its resources, and professionals in various disciplines ready and willing to collaborate in care of the mother and baby.

The most dangerous place for a preterm baby to be born is in a hospital with caregivers who believe that they have, but do not really have, the equipment and skills necessary to care effectively for these tiny babies before and after birth. Misplaced self-confidence can result in failure of timely referral to institutions where such facilities are available.

3.2 *Preparations for birth*

The woman at imminent risk of preterm delivery requires an immediate assessment as to whether transfer to a perinatal centre would be appropriate. The decision will depend on the gestational age of the fetus, on the facilities that are available (which may, for example, be sufficient for a baby born at 36 but not at 31 weeks), and on the imminence of the expected delivery.

If the woman is in labour, it is wise to inhibit labour with betamimetic drugs in order to postpone delivery at least until she arrives at the perinatal centre. Administration of corticosteroids should be commenced before transfer, unless gestational age has advanced beyond the stage at which respiratory distress syndrome is likely to be a problem.

Given the frequency with which preterm birth is associated with maternal disease, the institutions where preterm delivery is undertaken should be able to draw on the expertise not only of neonatologists or perinatologists, but also of other professionals who may be needed to provide counselling or advice. Ultrasound equipment should be available for every preterm labour or delivery, along with persons who can handle the equipment and interpret its findings correctly. Full laboratory facilities should be available at all times. Resuscitation equipment should be available on the spot, and the presence and proper working order of this equipment should be verified before each delivery.

Fetal assessment before birth should identify multiple pregnancy, assess whether or not the fetus is alive and well, differentiate the normally formed from the malformed fetus, and determine the fetal presentation. As mentioned earlier, careful review of the gestational data is necessary. All these are prerequisites for proper care for preterm birth. A policy of caesarean section for all is not an adequate substitute.

A professional who is skilled in resuscitation, and who can devote total attention to the infant, should be in attendance at all preterm births. This is readily achieved where labour and neonatal wards are adjacent (as they should be). Where distance presents a problem, every effort should be made first to ensure that these distances are phased out in the foreseeable future and second to increase staffing levels so that no preterm baby will be delivered without appropriate neonatal attention.

3.3 Prevention of intraventricular haemorrhage

Intraventricular haemorrhage is an important cause of mortality and morbidity in the very preterm infant. The risk of haemorrhage is inversely related to gestational age, ranging from more than 70 per cent for babies born under 26 weeks to less than 10 per cent for those born after 33 weeks. More than 90 per cent of the haemorrhages occur in infants below 35 weeks of gestation.

Trials have addressed the question of whether the incidence of intraventicular haemorrhage in preterm infants can be reduced by the administration of either phenobarbitone or vitamin K to the mother before delivery. Although the data are insufficient to make strong policy recommendations, they suggest that both these approaches may be effective. Further research is urgently required to confirm or refute these findings.

4 Route of delivery

One of the main decisions about care for preterm birth, and certainly the most controversial one, is the choice between vaginal delivery and caesarean section. From the volume of literature that has been published one would expect a wealth of evidence to be available for selecting the best approach. This is not the case. Calls for clinical trials to provide unbiased comparisons between vaginal and abdominal routes of delivery have been published for more than a decade, yet there have been very few attempts to put these recommendations into practice. Trials that have been attempted were abandoned before or soon after they started. To date, insufficient unbiased information is available to shed light on the question of when a caesarean section might add sufficient benefits to the infant to warrant the operation.

Breech presentation is far more common among preterm infants than among term infants, and breech presentation at birth carries a higher risk for the infant than does cephalic presentation. This increased risk has paved the way for 'prophylactic caesarean section' for the preterm baby presenting as a breech, as a presumably safer method of delivery. In some centres the supposed benefits of this approach have been extended to all preterm babies without adequate evidence that the assumed gain in safety is indeed a gain and not a loss.

Observational studies that have compared the outcome of preterm birth have usually found higher survival rates after caesarean section than after vaginal delivery. Unfortunately, even when they made genuine attempts to control for as many confounding factors as possible, these studies could not compare like with like. Infants born vaginally are more likely to be those who are considered to be too small, of too low gestational age, or too sick to receive sufficient benefit from caesarean section. They may be those whose mothers arrived too late in second stage labour to have a caesarean section. They are more likely to have foregone the benefits of antenatal corticosteroid treatment because they were born too quickly for these drugs to be administered or to have their full effect. Infants born vaginally are also more likely to have been born in the absence of a senior obstetrician and neo-natologist, and to have received less dedicated care at vaginal delivery. If the general policy favours caesarean section, vaginal delivery is likely to be assisted by a person without the necessary technical skills and experience of delivery of tiny infants.

In contrast, infants born by caesarean section are more likely to be those for whom delivery could be planned in advance. They are more likely to be born as the result of elective obstetric intervention to end pregnancy. Their birth is more likely to have been preceded by thorough assessment of the fetal condition, possibly including ascertainment of lung maturation, and after full preparations have been made for whatever special neonatal care is required. These infants are more likely to be born later in gestation and to have birthweights that are higher than those of infants born vaginally.

As infants born vaginally or by caesarean section vary in so many other ways in addition to the method of delivery, observational or descriptive studies provide no useful information. Properly controlled studies are required to address this important question.

4.1 *Abdominal delivery*

Although caesarean section is considered to be one of the safest of the major surgical procedures, it still carries an important risk of mortality and morbidity. Some special considerations apply to a preterm caesarean section.

A careful ultrasound examination is essential before caesarean delivery of a preterm infant. First, it is important to determine whether or not the fetus is normally formed; a caesarean to deliver a fetus with a lethal abnormality would be a double tragedy. Second, an exact diagnosis of the fetal presentation may allow correction of malpresentation before the uterus is incised, and this may make all the difference between an easy caesarean delivery and a traumatic extraction of a malpositioned infant through a poorly formed and thick lower uterine segment. Third, it is useful to know whether or not the placenta will be in the way. Trauma to the aftercoming head will not easily be avoided, for example, if a relatively large head must be delivered through a small incision in a thick uterine segment, with the placenta as well as the body of the baby bulging through it.

At caesarean section, particularly when it is performed for breech presentation, it is important to assess whether the lower uterine segment is sufficiently wide to permit easy delivery of the head. Often, and particularly in elective preterm delivery, there will be little lower uterine segment, and whatever incision is made will be made through the body of the uterus. Therefore some authors recommend a vertical rather than a transverse incision, but no controlled experiments have been conducted to evaluate the relative merits of the alternative policies.

If the presentation is not longitudinal, it should be corrected before the uterine incision is made, preferably to a vertex presentation. This is usually not too difficult, particularly when there is a normal volume of amniotic fluid and the uterine relaxant effect of betamimetic treatment can be maintained up to the moment of delivery with this aim in mind. It is just as important to strive for easy and gentle delivery of the fetal head at caesarean delivery as it is at vaginal delivery.

4.2 *Vaginal delivery*

The head of the preterm baby, with its soft bones and wide skull sutures, is more vulnerable than that of the term baby to compression by the maternal pelvic tissues, and to sudden decompression when the baby emerges from the birth canal. Several measures have been proposed to minimize the occurrence of these changes in intracranial pressure. These include liberal use of epidural analgesia to lower resistance in the birth canal; routine use of 'prophylactic forceps' delivery to counteract both compression before and decompression after birth, and routine use of early episiotomy to remove the resistance of rigid perineal tissues. There is little, if any, evidence to support the use of these measures.

4.2.1 *Epidural and other analgesia* Adequate analgesia is just as important for the mother giving birth preterm as for the mother delivering at term. The main difference may be that the woman in preterm labour may be less well prepared for birth, and may be overcome by the suddenness of it all and the increased risk of giving birth preterm. It is important to respond to her anxiety and discomfort in ways that do not depress the baby. When, as is often the case, pharmacological analgesia is required, an epidural block is probably safer than narcotics, although there have been no controlled studies to substantiate this recommendation.

The routine use of epidural block, particularly for the preterm breech, has also been suggested as a means of abolishing the urge to push before the cervix is fully dilated, and to reduce the resistance of the pelvic musculature. This is a reasonable hypothesis, but as yet there is no evidence to support it. Similar or greater protection might be gained by close communication with, and careful instructions to, the woman.

4.2.2 *Elective forceps delivery* Prophylactic forceps have become accepted practice for preterm vaginal delivery in many places, but there have been few attempts to assess the value, if any, of this practice. The postulated protection of the baby's head can be questioned on theoretical grounds. Forceps are effective levers, and any compression applied at the handles is transmitted directly to the blades and hence to the fetal head. In addition, a substantial part of the traction force during delivery is transmitted as compression force to the fetal head. This may be particularly damaging to a preterm baby with soft skull bones and wide skull sutures.

There is at present no unbiased evidence to suggest that routine use of forceps to deliver the preterm baby confers more benefit than harm.

4.2.3 *Routine use of early episiotomy* The few unbiased data on the question of whether or not routine use of episiotomy for delivery of the preterm baby improves neonatal outcome do not support a policy of routine episiotomy for infants to be delivered preterm.

5 Immediate care at birth

The physiological consequences of early versus late cord clamping have not been studied as well in the preterm infant as they have been at term. In the preterm infant, delayed cord clamping is associated with a 50 per cent increase in red cell volume and over half of this placental transfusion occurs within the first minute after birth. Proponents of early clamping suggest that the large transfusion may encourage pulmonary oedema and increase the risk of intracranial haemorrhage and hyperbilirubinaemia. Those who advocate delayed clamping point out that the placental transfusion may expand the pulmonary bed and prevent respiratory distress, prevent hypovolaemia and hypotension, and increase haemoglobin concentrations and total body iron stores.

It remains unknown whether alternative policies of cord clamping with preterm birth will have a significant impact on neonatal outcome. The available evidence from the few trials that have been reported show that early clamping reduces the incidence of hyperbilirubinaemia, but no clear effect is demonstrated on other outcomes. From the available evidence (admittedly gathered predominantly in infants who were not born preterm) there does not seem to be any justification for rushing to clamp the cord unless urgent paediatric attention is required.

A paediatrician should be present at all preterm deliveries. For the very preterm and very small infant, the paediatrician should be an experienced neonatologist, and decisions with regard to resuscitation and suctioning should be her or his prerogative, taken in harmony with the parents and the obstetrician.

The simple measure of providing adequate heat in the delivery room can contribute more to subsequent neonatal well-being than any of the other measures that, in many places, are routinely applied to the preterm infant.

6 Conclusions

The expression 'preterm birth' encompasses a variety of different clinical presentations. In some the risk is little different from that of delivery at term; in others the utmost sophistication of facilities and skills is necessary to allow the infant even a remote hope of intact survival. Many of these small babies are already compromised by other factors, such as congenital malformations, multiple pregnancy, or complicating maternal illness. Therefore care for preterm birth must be carefully individualized, taking all these factors into consideration.

The plan of care for the birth of a preterm baby should be governed by consideration of gestational age rather than estimated weight, because gestational age is a better indicator of prognosis.

The baby should be delivered in an institution that has all the necessary facilities and skilled personnel readily available. Transfer of the baby after birth is not as likely to be effective as birth in a centre that is adequately equipped and staffed to ensure that the fetus is alive and well, to rule out congenital malformation, to establish fetal presentation before birth, and to perform the skilled resuscitation that is necessary at the moment of birth. Even for the woman in active labour, labour may be inhibited with betamimetics to delay birth long enough to permit her transfer to such a centre.

The choice between vaginal delivery or caesarean section is not easy. Observational data on the differential effects of abdominal and vaginal delivery are all subject to such major biases that their results should be ignored. In the absence of guidance from controlled trials, caesarean section, with its known risks to the mother, should be the exception rather than the rule. This applies to the fetus presenting as a breech as well as to that presenting as a vertex.

A careful ultrasound examination is essential before caesarean delivery of a preterm infant to ensure that the baby is free of lethal congenital malformations, to determine fetal presentation and allow correction if necessary before the uterus is incised, and to determine the position of the placenta which may interfere with the extraction of the baby. The uterine incision must be adequate in size, even though this is sometimes difficult with the poorly developed lower segment characteristic of the preterm uterus.

The head of the preterm baby is more vulnerable to injury, from either compression or sudden expansion, than that of the baby at term. There is no evidence to suggest that either elective forceps delivery or performing an episiotomy reduces this risk. The routine use of both of these procedures should be abandoned, except in the context of controlled trials.

As the evidence in favour of early or late clamping of the umbilical cord is conflicting, decisions about when to clamp the cord should be based on the urgency of the need for resuscitation.

A paediatrician should be in attendance at all preterm deliveries, and for the very preterm or very small infant he or she should be an experienced neonatologist capable of making the decisions and performing the skilled resuscitation that may be necessary.

Meticulous attention to the features that should be present for all births, such as a warm environment and careful consideration, is even more important for the preterm baby than for the infant born at term without complications.

This chapter is derived from the chapter by Marc J.N.C. Keirse (74) in EFFECTIVE CARE IN PREGNANCY AND CHILDBIRTH

References to primary sources and more complete data for statements made in this chapter can be found in the source chapter and/or in the following reviews from the Cochrane pregnancy and childbirth database:

Elbourne, D.R.
— Early cord clamping in preterm infants. Review no. 05944.

Grant, A.M.
— Vitamin K prior to preterm delivery. Review no. 04748.
— Elective vs selective Caesarean delivery of the small baby. Review no. 06597.

Grant, A.M. and Crowther, C.A.
— Phenobarbital prior to preterm delivery. Review no. 04261.

---

38 Labour and delivery after previous caesarean section

1 Introduction   2 Results of a trial of labour   3 Risks of
caesarean section   3.1 *Risks to the mother*   3.2 *Risks to the
baby*   4 Factors to consider in the decision about a trial of
labour   4.1 *More than one previous caesarean section*
4.2 *Reason for the primary caesarean section*   4.3 *Previous
vaginal delivery*   4.4 *Type of previous incision in the uterus*
4.5 *Gestational age at previous caesarean section*
4.6 *Integrity of the scar*   5 Care during a trial of labour
5.1 *Use of oxytocics*   5.2 *Regional analgesia and anaesthesia*
5.3 *Manual exploration of the uterus*   6 Rupture of the
scarred uterus in pregnancy and labour   7 Gap between
evidence and practice   8 Conclusions

1 Introduction

There is wide variation internationally in the percentage of babies born
by caesarean section — almost 25 per cent in the United States and, by
way of contrast, 7 per cent in The Netherlands. It is highly unlikely that
such extremes of practice reflect major differences in the health of
women in these two affluent countries. Therefore one has to seek
differences in clinical philosophy for the explanation. One difference
has been in the greater readiness with which obstetricians in America
resort to repeat abdominal delivery after previous caesarean section.
Although the North American dogma 'once a caesarean, always a
caesarean' has come under professional and public scrutiny in recent

years, it still remains as stated policy in many institutions and is widely practised in other countries as well.

Two general propositions underlie the practice of repeat caesarean section: that trial of labour with its inherent risk of uterine rupture represents a significant hazard to the well-being of mother and baby, and that planned repeat caesarean operations are virtually free of risk. It is important to examine the truth of both these propositions.

2 Results of a trial of labour

No randomized controlled trials have compared the results of elective caesarean section with those of trial of labour for women who have had a previous caesarean section. In the absence of such trials, the best available data on the relative safety of a trial of labour come from prospective observational studies. In these studies, including many thousands of pregnant women with a history of one caesarean section, over two-thirds were allowed a trial of labour. Of these women almost 80 per cent gave birth vaginally. Thus, for the series for which total data are available, well over half of all women with a previous caesarean section gave birth vaginally.

Obstetricians' fear of uterine dehiscence (breakdown of the uterine scar) has had a major influence on clinical practice. The most dramatic consequences of this complication can be seen in developing countries in which pelvic contraction and cephalopelvic disproportion are common, and access to clinical facilities is often difficult. When obstructed labour occurs following previous caesarean section under these circumstances, dehiscence of the wound may extend into a rupture of other parts of the uterus and become a threat to the life of both mother and baby.

However, these are not the conditions in 'developed' countries in which the caesarean section rates are highest. In these countries, dehiscences that are encountered are usually slight, often representing so-called 'windows' in the uterus and carrying no sequelae. Indeed, the prospective observational studies found evidence of dehiscence in 0.5–2.0 per cent of women undergoing planned caesarean section before labour had even started. The corresponding figure among women undergoing a trial of vaginal delivery (successful or un-successful) was little different (0.5–3.3 per cent) although, because of lack of randomization, the two figures are not directly comparable. The important point is that serious wound dehiscence is a rare complication during labour after previous caesarean section.

Maternal morbidity was less in the trial of labour groups. Data from these prospective studies show that febrile morbidity rates were consistently and substantially higher in the groups of women who underwent elective caesarean section (range 11–38 per cent) than in the groups of women who had a trial of labour, including both those who had an emergency caesarean section and those who had a vaginal delivery (range 2–23 per cent). Although the febrile morbidity rates were higher among women who underwent caesarean section after a trial of labour, these were more than counterbalanced by the lower rate in the two thirds of women who give birth vaginally after a trial of labour.

Blood transfusions, endometritis, abdominal wound infections, thromboembolic phenomena, anaesthetic complications, pyelonephritis, pneumonia, and septicemia were also less common in women who had a vaginal delivery following low transverse caesarean section than in women who underwent a repeat caesarean section.

Perinatal mortality and morbidity rates were similar with trial of labour and elective caesarean section in the studies that report these data. However, such comparisons are of little value because the groups compared are not equivalent. The decision to perform a repeat caesarean section or to permit a trial of labour may have been made on the basis of whether or not the fetus was alive, dead, anomalous, or immature.

3 Risks of caesarean section

3.1 *Risks to the mother*

Large series of caesarean sections have been reported with no associated maternal mortality. One should not be lulled into a false sense of security by this. The risk of a mother dying with caesarean section is small, but it is still considerably higher than with vaginal delivery.

The rate of maternal death associated with caesarean section (approximately 40 per 100 000 births) is four times that associated with all types of vaginal delivery (10 per 100 000 births). The maternal death rate associated with elective repeat caesarean section (18 per 100 000 births), although lower than that associated with caesarean sections overall, is still almost twice the rate associated with all vaginal deliveries and nearly four times the mortality rate associated with normal vaginal delivery (5 per 100 000 births).

The rate of maternal mortality attributable to caesarean section *per se* is difficult to estimate, as some of the deaths are caused by the condition which prompted the caesarean section in the first place. While it is not possible to quantitate exactly the increase in risk of dying from elective caesarean section, the data available suggest that it is between two and four times that associated with vaginal delivery.

Most forms of maternal morbidity are higher with caesarean section than with vaginal delivery. In addition to the risks of anaesthesia attendant on all surgery, there are risks of operative injury, febrile morbidity, effects on subsequent fertility, and psychological morbidity.

3.2 Risks to the baby

The major hazards of caesarean section for the baby relate to the risks of respiratory distress contingent on either the caesarean birth itself or on preterm birth as a result of miscalculation of dates. Babies born by caesarean section have a higher risk of respiratory distress syndrome than babies born vaginally at the same gestational age.

The availability of more accurate and readily available dating with ultrasound should decrease the risk of unexpected preterm birth. Nevertheless, it is unlikely that errors in dating can ever be completely eliminated.

4 Factors to consider in the decision about a trial of labour

A mathematical utilitarian approach comparing the balance of risks and benefits of trial of labour with those of planned caesarean section will not always be the best way to choose a course of action. However, such an approach can provide important data that may be helpful in arriving at the best decision.

The technique of decision analysis has been used to determine the optimal birth policy after previous caesarean section. The probabilities and utilities of a number of possible outcomes, including the need for hysterectomy, uterine rupture, iatrogenic preterm birth, need for future repeat caesarean sections, prolonged hospitalization and recovery, additional cost, failed trial of labour, discomfort of labour, and inconvenience of awaiting labour, can be put into a mathematical model comparing different policies. Over a wide range of probabilities and utilities, which included all reasonable values, trial of labour proved to be the safer choice.

4.1 More than one previous caesarean section

Data on the results of trials of labour in women who have had more than one previous caesarean section tend to be buried in studies of trial of labour after previous caesarean section as a whole. The available data on outcomes after a trial of labour in women who have had more than one previous caesarean section show that the overall vaginal delivery rate is little different from that seen in women who have had only one previous caesarean section. Successful trials of labour have been carried out on women who have had three or more previous caesarean sections.

The rate of uterine dehiscence in women who have had more than one previous caesarean section is slightly higher than the dehiscence rate in women with only one previous caesarean, but all dehiscences in the reported series were without symptoms and without serious sequelae. No maternal or perinatal mortality associated with any of the trials of labour after more than one previous caesarean section was reported in these series. No data have been reported on other maternal or infant morbidity specifically associated with multiple previous caesarean sections.

While the number of cases reported is still small, the available evidence does not suggest that a woman who has had more than one previous caesarean section should be treated any differently from the woman who has had only one caesarean section.

4.2 Reason for the primary caesarean section

The greatest likelihood of vaginal birth following previous section is seen when the first caesarean section was done because of breech presentation; vaginal delivery rates are lowest when the initial indication was failure to progress in labour, dystocia, or cephalopelvic disproportion. Even when the indication for the first caesarean section was disproportion, dystocia, or failure to progress, successful vaginal delivery was achieved in more than 50 per cent of the women in most published series, and the rate was over 75 per cent in the largest series reported. It is clear that a history of caesarean section for dystocia is not a contraindication to a trial of labour. It has only a small effect on the chances of vaginal birth when a trial of labour is permitted.

4.3 Previous vaginal delivery

Mothers who have had a previous vaginal birth in addition to their previous caesarean sections are more likely to deliver vaginally after trial of labour than mothers with no previous vaginal births. This advantage is increased even further in those mothers whose previous vaginal birth occurred after, rather than before, the caesarean section.

4.4 Type of previous incision in the uterus

Modern experience with operative approaches other than the lower segment operation for caesarean section is limited. However, there is a growing trend towards the use of vertical incisions in preterm caesarean sections. This, and the inverted-T incision sometimes necessary to allow delivery, shows that consideration of the type of uterine scar is still relevant.

The potential dangers of uterine rupture are related to the rapid 'explosive' rupture which is most likely to be seen in women who have a classical mid-line scar. The majority of dehiscences after lower segment transverse incisions are 'silent', 'incomplete', or incidentally discovered at the time of repeat caesarean section.

Rupture of the scar after a classical caesarean section is not only more serious than rupture of a lower-segment scar, but is also more likely to occur. Rupture may occur suddenly during the course of pregnancy, prior to labour, and before a repeat caesarean section can be scheduled. A review of the literature at a time when classical caesarean section was still common showed a 2.2 per cent rate of uterine rupture with previous classical caesarean sections, and a rate of 0.5 per cent with previous lower-segment caesarean sections. That is, the scar of the classical operation was more than four times more likely to rupture in a subsequent pregnancy than was that of the lower-segment incision.

Unfortunately, even in the older literature, there are very few data on the risk of uterine rupture of a vertical scar in the lower-segment. One 1966 study reported an incidence of rupture of 2.2 per cent in classical incision scars, 1.3 per cent in vertical incision lower-segment scars, and 0.7 per cent in transverse incision lower–segment scars. The distinction between the risk of rupture of vertical and transverse lower-segment scars may be related to extension of the vertical incision from the lower-segment into the upper segment of the uterus.

The uncertain denominators in the reported series make it difficult to quantitate the risk of rupture with a previous classical or vertical incision lower–segment scar. However, it is clear that the risk that rupture may occur, that it may occur prior to the onset of labour, and that it may have serious sequelae is considerably greater with such scars than with transverse incision lower-segment scars. It would seem reasonable that women who have had a hysterotomy, a vertical uterine incision, or an inverted-T incision be treated in subsequent pregnancies in the same manner as women who have had a classical caesarean section, and that trial of labour, if permitted at all, should be carried out with great caution and with acute awareness of the increased risks that are likely to exist.

4.5 Gestational age at previous caesarean section

During the past decade improved neonatal care has increased the survival rate of preterm babies. This in turn has led to a reduction in the stage of gestation at which obstetricians are prepared to perform caesarean sections for fetal indications. It has resulted in caesarean sections being used to deliver babies at, or even before, 26 weeks. At these early gestations the lower segment is poorly formed, and so-called 'lower-segment' operations at this period of gestation are, in reality, transverse incisions in the body of the uterus. Whether or not such an incision confers any advantage over a classical incision remains in doubt. Indeed, some obstetricians now recommend performing a classical incision in these circumstances.

Whichever of these incisions is used at these early gestational ages, their consequences for subsequent pregnancies are currently unknown. It is quite possible, in theory at least, that they may result in a greater morbidity in future pregnancies than that associated with the lower segment operation at term.

4.6 *Integrity of the scar*

The decision to advise for or against a trial of labour may be influenced by an assessment of the integrity of the scar. This assessment may be helped by knowledge of the operative technique used for the previous caesarean section, the operative findings at the time of surgery, whether an extension of the operative incision had occurred, and the nature of the postoperative course.

5 Care during a trial of labour

5.1 *Use of oxytocics*

The use of oxytocin or prostaglandins for induction or augmentation of labour in women with a previous caesarean section has remained controversial because of speculation that there might be an increased risk of uterine rupture or dehiscence. This view is not universally held, nor is it strongly supported by the available data. A number of series have been reported in which oxytocin or prostaglandins were used for the usual indications with no suggestion of increased hazard. Review of the reported case series shows that any increased risk of uterine rupture with the use of oxytocin or prostaglandins is likely to be extremely small. When dehiscences occur, they are more likely to be in women who have received more than one oxytocic agent rather than a single agent, used in an appropriate manner.

Such comparisons are rendered invalid by the fact that the cohorts of women who received or did not receive oxytocics may have differed in many other respects in addition to the use of oxytocic agents. Nevertheless, the high vaginal delivery rates and low dehiscence rates noted in these women suggest that oxytocics can be used for induction or augmentation of labour in women who have had a previous caesarean section, with the same precautions that should always attend the use of oxytocic agents.

5.2 *Regional analgesia and anaesthesia*

The use of regional (caudal or epidural) analgesia in labour for the woman with a previous caesarean section has been questioned because of fears that it might mask pain or tenderness, which are considered to be early signs of rupture of the scar. The extent of the risk of masking a catastrophic uterine rupture is difficult to quantitate. It must be minuscule as only one case report of this having occurred was located. In a number of reported series regional block is used whenever requested by the woman for pain relief, and no difficulties were encountered with this policy.

There does not appear to be any increased hazard from uterine rupture associated with the use of regional anaesthesia for women who have had a previous caesarean section. It is sensible, safe, and justified to use analgesia for the woman with a lower-segment scar in the same manner as for the woman whose uterus is intact.

5.3 Manual exploration of the uterus

In many reported series of vaginal births after previous caesarean section, mention is made of the fact that the uterus was explored postpartum in all cases in a search for uterine rupture or dehiscence without symptoms. The wisdom of this approach should be seriously challenged.

Manual exploration of a scarred uterus immediately after a vaginal delivery is often inconclusive. It is difficult to be sure whether or not the thin soft lower segment is intact. In any case, in the absence of bleeding or systemic signs, a rupture without symptoms discovered postpartum does not require any treatment and so the question of diagnosis would be academic. In the absence of epidural or general anaesthesia, it is also very painful to the woman.

No studies have shown any benefit from routine manual exploration of the uterus in women who have had a previous caesarean section. There is always a risk of introducing infection by the manual exploration, or of converting a dehiscence into a larger rupture. A reasonable compromise consists of increased vigilance in the hour after delivery of the placenta, reserving internal palpation of the lower segment for women with signs of abnormal bleeding.

6 Rupture of the scarred uterus in pregnancy and labour

Complete rupture of the uterus can be a life-threatening emergency. Fortunately, the condition is rare in modern obstetrics despite the increase in caesarean section rates, and serious sequelae are even more rare. Although often considered to be the most common cause of uterine rupture, previous caesarean section is a factor in less than half the cases.

Excluding symptomless wound breakdown, the rate of reported uterine rupture ranges from 0.09 per cent to 0.22 per cent for women with a singleton vertex presentation who had undergone a trial of labour after a previous transverse lower-segment caesarean section. To put these rates into perspective, the probability of requiring an emergency caesarean section for other acute conditions (fetal distress, cord prolapse, or antepartum haemorrhage) in any woman giving birth is approximately 2.7 per cent, or 30 times as high as the risk of uterine rupture with a trial of labour.

Treatment of rupture of a lower-segment scar does not require extraordinary facilities. Hospitals whose capabilities are so limited that they cannot deal promptly with problems associated with a trial of labour are also incapable of dealing appropriately with other obstetrical emergencies. Any obstetrical department that is prepared to look after women with much more frequently encountered conditions such as placenta praevia, abruptio placentae, prolapsed cord, and acute fetal distress should be capable of the safe management of a trial of labour after a previous lower-segment caesarean section.

7 Gap between evidence and practice

Obstetric practice has been slow to adopt the scientific evidence confirming the safety of trial of labour after previous caesarean section. The degree of opposition to vaginal birth after caesarean section, in North America in particular, is difficult to explain, considering the strength of the evidence that, under proper circumstances, trials of labour are both safe and effective. Two national consensus statements and two national professional bodies in Canada and the United States have recommended policies of trial of labour after previous caesarean section.

Increasing numbers of pregnant women, as well as professionals, are vehemently protesting against the status quo. For a variety of reasons many women prefer to attempt a vaginal birth after a caesarean section. Their earlier caesarean experience may have been emotionally or physically difficult. They may be unhappy because they were separated from their partners or their babies. They may wonder if it was all necessary in the first place. They may be aware of the accumulated evidence on the relative safety and advantages of trial of labour, and simply be looking for a better experience this time.

In recent years a number of consumer 'shared predicament' groups have appeared with the expressed purposes of demythologizing caesarean section, of combating misinformation, and of disseminating both accurate information and their own point of view. In some countries special prenatal classes are available for parents who elect to attempt a vaginal birth after caesarean.

8 Conclusions

A trial of labour should be recommended for women who have had a previous caesarean section by lower-segment transverse incision and who have no other indication for caesarean section in the present pregnancy. The likelihood of vaginal birth is not significantly altered by the indication for the first cesarean section (including 'cephalopelvic disproportion' and 'failure to progress'), nor by a history of more than one previous caesarean section.

A history of classical, low-vertical, or unknown uterine incision, or hysterotomy, carries with it an increased risk of uterine rupture, and in most cases is a contraindication to trial of labour.

The care of a woman in labour after a previous lower-segment caesarean section should be little different from that of any woman in labour. Oxytocin induction or stimulation and epidural analgesia may be used for the usual indications. Careful monitoring of the condition of the mother and fetus is required, as for all pregnancies. The hospital facilities required do not differ from those that should be available for all women giving birth, irrespective of their previous history.

This chapter is derived from the chapter by Murray Enkin (70) in EFFECTIVE CARE IN PREGNANCY AND CHILDBIRTH.

References to primary sources and more complete data for statements made in this chapter can be found in the source chapter.

TECHNIQUES OF INDUCTION AND OPERATIVE DELIVERY

39 Preparing for induction of labour

1 Introduction

The decision to bring pregnancy to an end before the spontaneous onset of labour is one of the most drastic ways of intervening in the natural history of pregnancy and childbirth. The reasons given for elective delivery (which may be achieved either by inducing labour or by elective caesarean section) range from the life-saving to the trivial. There has been very little controlled research on the indications for elective delivery; most research has been concerned with methods to implement elective delivery. Although comparisons of these methods are secondary to the more fundamental question of when or whether an elective delivery is required, once the decision for an elective delivery is made, the method chosen becomes important. If induction of labour and vaginal delivery is planned, then attention to the state of the cervix is essential.

2 Assessing the cervix

The state of the cervix at the time that induction of labour is attempted is the most important determinant of the subsequent course of events. An 'unripe' cervix (one which is not ready for induction of labour) fails to dilate adequately in response to uterine contractions. Attempted induction when the cervix is not ripe may result in high rates of induction failure, protracted and exhausting labours, a high caesarean section rate, and other complications. These include intrauterine infection and pyrexia when amniotomy is employed, as well as uterine hypertonus and drug-induced side-effects (due to the high doses needed) with the use of oxytocics.

Assessment of the state of the cervix is highly subjective, and even experienced examiners may differ in their appraisal of cervical features. Several scoring systems have been developed in attempts to establish more comparable guidelines for cervical assessment. The best known of these is the score proposed by Bishop, which rates five different qualities: effacement, dilatation, and consistency of the cervix, position of the cervix relative to the axis of the pelvis, and descent of the fetal presenting part. Most other scoring systems use the same components, although with different weighting.

3 Prostaglandins for cervical ripening

Doses of prostaglandins which by themselves are insufficient to induce labour successfully produce a marked softening of the uterine cervix. This softening results more from an effect on the connective tissue of the cervix than from uterine contractions.

Prostaglandins can effectively ripen the cervix and facilitate induction of labour. Labour commences before the start of induction (during the period allocated for cervical ripening) more often in women receiving prostaglandins for cervical ripening than in women who receive placebo or no specific treatment. Prostaglandin ripening of the cervix increases the likelihood of a successful induction of labour, and of achieving vaginal delivery within 12 or 24 hours.

The effect of prostaglandin ripening on the pain experienced during labour is not clear. The few reports that provide data on the use of pharmacological or epidural analgesia show that epidural analgesia is less frequently used among prostaglandin-treated women than among women in the control groups.

'Uterine hypertonus' or 'uterine hyperstimulation' (excessive uterine contractility), either during the period of cervical ripening or during the subsequent induction of labour, occurs more often in prostaglandin-treated women than among women who receive placebo or no treatment prior to induction. Whether for this or other reasons, fetal heart rate abnormalities also tend to occur more frequently with prostaglandin treatment.

Neither of these trends lead to an increased incidence of operative delivery. On the contrary, prostaglandin treatment results in a major reduction in the operative delivery rate, which is based on a modest, but statistically significant, decrease in the caesarean section rate and a more marked decrease in the rate of instrumental vaginal delivery.

Relatively few trials provide data on the incidence of postpartum haemorrhage and/or the use of blood transfusion. The available data do not suggest that these outcomes are influenced by the use of prostaglandins for cervical ripening, but the estimates are not precise.

None of the data available suggest any influence, good or bad, of cervical ripening on neonatal outcomes. There is a trend towards fewer low Apgar scores in babies born to mothers who received prostaglandins, but few trials provide data on more substantive infant outcome measures such as resuscitation of the newborn, admission to a special care nursery, and perinatal death.

There is no convincing evidence that prostaglandins PGE_2 and $PGF_{2\alpha}$ have clinically important differential effects on cervical ripening when used in equipotent doses. However, the much lower drug dose needed for PGE_2 makes this compound preferable to $PGF_{2\alpha}$.

3.1 Oral prostaglandins

Oral prostaglandins have shown little or no beneficial effect when compared with either placebo or no treatment for cervical ripening. The drug must be administered repeatedly over a period of several hours in order to have effects that, on the whole, are not very impressive. Overall, the data suggest that oral prostaglandins are not a suitable approach to increasing the readiness of the cervix for induction.

3.2 Vaginal prostaglandins

Vaginal administration of prostaglandins for cervical ripening has been studied more extensively than oral administration. The results show an increase in the frequency of labour onset during the ripening period, a decrease in the incidence of failed induction, and an increase in the likelihood of 'hypertonus' or 'hyperstimulation'. The rate of both instrumental vaginal delivery and caesarean section is decreased with the use of vaginal prostaglandins. No effect has been demonstrated on any infant outcomes.

Women receiving vaginal prostaglandins for cervical ripening are more likely to undergo caesarean section during the time interval allowed for cervical ripening than women receiving the control treatment. This higher incidence of caesarean section during cervical ripening is more than compensated for by a lower caesarean section rate during induced labour.

Of the many preparations used for vaginal administration, the modern gel preparations are more effective than the tablet forms previously used.

3.3 Endocervical prostaglandins

Evaluations of prostaglandins administered in a viscous gel in the cervical canal have been conducted mostly with 0.5 mg PGE_2, a dose that is much smaller than is used with either the oral or vaginal routes of administration. Endocervical prostaglandin administration is more likely than either placebo or no treatment to result in uterine activity, in the onset of labour, in a reduction of the need for formal induction of labour at the end of the ripening period, and in delivery during the ripening period. As with vaginal prostaglandins, more women undergo caesarean section during the ripening period, but this is again compensated for by a reduction in the rate of caesarean sections during labour. The likelihood of failed induction is reduced, as is the rate of operative delivery.

3.4 *Extra-amniotic prostaglandins*

Few placebo controlled trials have evaluated extra-amniotic administration of a prostaglandin for cervical ripening. The few data available do not permit adequate judgement about the relative merits or hazards of this specific route of administration.

3.5 *Direct comparisons between different routes*

Controlled comparisons of endocervical with vaginal PGE$_2$ gel for cervical ripening suggest that the endocervical approach results in more women going into labour or delivering during ripening (if this is considered to be a desirable outcome). No statistically significant differences are found in any other outcomes. The choice between these approaches may be best determined by clinical preference.

Labour is more likely to occur during ripening with the use of extra-amniotic rather than vaginal or endocervical prostaglandins. No differential effects on other outcome measures have been found in the few controlled trials that have been conducted, but the trials were too small to detect modest differences.

4 Other methods for cervical ripening

4.1 *Oestrogens*

The use of oestrogens for cervical ripening has been suggested on the theoretical grounds that these agents might ripen the cervix without concomitant effects on uterine contractility. Data from controlled trials with a variety of oestrogenic preparations failed to show any beneficial effects.

4.2 *Oxytocin*

Oxytocin infusions for prolonged periods of time (as is usually the case when the aim is to ripen the cervix) are unpleasant, limit the woman's mobility, and may lead to water intoxication when administered in large doses. Oxytocin administration to ripen the cervix (rather than to induce labour) serves no useful purpose and should be abandoned.

4.3 *Mechanical methods*

Mechanical devices, such as laminaria tents, catheters, and synthetic hydrophylic materials, have been shown to increase cervical ripeness scores and may increase the proportion of women going into labour. No effect has been shown on the incidence of operative delivery, puerperal fever, or low Apgar scores.

These devices are often painful to insert. Overall the data do not suggest that the insertion of mechanical devices into the cervix or the extra-amniotic space is a useful approach to ripening the cervix before induction of labour.

4.4 *Relaxin*

The use of porcine relaxin to soften the cervix and shorten labour had a brief vogue of popularity in the 1950s. Placebo controlled trials failed to show any benefit.

Recent trials with a purer preparation have also failed to demonstrate any useful effects. Relaxin should not be adopted into clinical practice unless large properly controlled trials show evidence of benefit.

4.5 *Breast stimulation*

In an attempt to explore 'natural' methods for ripening the cervix, two groups of investigators have evaluated the effects of breast stimulation. Both reported that women allocated to breast stimulation were more likely to go into labour during the intervention period than those allocated to the control group, but there was no indication as to whether this resulted in easier labour or delivery.

5 Prostaglandins versus other methods

A number of small trials have compared prostaglandins with alternative methods of cervical ripening, including oxytocin, oestrogens, and mechanical devices. None of these trials were large enough to provide useful estimates of effect, either individually or by combining their data. The general conclusion derived from these studies suggests that prostaglandins are more effective than any of the other approaches.

6 Hazards of cervical ripening

Increasing the readiness of the cervix for induction is not a trivial intervention. Depending on the method used, the risks of the ripening include a (small) danger of intrauterine infection with mechanical procedures and extra-amniotic drug administration, an increased likelihood of uterine hypertonus and fetal heart rate abnormalities, and a certain amount of discomfort and inconvenience for the mother. Some of these hazards, uterine hypertonus and fetal heart rate abnormalities in particular, are ill defined and of unclear significance, but they have sometimes prompted caesarean sections during cervical ripening.

The risks of cervical ripening are not limited to those related to intervention itself, but include those associated with induction of labour. The greatest hazard is that the ease of cervical ripening may result in unnecessary induction of labour in women for whom an artificial ending of pregnancy would not otherwise have been contemplated.

7 Conclusions

No attempts should be made to ripen the cervix unless there are valid grounds for ending pregnancy artificially. None of the successful methods of cervical ripening act exclusively on the cervix, and all of them tend to increase myometrial contractility.

When it is necessary to induce labour in the presence of an 'unripe' cervix, the method used should not only increase cervical readiness, but must also increase the likelihood of spontaneous vaginal delivery of a healthy baby within a reasonable period of time and involve minimal inconvenience or discomfort for the mother.

Of the various interventions used, only prostaglandins have so far approached this goal. Use of prostaglandins decreases the likelihood of 'failed induction', decreases the incidence of prolonged labour, and increases the chances of a spontaneous vaginal delivery. There are still insufficient data to allow any confident conclusions about the effects on the baby.

Oral administration of prostaglandins has no value for ripening the cervix. Extra-amniotic administration, if used at all, should be reserved for women with very 'unripe' cervices, and for those who are unlikely to respond adequately to vaginal or endocervical administration.

Vaginal and endocervical administration of prostaglandin PGE_2 gel are the current methods of choice. The two forms of administration seem equally effective. At present there is no firm basis for considering one of the two approaches to be preferable, but the dose used must be appropriate for the route chosen.

This chapter is derived from the chapters by Iain Chalmers and by Marc J.N.C. Keirse (60) and by Marc J.N.C. Keirse and A. Carla C. Van Oppen (61) in EFFECTIVE CARE IN PREGNANCY AND CHILDBIRTH

References to primary sources and more complete data for statements made in this chapter can be found in the source chapters and/or in the following reviews from the *Cochrane pregnancy and childbirth database*:

Keirse, M.J.N.C.
— Vaginal PGE$_2$ for cervical ripening. Review no. 03774.
— Oral PGE$_2$ for cervical ripening. Review no. 03778.
— Vaginal PGE$_1$ for cervical ripening. Review no 03780.
— Endocervical PGE$_2$ for cervical ripening. Review no. 03781.
— Extra-amniotic vs endocervical PGE$_2$ for cervical ripening. Review no. 03784.
— Extra-amniotic vs vaginal PGE$_2$ for cervical ripening. Review no. 03785.
— Vaginal PGE vs PGF for cervical ripening. Review no. 03787.
— Endocervical vs vaginal PGE$_2$ for cervical ripening. Review no. 03788.
— Extra-amniotic PGE$_2$ for cervical ripening. Review no. 03792.
— Extra-amniotic/cervical vs vaginal PGE$_2$ for cervical ripening. Review no. 03793.
— PGF$_{2\alpha}$ (any route) for cervical ripening. Review no. 03870.
— Any prostaglandin/any route for cervical ripening. Review no. 04534.
— Vaginal prostaglandins for cervical ripening. Review no. 04535.
— Extra-amniotic prostaglandins for cervical ripening. Review no. 04537.
— Endocervical prostaglandins for cervical ripening. Review no. 04539.
— Endocervical PGE$_2$ in triacetin gel for cervical ripening. Review no. 05532.
— Endocervical vs vaginal PGs for cervical ripening/induction. Review no. 06196.
— Adjuvant betamimetics in PG-mediated cervical ripening. Review no. 06657.
— Vaginal PGF$_{2\alpha}$ for cervical ripening. Review no. 03864.

1 Introduction

For practical purposes, modern obstetric practice uses only three broad approaches to the induction of labour: stripping (sweeping) of the membranes, amniotomy (artificial rupture of the membranes), and oxytocic drugs (oxytocin or a prostaglandin). Other methods, although still occasionally reported, have generally been abandoned. Traditional practices such as the use of castor oil have never been formally evaluated.

2 Sweeping (stripping) the membranes

Sweeping the membranes (digital separation of the fetal membranes from the lower uterine segment) has been used for many years to induce labour. There are good theoretical reasons to suggest that it may be effective in that it stimulates intrauterine prostaglandin synthesis. Although it is now rarely considered as a formal method of induction, the procedure is still frequently carried out at term in the hope that it will circumvent the need to later induce labour with amniotomy or oxytocic drugs.

The few formal attempts to assess the effects of digital stripping of the membranes at term have been marred by their large potential for bias. They suggest that women who had the membranes stripped were more likely to go into labour within the next few days, and less likely to have post-term pregnancies, than were the women who served as controls. However, there is too little good evidence to assess the effectiveness of this technique, and the potential maternal and perinatal morbidity that it may entail.

It is remarkable that stripping the membranes has been the subject of so little controlled research. Randomized trials of adequate size could easily be mounted to assess the effects of this commonly performed and potentially painful procedure. Such trials should include an assessment of women's views.

3 Amniotomy

3.1 *Amniotomy used alone*

Amniotomy (rupturing the membranes) can induce labour, but its use implies a firm commitment to delivery; once the membranes have been ruptured, there is no turning back. The main disadvantage of amniotomy when used alone for the induction of labour is the unpredictable and occasionally long interval to the onset of uterine contractions and thus to delivery.

3.2 *Amniotomy with oxytocic drugs versus amniotomy alone*

In order to shorten the interval between amniotomy and delivery, oxytocic drugs are usually used either at the time that the membranes are ruptured or after an interval of a few hours if labour has not started. Evidence from controlled trials shows that women who receive oxytocics from the time of amniotomy are more likely to be delivered within 12 hours or within 24 hours, and less likely to be delivered by caesarean section or forceps, than those who have had amniotomy alone.

Women who receive early oxytocin use less analgesia than those receiving oxytocin later. This does not necessarily mean that early oxytocin results in a less painful labour for these women; it may simply reflect the shorter interval between amniotomy and birth. The trials for which data are available also suggest a lower incidence of postpartum haemorrhage when amniotomy is combined with early oxytocin administration.

Low Apgar scores are seen less frequently with a policy of using oxytocin from the time of amniotomy. No other differential effects on the baby have been noted in controlled trials.

3.3 Amniotomy with oxytocic drugs versus oxytocic drugs alone

When compared with a policy in which the membranes are left intact, routine amniotomy at the time of starting oxytocic drugs to induce labour is more likely to result in established labour within hours of starting the induction. The limited amount of controlled trials precludes firm conclusions on other outcome measures, such as the likelihood of birth within 24 hours, caesarean section, or perinatal morbidity and mortality.

Observational data derived from studies conducted in the 1960s suggest that about a third of women in whom induction of labour is attempted with oxytocin administration but without concurrent amniotomy will remain undelivered two to three days after the beginning of the induction attempt. Not surprisingly, in the light of these observations amniotomy has come to be used routinely at the time that oxytocin is started to induce labour. Only with the development of prostaglandin preparations has the choice of leaving the membranes intact during induction of labour become a reasonable option.

3.4 Hazards of amniotomy

A number of undesirable consequences have been attributed to artificial rupture of the membranes. These include pain and discomfort, intrauterine infection (occasionally leading to septicaemia), early decelerations in the fetal heart rate, umbilical cord prolapse, and bleeding from fetal vessels in the membranes, from the cervix, or from the placental site. Fortunately, serious complications are rare.

Any instrument (or a finger) passing up the vagina in order to rupture the amniotic sac will carry some of the vaginal bacterial flora with it. The risk of clinically significant intrauterine infection ensuing from these procedures is largely dependent on the interval between amniotomy and delivery.

The view that amniotomy predisposes to fetal heart rate decelerations is largely based on potential cord compression due to diminished amniotic fluid volume, but there is no evidence that this risk is important enough to be a main determinant in choosing a method for the induction of labour.

4 Oxytocin

4.1 Routes and methods of administration

Intravenous infusion of oxytocin is probably the most widely used method for inducing labour. No formal comparisons between the intravenous route and other routes of administration have been reported.

Intravenous oxytocin has been administered in different ways ranging from simple manually adjusted gravity-fed systems, through mechanically or electronically controlled infusion pumps, to fully automated closed-loop feedback systems in which the dose of oxytocin is regulated by the intensity of uterine contractions. Gravity-fed systems have the disadvantage that the amount of oxytocin infused may be difficult to regulate accurately and may vary with the position of the woman. A further disadvantage is that the amount of fluid administered intravenously may be large, and thus may increase the risk of water intoxication. Automatic oxytocin infusion equipment, in contrast, delivers oxytocin at a well-regulated rate in a small volume of fluid. In theory it should optimize efficacy and safety during oxytocin administration, but there is no evidence that these theoretical advantages confer any benefit in practice.

The only formal comparisons of different methods for administering oxytocin to induce labour consist of trials comparing automatic oxytocin infusion systems with 'standard regimens'. These trials have been too small to detect differences in substantive outcomes. The merits and risks of automated infusion systems and alternative dose regimens must be more thoroughly evaluated before their place, if any, in clinical practice can be determined.

4.2 *Hazards of oxytocin administration*

The possible hazards of oxytocin *per se* must be distinguished from the hazards associated with any attempt to induce labour, and those associated with any artificial stimulation of uterine contractions.

The antidiuretic effect of oxytocin can result in water retention and hyponatraemia, and may lead to coma, convulsions, and even maternal death. These risks are mainly associated with oxytocin infusions at early stages of pregnancy, when uterine sensitivity to oxytocin is far less than it is at term and when much larger doses are required to stimulate uterine contractions than are needed at term. In women with an already reduced urinary output the danger of water intoxication is an important consideration at any stage of gestation.

Any agent that causes uterine contractions, whether it be a drug, such as oxytocin or a prostaglandin, or a practice such as nipple stimulation, may also cause excessive uterine contractility. If the myometrial contraction force exceeds the levels of venous pressure, the venous circulation through the myometrium will become compromised. Blood flow from and to the placenta will be affected, which will in turn reduce fetal oxygenation. Uterine rupture is a further, though much rarer, consequence of excessive stimulation of uterine activity. The balance of evidence suggests that induction of labour with oxytocin increases the incidence of neonatal hyperbilirubinaemia.

5 Prostaglandins

5.1 *Comparisons with placebo*

Prostaglandins are the only agents which have been evaluated against placebo for the induction of labour. Not surprisingly, the rates of 'failed induction' and the proportions of women needing a second induction attempt are lower with prostaglandin administration (in various doses, formulations, and routes) than with placebo treatments. There were fewer caesarean sections in the prostaglandin than in the placebo groups in the reported trials, but the rates of instrumental vaginal delivery rates were similar. Most of the trials mentioned specifically that 'uterine hypertonus' and/or 'uterine hyperstimulation' were not observed, and several commented on the low incidence of gastrointestinal side-effects encountered.

Very few infant outcomes were reported in any of these trials. Among those in which they were reported, none showed any differences between the prostaglandin and placebo groups.

5.2 *Prostaglandin E versus prostaglandin F*

Both E and F prostaglandins stimulate uterine contractility. Among these two series of prostaglandins, only PGE_2 and $PGF_{2\alpha}$ (which are also naturally formed during spontaneous labour) have been used to any extent for the induction of labour.

In order to achieve a similar effect on uterine contractility, $PGF_{2\alpha}$ must be administered in a dose eight to ten times as large as that needed when PGE_2 is used. This difference in potency applies to the stimulating properties of these compounds on the myometrium. It does not apply to the same extent to their effects on other organ systems, such as the gastrointestinal tract. Consequently, for a comparable uterotonic effect the incidence of side-effects tends to be larger with $PGF_{2\alpha}$ than with PGE_2.

There are two exceptions to the general rule that PGE_2 is less likely to cause side-effects than $PGF_{2\alpha}$: hyperthermia and venous erythema are more often observed with PGE_2 than with $PGF_{2\alpha}$ administration. Venous erythema, which relates to dilatation of the vasa vasorum of the venous system, is seen virtually only with intravenous PGE_2 infusions. Although on occasions alarming from a visual point of view, venous erythema is painless and disappears after the end of the infusion period. Because of the difference in potency and side-effects, PGE_2 has become the only natural prostaglandin used for induction of labour.

5.3 *Routes and methods of administration*

Early studies of prostaglandins for the induction of labour used the intravenous route of administration. These studies showed few, if any, advantages of prostaglandins over other methods. Compared with oxytocin they offered no real benefit and were considerably more expensive.

Oral administration of PGE_2 (in repeat doses increasing from 0.5 to 2 mg) became widely used as an alternative to intravenous infusions of prostaglandins for inducing labour, particularly when combined with amniotomy and in women with a favourable cervix. Gastrointestinal side effects were common, and oral administration has been almost entirely replaced by vaginal administration.

The most widely adopted mode of administration of PGE_2 has become the vaginal route, particulary since the newer formulations using viscous gel became available.

Because intravenous, oral, and, to some extent, vaginal administration of prostaglandins leads to high levels of these drugs in the blood, the gastrointestinal tract, or both, intrauterine (extra-amniotic) routes of administration have been used in attempts to reduce the side-effects associated with the other routes. Continuous or intermittent extra-amniotic infusion of a PGE_2 solution and extra-amniotic injection of a PGE_2 gel suspension have been used for this purpose.

There is a limited amount of controlled data comparing the extra-amniotic route with other routes of prostaglandin administration. Although the data are too limited for a precise estimate, they show no advantage for the more invasive extra-amniotic route which is both cumbersome and inconvenient for the mother.

Another route of local administration, endocervical injection of PGE_2 in a viscous gel, has been used mainly for ripening the cervix rather than for induction. The relative merits and hazards of endocervical versus vaginal administration have been assessed in a few trials. These have not indicated that either of these approaches is clearly superior to the other in terms of substantive outcome measures.

5.4 *Hazards of prostaglandin administration*

The specific hazards attributable to prostaglandins *per se* relate mainly to their effects on the gastrointestinal tract (nausea, vomiting, and diarrhoea). These effects are minimal when the drugs are administered endocervically or extra-amniotically, and maximal when routes of administration (intravenous, oral) are used that lead to high levels of the drugs in either the blood or the gastrointestinal tract.

Fever may result from a direct effect of prostaglandins on thermoregulating centres in the brain. This is particularly a problem with systemic prostaglandin E_2 administration, and may give rise to concern that intra-uterine infection has supervened. This concern may be further fuelled by a rise in the leucocyte count, which can also be stimulated by prostaglandin administration. Fever is rarely observed with the newer vaginal and endocervical preparations.

More worrying than the specific hazards associated with prostaglandins are concerns that the simplicity of their administration may encourage their use for trivial indications or without adequate surveillance of mother and fetus.

6 Prostaglandins versus oxytocin for inducing labour

Oxytocin administration, combined with amniotomy, remains the most widely used approach to the induction of labour. In recent years prostaglandins have become more popular, particularly when the cervix is 'unripe'. The important question is whether prostaglandins are, on balance, superior to oxytocin for the induction of labour.

6.1 *Effects on time and mode of delivery*

The total amount of uterine work required to achieve delivery is lower with prostaglandins than with oxytocin, presumably because the former also influence connective tissue compliance (cervical ripening) whereas the latter does not. The proportions of women who deliver within 12 hours of the start of induction are similar for women induced with prostaglandins and for those induced with oxytocin. By 24 hours, however, more women have given birth after induction with prostaglandins, and after 48 hours the proportion of women who have given birth shows an even larger difference in favour of prostaglandins. When only women who give birth vaginally are considered, this advantage of prostaglandins becomes even more pronounced.

There is no clear evidence of a differential effect of prostaglandins and oxytocin on the caesarean section rate. The rate of instrumental vaginal delivery is lower in the women induced with prostaglandins, as is the incidence of operative delivery overall. This may be due partly to the influence of prostaglandins on connective tissue and partly to the greater freedom of movement allowed because it is not administered intravenously.

6.2 *Effects on the mother*

There are some major differences between the effects of the oxytocin and prostaglandin classes of drugs on organ systems other than the uterus. More women experience gastrointestinal side-effects, such as nausea, vomiting, and diarrhoea, when prostaglandins rather than oxytocin are used for the induction of labour. Fever during labour is more likely to occur with prostaglandins than with oxytocin, although the differential effect is limited to the earlier studies of intravenous PGE_2.

Uterine hyperstimulation occurs more frequently with prostaglandin than with oxytocin administration. This complication is seen mainly in institutions with little experience in the use of prostaglandin and was not observed in many trials. A diagnosis of hyperstimulation may lead to a variety of interventions ranging from changes in position, through fetal scalp blood sampling and administration of betamimetic agents, to caesarean section. Thus hyperstimulation is important to the mother, irrespective of whether or not it directly jeopardizes her or the fetus.

Data on the incidence of retained placenta, postpartum haemorrhage, and fever during the puerperium show no difference in the effects of prostaglandins and oxytocin.

Few data on mothers' views of induction have been reported, but they are consistently in favour of prostaglandin administration which is considered to be more agreeable, more natural, and less invasive than intravenous administration of oxytocin.

6.3 *Effects on the infant*

In view of the increased incidence of uterine hyperstimulation associated with induction using prostaglandins, it is reassuring to note that the incidence of fetal heart rate abnormalities is similar in labour induced with prostaglandins and among fetuses of women receiving oxytocin.

Unfortunately, few trials provide data on substantive infant outcomes, such as resuscitation of the newborn, admission to a special care nursery, or early neonatal convulsions. Even data on perinatal death are only available from half of the trials. No differential effects of prostaglandins and oxytocin emerge from those trials that provide data, but the precision of these estimates is extremely low.

Rather more data are available on the incidence of low 1 minute and 5 minute Apgar scores, but these show no statistically significant differences between prostaglandin and oxytocin inductions.

The incidence of neonatal hyperbilirubinaemia (jaundice) is lower among infants born after induction of labour with prostaglandins than among those born after induction with oxytocin, but the difference is compatible with chance.

7 Conclusions

The most important decision to be made when considering the induction of labour is whether or not the induction is justified, rather than how it is to be achieved. Whatever method is chosen to implement a decision to induce labour, uterine contractility and maternal and fetal well-being must be monitored carefully.

Amniotomy alone is often inadequate to induce labour. When amniotomy is used to induce labour and fails to result promptly in adequate uterine contractility, oxytocic drugs should be administered. The administration of oxytocin without amniotomy is also associated with an unacceptable failure rate.

Prostaglandins are more likely than oxytocin to result in vaginal birth within a reasonable length of time after the start of induction, and to lower the rate of operative delivery associated with induction of labour. The extent to which this may reflect the greater mobility possible with some forms of prostaglandin administration than with intravenously administered oxytocin is unknown. These positive effects of prostaglandins must be balanced against their negative effects, troublesome gastrointestinal symptoms or fever, although these are rarely seen with the newer formulations of prostaglandin E_2 that are now available.

If the cervix is ripe and a decision has been made to use prostaglandins to induce labour, the best option appears to be vaginal administration of prostaglandin E_2 in a viscous gel. $PGF_{2\alpha}$ should no longer be used.

There is too little evidence to allow any judgement about whether prostaglandins are more or less safe for the baby than oxytocin.

This chapter is derived from the chapters by Michel Thiery, Cornelia J. Baines, and Marc J.N.C. Keirse (59), Marc J.N.C. Keirse and Iain Chalmers (62), and Marc J.N.C. Keirse and A. Carla C. van Oppen (63) in EFFECTIVE CARE IN PREGNANCY AND CHILDBIRTH.

References to primary sources and more complete data for statements made in this chapter can be found in the source chapters and/or in the following reviews from the *Cochrane Collaboration pregnancy and childbirth database*:

Keirse, M.J.N.C.
— Any PG vs placebo for induction of labour. Review no. 03319.
— Vaginal PGE_2 vs placebo for induction of labour. Review no. 03323.
— Oral PGE_2 vs intravenous oxytocin for induction of labour. Review no. 03327.
— Oral PGE_2 vs buccal oxytocin for induction of labour. Review no. 03331.
— Oxytocin with amniotomy vs oxytocin without amniotomy for induction of labour. Review no. 03855.
— Oral PGE_2 with or without amniotomy for induction of labour. Review no. 03856.
— Extra-amniotic PGE_2 vs other route for induction of labour. Review no. 03866.

— Oxytocics with amniotomy vs oxytocics alone for induction of labour. Review no. 03857.
— Amniotomy plus early vs late oxytocin infusion for induction of labour. Review no. 03858.
— Automated vs standard oxytocin infusion for induction of labour. Review no. 03860.
— Intravenous PGE_2 vs oxytocin for induction of labour. Review no. 04529.
— Intravenous $PGF_{2\alpha}$ vs oxytocin for induction of labour. Review no. 04530.
— Intravenous PGs vs oxytocin for induction of labour. Review no. 04533.
— Any prostaglandin (by any route) vs oxytocin (any route) for induction of labour. Review no. 04536.
— Vaginal prostaglandins vs oxytocin for induction of labour. Review no. 04538.
— Oral PGE_2 vs oral $PGF_{2\alpha}$ for induction of labour. Review no. 04540.
— Stripping/sweeping membranes at term for induction of labour. Review no. 05090.
— Corticosteroids for induction of labour. Review no. 05530.
— Lamicel as an adjunct to oxytocics for inducing labour. Review no. 05557.
— Pulsatile oxytocin for induction of labour. Review no. 06193.
— Breast pump vs oxytocin for induction of labour. Review no. 06197.
— Timing of amniotomy in relation to oxytocics for labour induction. Review no. 06438.
— Mifepristone for inducing labour at term. Review no. 06813.

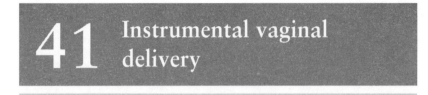

1 Introduction

When there is a valid indication for expediting the birth of the baby, instrumental vaginal delivery rather than caesarean section may be selected on the basis of a number of factors. These include the condition of the fetus and mother, progress in labour, dilatation of the cervix, the station of the presenting part, position and moulding of the fetal head, comfort, morale, and cooperation of the mother, experience and attitudes of the operator, and the availability of the necessary equipment.

Few indications for instrumental delivery are absolute, and there are considerable regional and international differences in the rate of instrumental deliveries. In the English-speaking world, in general, forceps are the preferred instruments and adequate familiarity with the vacuum extractor is rare. The situation is the reverse in many European countries where forceps are less frequently used.

2 Conditions for instrumental delivery

The operator is a major determinant of the success or failure of instrumental delivery. Unfavourable results are almost always caused by the user's unfamiliarity with either the instrument or the basic rules governing its use.

A fully dilated cervix is a prerequisite for instrumental vaginal delivery. Moreover, use of oxytocin may be better than too early instrumental delivery for dealing with a delay in second-stage labour before the baby's head reaches the pelvic floor. The station and degree of moulding of the head must be carefully assessed and its position accurately known.

The common indications for instrumental delivery, such as fetal distress or delay in the second stage of labour, are likely to create anxiety in the mother and her partner. Some of this anxiety can be relieved by keeping them fully informed of the reasons for and the nature of the procedures that are undertaken.

Proper and effective analgesia should be provided before instrumental delivery is commenced. Less pain relief is needed, as a rule, for vacuum extraction than for forceps delivery. Vacuum extraction or outlet forceps delivery can usually be accomplished comfortably with local infiltration of the perineum or pudendal nerve block. Rotational forceps deliveries will often require a more profound form of anaesthesia, such as epidural or spinal block.

The total force exerted on the fetal head during instrumental delivery will depend on the type of instrument, the duration of the procedure, and the number and strength of pulls. Some descent of the head should occur with each pull. Absence of descent with traction on a correctly positioned instrument should be regarded as a reason to abandon the procedure in favour of caesarean section.

Elective instrumental vaginal delivery has been compared with spontaneous vaginal delivery in only a few small trials. Instrumental delivery results in significantly more perineal trauma (both episiotomy and laceration) than spontaneous birth. The alleged benefits of a 'lift out' forceps in terms of slightly fewer babies with low cord pH values have to be weighed against the more frequent problem of maternal vaginal and perineal trauma.

3 Equipment and techniques

3.1 Forceps

Since the introduction of forceps, numerous modifications have been made in attempts to improve their efficiency and safety. Forceps can be grouped on a functional basis into those whose primary function is to exert traction and those whose primary function is to correct malposition. No controlled trials of the use of different types of forceps have been reported. One trial evaluated the use of a forceps pad designed to reduce infant trauma; the use of the pad resulted in fewer babies having craniofacial markings.

3.2 Vacuum extraction

Various modifications of the vacuum extractor have been made in attempts to increase manoeuvrability of the cup and to achieve a more correct application on the fetal head.

All the vacuum cups are satisfactory for outlet and non-rotational mid-pelvic operations. Operators would be well advised first to develop confidence in outlet and non-rotational mid-pelvic procedures. The basic technique is similar in all positions of the occiput, and experience gained with non-rotational procedures will prove invaluable when the more difficult rotational operations are attempted. The few trials that have been carried out comparing the various rigid cup designs with one another have not demonstrated any differences in outcome. Direct comparisons of soft with hard vacuum extraction cups suggest that soft cups are less likely to achieve vaginal delivery, but result in significantly fewer fetal scalp injuries than metal cups.

The success rate with a metal cup is better than with a soft cup, particularly for delivering a baby in an occiput posterior position, where the 'OP' cup is very useful. Because soft cups are also more likely to fail with a large baby, a high head, or a large amount of caput, it is reasonable to limit their use to more straightforward deliveries. Despite these disadvantages, it is worth continuing to use soft cups when feasible because they are associated with less neonatal trauma.

The risk of injury to the infant is directly related to the number of pulls with the vacuum extractor. Sudden cup detachments may cause injury to the scalp of the infant.

4 Comparison of vacuum extraction and forceps

4.1 *Efficiency*

With adequate experience and proper placement of the vacuum cup, most deliveries that require instrumental rotation of the head can be accomplished by vacuum extraction, thus obviating the need for painful and potentially traumatic forceps rotations. This experience should not be difficult to obtain and should form part of all specialty training programmes in obstetrics. Although the vacuum extractor is less likely than forceps to achieve a vaginal delivery with the chosen instrument, vacuum extraction with forceps back-up when required, is associated with a lower overall caesarean section rate.

The mean time between the decision to deliver and delivery itself is similar for forceps and vacuum extraction, although the range of the decision-to-delivery interval is greater for forceps. This is at least in part due to the time required to institute the more complex forms of analgesia used for forceps delivery. The widely held belief that vacuum extraction is too slow to be useful when rapid delivery is required for fetal distress can be firmly laid to rest.

On balance, for most instrumental vaginal deliveries vacuum extraction is to be preferred over forceps. Reserving one instrument for routine applications and the other for particularly difficult situations would be ill advised. Difficult extractions, whether by forceps or by vacuum extraction, should not be undertaken unless the operator has considerable expertise with the instrument chosen. In the absence of such expertise, delivery by caesarean section should be considered as the alternative.

4.2 *Effects on the mother*

The vacuum extractor is significantly less likely to cause serious maternal injury than the forceps. Its application is associated with a lower usage of regional and general anaesthesia, and with significantly less pain to the mother, both at delivery and in the puerperium.

4.3 *Effects on the infant*

Vacuum extraction is more likely to cause cephalhaematoma than forceps, but forceps are more likely to cause other kinds of scalp and facial injuries. No significant differences between the instruments have been found in the number of babies requiring phototherapy. Perhaps because of the chignon caused by the vacuum cup, mothers tend to be more worried about the baby delivered by vacuum extraction than with forceps.

The vacuum extractor appears to be associated with an increased incidence of retinal haemorrhages (although this latter result is largely influenced by a single study which was methodologically the least sound of all the trials reviewed).

There is not enough information available to judge the relative effects of the two instruments on the risk of perinatal death or the long term condition of the infant. In the only follow-up study of cohorts randomized to the two instruments, the incidence of problems was similar in the vacuum and forceps groups but the numbers of infants studied was too small to exclude anything other than very dramatic differences in outcome.

Follow-up studies showed no significant differences in mothers' attitudes to the instruments or in infant readmissions to hospital.

5 Conclusions

Both a valid indication and the necessary conditions must be met before instrumental delivery is undertaken. The cervix must be fully dilated, effective analgesia must be in place, and the operator must be familiar with the chosen instrument. There is no justification for a 'difficult instrumental delivery'. Caesarean section would be preferable.

Shortening of the second stage with elective instrumental delivery can result in a clinically unimportant gain in umbilical cord pH, but, at least with forceps, it leads to a considerable increase in maternal vaginal and perineal trauma.

Forceps delivery and vacuum extraction are to a large extent interchangeable procedures. The available evidence indicates that the use of forceps is more likely to result in maternal injury and is more dependent on extensive analgesia or anaesthesia than is vacuum extraction. There is no evidence of any compensating benefits that could support the choice of forceps over vacuum extraction.

This chapter is derived from the chapter by Aldo Vacca and Marc J.N.C. Keirse (71) in EFFECTIVE CARE IN PREGNANCY AND CHILDBIRTH.

References to primary sources and more complete data for statements made in this chapter can be found in the source chapter and/or in the following reviews from the *Cochrane pregnancy and childbirth database*:

Johanson, R.
— Forceps vs spontaneous vaginal delivery. Review no. 07087.
— Soft vs hard vacuum extractor cups. Review no. 05451.
— Obstetric forceps pad designed to reduce trauma. Review no. 07086.
— O'Neil vs Malmstrom vacuum extraction. Review no. 03795.
— New Generation vs original Bird vacuum extraction. Review no. 03259.
— Silastic vs Mityvac vacuum extraction. Review no. 03258.
— Vacuum extraction vs forceps delivery. Review no. 03256.

Renfrew, M.J.
— Vacuum extraction compared to normal delivery. Review no. 06517.

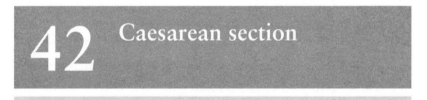

42 Caesarean section

1 Introduction   2 Anaesthesia for caesarean section
3 Surgical technique   4 Conclusions

1 Introduction

The term 'caesarean section' refers to the operation of delivering the baby through incisions made in the abdominal wall and uterus. It has an enormous potential for the preservation of life and health, probably greater than that of any other major surgical operation.

The rate of caesarean section varies considerably among countries, from about 5 per cent to over 25 per cent of all deliveries. The optimal rate is not known, but from national data available, little improvement in outcome appears to occur when rates rise above about 7 per cent. Despite this, rates of caesarean section well above this level exist in many parts of the world.

Many caesarean sections are carried out for unequivocal indications, such as placenta praevia or transverse lie. However, the majority of the operations are carried out for rather ambiguous indications. Criteria for the diagnoses of dystocia (prolonged labour) and fetal distress, two of the most commonly given reasons for performing a caesarean section, are by no means clear. No data are available to suggest what proportion of babies presenting as a breech would benefit from delivery by caesarean section. Previous caesarean section is rarely an adequate indication by itself.

The extent to which obstetricians differ in the use of this major operation to deliver babies suggests that the obstetrical community is uncertain as to when caesarean section is indicated. It also suggests that other factors, such as the socio-economic status of the woman, the influence of malpractice litigation, women's expectations, financial considerations, and convenience, may sometimes be more important than obstetrical factors in determining the decision to operate.

2 Anaesthesia for caesarean section

When caesarean section is required, its safety depends on the care with which the anaesthetic is administered and the operation performed.

Caesarean section can be carried out under either regional (epidural or spinal block) or general anaesthesia. Regional anaesthesia has many advantages. It largely avoids the risk of regurgitation and aspiration of stomach contents associated with general anaesthesia, allows the mother to remain awake, and permits early contact between mother and baby at birth. The main disadvantages of using regional anaesthesia for caesarean section relate to the extensive block required for the operation. This may result in a drop in blood pressure which may require treatment with intravenous fluids and vasopressors. The limited data available from controlled trials show only minor differences in the effects of epidural or spinal anaesthesia for caesarean section, with spinal resulting in a quicker onset of anaesthesia, less shivering with the onset of block, more frequent hypotensive episodes, and no differences in other substantive outcomes.

Despite the increasing popularity of regional anaesthesia, general anaesthesia is sometimes required. Regional anaesthesia is contraindicated if the mother has a coagulation disorder. If the reason for the caesarean section relates to a bleeding complication in the mother, the drop in blood pressure with regional anaesthesia can be particularly dangerous. General anaesthesia can be more rapidly administered, and is of value when speed is important such as when the fetus is severely distressed. Some women prefer to be asleep for the operation.

The disadvantages of general anaesthesia relate to the serious problems that are sometimes associated with it, such as pulmonary aspiration, inadequate airway control, and neonatal depression. If improperly managed, these can lead to significant maternal and fetal morbidity, and sometimes death. Aspiration of acidic stomach contents can cause acid pneumonitis (Mendelson's syndrome), the severity of which depends on the acidity of the aspirate (see Chapter 29).

The most important measure in preventing pulmonary aspiration is occlusion of the oesophagus by cricoid pressure. The manoeuvre requires an assistant who is knowledgeable and capable. It loses its effectiveness unless skilled help is available. Deaths from Mendelson's syndrome usually result from not applying cricoid pressure, relaxing the pressure before intubation, or applying pressure inefficiently. An equally important problem with general anaesthesia arises when the anaesthetist is unable to intubate the trachea. Deaths due to aspiration or failed intubation are largely preventable, and can be almost eradicated by improvements in clinical care.

Whatever form of anaesthesia is chosen, the woman should be in a 15°–20° lateral tilt position rather than lying flat on her back. This results in improved umbilical artery pH and neonatal Apgar scores, probably as a result of relieving the pressure of the pregnant uterus on the vena cava.

3 Surgical technique

Details of operative technique vary from surgeon to surgeon, and few of these techniques have been evaluated in controlled trials.

Pulmonary embolism is an important cause of maternal morbidity and mortality with caesarean section. Evidence from other surgical fields shows that this risk can be reduced by the preoperative use of low dose heparin, and the available data show no evidence that the use of heparin increases the risk of bleeding with caesarean section.

One small trial compared the use of an antiseptic impregnated film with the usual antiseptic scrub technique for skin preparation prior to caesarean section. No differences in infection rates were demonstrated.

When a transverse rather than a vertical skin incision is used, average operating time is longer and more women require blood transfusions. However, febrile morbidity occurs somewhat less frequently with the transverse skin incision, and most women find the transverse scar more acceptable cosmetically.

The uterine incision should be made transversely in the lower uterine segment, except in extremely rare circumstances. The initial incision is made with a scalpel. Whether scissors or fingers should be used to extend the incision, and whether the suture material or technique used for closure influences the post-operative result, has not been adequately evaluated. No benefits have been demonstrated for the use of a haemostatic stapler for incising the uterus, although the number of women involved in the trials has not been large enough to rule out significant benefits or adverse effects of the technique.

Elective manual removal of the placenta during the caesarean should be avoided, particularly in rhesus-negative women or others where transplacental bleeding might increase the risk of isosensitization. The available information from controlled trials suggests that manual removal increases maternal blood loss.

The results of single-layer or two-layer closure of the uterine incision are similar, and the time saved by use of a single-layer closure may be worthwhile. A policy of repairing the uterine incision after bringing the uterus out through the abdominal wound results in somewhat lower blood loss than one of repairing it within the pelvis, although it may increase the risk of infection. It would seem sensible to exteriorize the uterus for repair if there is difficulty with intra-abdominal exposure.

The few small trials that have been carried out on other minutiae of operative technique do not provide sufficient data to guide clinical policy.

4 Conclusions

Caesarean section is a major operation with great potential benefit, but also with substantial risks for both mother and baby. The hazards can be kept to a minimum first by avoiding unnecessary use of the operation, and second by meticulous attention to proper anaesthetic and surgical techniques.

This chapter was derived from the chapters by Jonathan Lomas and Murray Enkin (69), and by Jim Pearson and Gareth Rees (72) in EFFECTIVE CARE IN PREGNANCY AND CHILDBIRTH.

References to primary sources and more complete data for statements made in this chapter can be found in the source chapters and/or in the following reviews from the *Cochrane pregnancy and childbirth database:*

Enkin, M.W.
— Lateral tilt during caesarean section. Review no. 05527.
— Adhesive film as skin preparation for caesarean section. Review no. 05539.
— Haemostatic stapler for incising uterus at LSCS. Review no. 05543.
— Manual removal of placenta at caesarean section. Review no. 05528.

- Single-layer closure of uterine incision at caesarean section. Review no. 05542.
- Uterine exteriorisation vs intraperitoneal repair at caesarean section. Review no. 03882.
- Low-dose heparin before caesarean section. Review no. 05540.
- Hydroxyethyl starch vs low-dose heparin with caesarean section. Review no. 06658.
- Leg compression during spinal anaesthesia for caesarean section. Review no. 05529.
- Non-closure of peritoneum at caesarean section. Review no. 06655.
- Closed suction wound drainage at caesarean section. Review no. 05541.
- Wound edge infiltration with local anaesthetic at caesarean section. Review no. 06659.
- Bladder drainage after caesarean section under epidural. Review no. 05537.

Howell, C.J.
- Spinal vs epidural block for caesarean section. Review no. 07057.

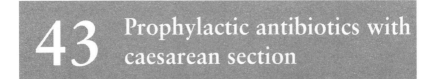

43 Prophylactic antibiotics with caesarean section

1 Introduction   2 Effects on infection and febrile morbidity
3 Choice of antibiotic preparation   4 Timing, dose, and
frequency of administration   5 Route of administration
6 Potential adverse consequences of antibiotic prophylaxis
7 Conclusions

1 Introduction

Maternal morbidity after caesarean section has not been studied as systematically as the maternal mortality associated with the operation, but the problem is undoubtedly substantial. Febrile morbidity, caused by postoperative infection or other factors, appears to follow caesarean section in at least one in five women. Serious infections, such as pelvic abscess, septic shock, and septic pelvic vein thrombophlebitis, are not rare.

Labour and ruptured membranes are the most important factors associated with an increased risk of infection, with the risk rising with increased duration of each. Obesity appears to be a risk factor of particular importance for wound infection. At one time, extraperitoneal caesarean section was proposed to reduce infectious morbidity in women at high risk of infection, but this approach is now of historical interest only.

The potential for prophylactic antibiotics to reduce maternal morbidity after caesarean section has now been investigated systematically. The benefits have been unequivocally demonstrated. Although the extent to which toxic or allergic effects of antibiotics may cause maternal morbidity is not well established, the information that is available provides clear guidelines for practice.

2 Effects on infection and febrile morbidity

Antibiotic prophylaxis markedly reduces the risk of serious postoperative infection, such as pelvic abscess, septic shock, and septic pelvic vein thrombophlebitis. A protective effect of the same order of magnitude is seen for endometritis. The degree of reduction in the risk of wound infection is somewhat less, but is still substantial. The evidence for these benefits is overwhelming.

Prophylactic antibiotics reduce the relative risk of both endometritis and wound infection to a similar extent for women having planned (elective) caesarean sections as for those having emergency procedures. The absolute numbers of serious infections avoided by prophylactic administration are greater with emergency caesarean sections because the rates of infection are higher. Postoperative febrile morbidity has fewer sequelae than the more serious infections, but is important because of its higher incidence (which makes it possible to distinguish differential effects of different forms of antibiotic prophylaxis).

3 Choice of antibiotic preparation

The risk of postoperative febrile morbidity is reduced to a comparable extent by broad spectrum penicillins such as ampicillin, cephalosporins, and metronidazole. The evidence from direct comparisons between broad spectrum penicillins and cephalosporins suggests that they have similar effects on the risk of postoperative febrile morbidity. There is no convincing evidence that antibiotics with a broader spectrum of activity, such as second- and third-generation cephalosporins, are more efficacious than a first-generation cephalosporin, although there is a trend towards better outcome with second- or third-generation cephalosporins.

Trials comparing a combination of ampicillin and aminoglycosides with ampicillin alone suggest that the combination might be a more effective prophylactic. However, the evidence is not sufficiently strong to warrant the combined regimen, which is more complicated, more costly, and carries a risk of ototoxicity and nephrotoxicity from the aminoglycosides.

4 Timing, dose, and frequency of administration

The use of three to five doses, rather than a single dose, of antibiotics for prophylaxis of infection with caesarean section does not appear to confer any additional benefits, although there is some evidence that considerably longer courses (three to five days) offer additional protection over one day courses. Whether or not this improved outcome is sufficient to justify the additional costs and risks remains to be determined.

5 Route of administration

Intra-operative irrigation with antibiotics has been shown to be more effective than irrigation with placebo in reducing the risk of post-operative febrile morbidity, and indirect comparisons suggest that antibiotic irrigation may be as effective (and possibly more effective) than systemic administration. Evidence from direct comparisons of the two routes of administration provides no strong evidence that one route of administration is superior to the other.

6 Potential adverse consequences of antibiotic prophylaxis

Only a minority of the reports of controlled trials included information about adverse effects of the prophylactic agents used, and even in these the reference was usually rather casual. Thus it is not surprising that the reported incidence of adverse reactions was very low (1 per cent or less). This is well below the rate of adverse reactions that one would expect of antibiotics, particularly broad-spectrum antibiotics given intravenously.

Drug effects on the infant (which might include protective as well as unwanted effects) have not been studied systematically by the majority of investigators. Preventing exposure of the baby to antibiotics by starting them after the umbilical cord has been clamped, as was done in most of the trials reported, would seem to be a sensible precaution even if there is some slight, as yet undetected, loss of prophylactic efficacy.

Antibiotics received by the mother can also reach the baby through breast milk. The drug levels involved seem likely to be very low, particularly if the course of prophylactic antibiotics has been relatively short.

An important argument of those who have objected to routine antibiotic prophylaxis has been their concern about the effects of this practice on the bacterial flora, namely replacement of non-pathogenic with pathogenic bacteria, and a rise in resistance of bacteria in the women and in the hospital environment generally. At least some antibiotics appear to cause these changes with relatively few doses. There is some suggestion that certain prophylactic regimens, for example trimethoprim and sulphamethoxazole, may be less disruptive of flora, yet remain effective.

Adverse ecological effects on bacterial flora are difficult to quantify and predict, but are potentially of greater concern than adverse drug reactions in individual mothers and babies. If prophylactic antibiotics are used routinely, genital tract cultures and drug-sensitivity studies should be performed in all women who become infected despite receiving prophylaxis, and the hospital bacteriology laboratory should conduct a periodic review of the susceptibility patterns of commonly isolated organisms to detect gradual changes in antibiotic resistance.

7 Conclusions

The first step towards reducing the infectious morbidity that is so common after caesarean section is to minimize the number of unnecessary operations. The second step requires attention to the many factors that reduce the risk of infection when the operation is justified, such as minimizing the length of hospital admission before surgery, delaying shaving of the operation site until immediately before the operation, sterilizing swabs, instruments, and the gloves worn by the operating team, cleaning the skin of the woman and the air of the operating theatre, and paying attention to good surgical technique.

Antibiotic prophylaxis can reduce the risk of serious infections. If the level of post-caesarean infectious morbidity is very low without a policy of antibiotic prophylaxis, the ratio of benefits to cost in absolute terms might argue against instituting such a policy. Such circumstances are rare, and the evidence justifies far wider adoption of antibiotic prophylaxis than currently exists. Although the incidence of adverse drug effects among women receiving prophylactic antibiotics has probably been underestimated, it is inconceivable that it could outweigh the reduction in serious maternal morbidity that can be achieved by a policy of antibiotic prophylaxis. Adverse drug effects in the baby can be lessened by beginning prophylaxis after the umbilical cord has been divided.

The risk of adverse ecological effects is likely to be reduced if the total load of antibiotics is reduced. The disadvantages of longer courses of antibiotics in terms of an increase in the total antibiotic load and in the number of women experiencing side-effects, and the additional financial cost, may outweigh the advantages of greater prophylactic efficacy compared with shorter or single dose regimens.

With regard to choice of antibiotic, the broad-spectrum penicillins are as effective as the cephalosporins. No strong case can be made for using a second- or third-generation cephalosporin or for adding aminoglycosides to broad-spectrum penicillins.

Withholding prophylactic antibiotics from women having caesarean section will increase the chances that they will experience serious morbidity. Further trials which include no-treatment controls would be unethical.

This chapter is derived from the chapter by Murray Enkin, Eleanor Enkin, Iain Chalmers, and Elina Hemminki (73) in EFFECTIVE CARE IN PREGNANCY AND CHILDBIRTH.

References to primary sources and more complete data for statements made in this chapter can be found in the source chapter and/or in the following reviews from the *Cochrane pregnancy and childbirth database*:

Smaill F.
— Prophylactic antibiotics in caesarean section (all trials). Review no. 03690.
— Prophylactic antibiotics for elective caesarean section. Review no. 03775.
— Cephalosporins vs placebo for caesarean section. Review no. 03240.
— Broad spectrum penicillin vs placebo for caesarean section. Review no. 03241.
— Broad spectrum penicillin + aminoglycoside vs placebo for caesarean section. Review no. 03243.
— Metronidazole vs placebo for caesarean section. Review no. 03242.
— Aminoglycoside + broad spectrum penicillin vs broad spectrum penicillin alone for caesarean section. Review no. 03248.
— Broad spectrum penicillin vs cephalosporins for caesarean section. Review no. 03247.
— Penicillin/aminoglycoside vs cephalosporin for caesarean section. Review no. 06654.
— Extended spectrum vs 1st generation cephalosporins with caesarean section. Review no. 06653.
— Trimethoprim sulfa vs placebo for caesarean section. Review no. 03244.
— One vs 3–5 doses of antibiotics for caesarean section. Review no. 03249.

— 1-day vs 3–5 day courses of antibiotics for caesarean section. Review no. 03250.
— Antibiotic irrigation vs placebo at caesarean section. Review no. 03245.
— Extraperitoneal caesarean section vs prophylactic antibiotics. Review no. 05538.
— Antibiotic peritoneal irrigation vs systemic antibiotics for caesarean section. Review no. 03246.

CARE AFTER CHILDBIRTH

44 Immediate care of the newborn infant

1 Introduction

Unless specific problems need urgent attention, babies should be given to their mothers as soon after birth as the mother is ready. The vast majority of newborn babies require little more than a clear airway and adequate warmth to support the first few minutes of adaptation to extrauterine life. The success of human evolution and the exceptionally high survival rate of human infants attest to this. Despite this truth, there are striking variations in the patterns of care that newborn babies receive in the immediate postdelivery period. These variations reflect both the inherent resilience and adaptability of the newborn infant and the lack of consensus among caregivers as to what constitutes appropriate care.

2 Immediate care of the normal newborn infant

2.1 *Welcoming the newborn infant*

In his book *Birth without violence,* Frederick Leboyer described a number of measures designed to minimize 'the shock of the newborn's first separation experiences': the use of a dark delivery room, delayed clamping of the umbilical cord, gentle massage, and a warm bath for the infant. Controlled trials of these specific measures have not shown any effects, either adverse or beneficial, on infant health, neurobehavioural status in the first few days of life, or subsequent development. As no adverse effects have been noted, there is no reason to refuse a mother's request for specific practices.

The fact that no long-term advantages of the specific measures advocated by Leboyer have been demonstrated does not obviate the need for treating the newborn with the regard and respect due to any human being, including gentleness and avoidance of excessive noise in the environment.

2.2 *Ensuring a clear airway*

The common practice of routine suctioning to remove secretions from the newborn infant's oral and nasal passages has not been assessed in any clinical trials, and its value is uncertain. Possible benefits of the practice include improved air exchange, reduced likelihood of aspiration of secretions, and, perhaps, reduced acquisition of any pathogens present in the amniotic fluid or birth canal. Potential hazards include cardiac arrhythmias, laryngospasm, and pulmonary artery vasospasm.

If nasal and pharyngeal suctioning is used, care should be taken to minimize pharyngeal stimulation. Suction bulbs rather than catheters should be used, because suction bulbs are less likely to induce cardiac arrhythmias.

The practice of routine suctioning of the stomach was introduced following an untested suggestion that the respiratory distress of infants of diabetic women often resulted from regurgitation and aspiration that might have been prevented by gastric suctioning. As the passage of the tube during the immediate neonatal period may produce bradycardia or laryngospasm and disruption of prefeeding behaviour, there is no justification for routine gastric suctioning in the delivery room.

2.3 *Maintaining body temperature*

The recommendation that all babies be kept warm immediately after birth is based on a large body of evidence about thermal physiology of both newborn animals and humans. Newborns maintain their body temperature in a cool environment at the metabolic cost of greatly increased energy expenditure. Even vigorous newborns exposed to cold delivery rooms may experience marked drops in body temperature and develop metabolic acidosis during the first two hours of life. The post-natal fall in infant temperature can be reduced by skin-to-skin contact between baby and mother, or by the use of an incubator or a radiant warmer. Babies should be dried with prewarmed towels, giving particular attention to drying the head. They should then be covered with a dry warm blanket and placed in the mother's arms or under a radiant warmer if the mother is unable to hold her baby.

2.4 Prophylactic administration of vitamin K to prevent haemorrhagic disease

Most textbooks recommend routine parenteral administration of vitamin K to all newborn infants, yet there is a continuing debate about whether or not this is necessary for healthy formula-fed infants. The concentration of vitamin K in cow's milk or infant formula is consider-ably greater than that in human milk, and prothrombin levels in infants given parenteral vitamin K at birth are similar to those in comparable infants fed on cow's milk for 24 hours.

The usually quoted incidence rates of haemorrhagic disease of the newborn in the absence of vitamin K administration range from 0.25 to 0.50 per cent. These rates are derived from studies performed many years ago, at a time when traumatic deliveries were more common and first feeds were often delayed. The risk of healthy formula-fed infants developing haemorrhagic disease today is likely to be considerably smaller. These considerations have led some paediatricians to abandon routine prophylaxis in normal formula-fed infants.

Breastfed infants are reported to be at greater than average risk of developing haemorrhagic disease of the newborn. Further research is needed to clarify this issue; much of the research on this topic was carried out at a time when normal breastfeeding practices included restricting breastfeeds, separating mothers and babies, and giving sup-plements. As a result, babies did not receive their full complement of colostrum which has a high vitamin K content. Until more evidence is available, breastfed infants should receive vitamin K. Failure to admin-ister vitamin K to breastfed infants may predispose them to serious bleeding, such as intracranial haemorrhage. Questions have been raised about a link between the parenteral administration of vitamin K and the development of childhood cancer but further observational studies have not confirmed this association.

2.5 Prophylactic measures to prevent eye infections

The Credé procedure of instilling silver nitrate routinely into the eyes of all newborn babies, introduced in 1881, was credited with the control of gonococcal ophthalmia of the newborn in the last century. As a result, many countries have a legal requirement that one of a list of approved chemical agents be instilled routinely into the eyes of all newborn infants, with the aim of preventing infectious conjunctivitis. No controlled trials have been carried out to ascertain whether or not this is a more effective means of preventing blindness than careful observation of the newborn, followed by adequate treatment of any conjunctivitis that should appear. In circumstances where the incidence of bacterial ophthalmia is high, routine chemical prophylaxis may be useful. In these circumstances, the next question concerns the choice of the most effective and least harmful agent. Silver nitrate results in more chemical conjunctivitis, and provides no greater protection against gonococcal ophthalmia than tetracycline or erythromycin. It is also ineffective against *Chlamydia* (which in many areas is the most common cause of neonatal ophthalmia) and should no longer be used. Both tetracycline and erythromycin provide protection against *Chlamydia*, as well as gonococcal conjunctivitis.

Topical agents applied to the eyes of newborn infants may decrease eye openness and inhibit visual responses. This may disrupt the visual interaction between mother and baby during the first hour of life. If topical agents are necessary, they should be delayed for an hour after birth. Mothers and babies should be able to enjoy the immediate closeness of the first hour or so after birth before chemical agents are applied.

3 Prophylactic measures in newborns considered to be at above average risk

3.1 *Suctioning of infants who have passed meconium before birth*

For infants who have passed meconium before birth, suctioning the nostrils, mouth, and pharynx before delivery of the chest may prevent postnatal aspiration of meconium in the pharynx. This procedure is sufficiently safe to be recommended even though its effectiveness in preventing severe meconium aspiration is unproven.

Tracheal suctioning for infants who passed meconium before birth is a more hazardous procedure which has been introduced into practice without testing in randomized trials. The apparently low rate of morbidity following intubation reported by very experienced resuscitators is not likely to be achieved by persons who only occasionally perform the procedure. The potential hazards of intubation include trauma, hypoxia, and bradycardia. The increase in blood pressure that regularly occurs during intubation may increase the risk of intracranial haemorrhage in preterm infants. Severe pulmonary artery vasospasm may occur in babies with pre-existing pulmonary artery hypertension, particularly in meconium-stained babies. Other risks include cross-contamination of bacteria or viral infection between newborns and caregivers.

Because of these risks, and because there are no demonstrated benefits, it would seem unwise to perform tracheal intubation for infants who are not depressed simply because they have been born following meconium passage *in utero*. Careful tracheal suction should be carried out only for infants who are depressed at birth (heart rate less than 80–100 beats minute at birth) and have meconium in the pharynx.

3.2 *Elective tracheal intubation for very-low-birthweight infants*

Although some neonatologists advocate immediate intubation of all very-low-birthweight infants, whether or not signs of respiratory depression or respiratory distress are present, the available evidence does not warrant such a policy. Because of the potential hazards of intubation, routine delivery room intubation of infants below 1500 grams with no signs of respiratory distress or respiratory depression is unjustified on the basis of current evidence.

3.3 *Prophylactic administration of surfactant to immature infants*

Prophylactic administration of surfactant to preterm newborns at high risk of developing respiratory distress syndrome reduces the incidence of pneumothorax, bronchopulmonary dysplasia, or death at 28 days, and reduces overall mortality in treated newborns. Surfactant products are rapidly becoming available world-wide. Either synthetic surfactants or natural surfactant extract products will probably become used routinely in the care of very preterm newborns in the next few years.

The quality of the evidence, which supports the administration of surfactant to high-risk newborns, is higher than that for any other immediate prophylactic treatment for newborn infants or indeed for any other method of delivery room care.

4 Immediate resuscitation of ill newborn infants

The availability of professionals skilled in neonatal resuscitation has increased with the growth of neonatology as a specialty. This means that the birth of a very preterm asphyxiated or otherwise high-risk neonate is now more likely to be attended by someone who is experienced in giving care to such infants. However, a proportion of ill and high-risk infants will continue to present as unpredicted emergencies, and it will often fall to a midwife, nurse, general practitioner, or trainee obstetric specialist to initiate and continue neonatal resuscitation.

Whenever possible, a person skilled in resuscitation who can devote all of his or her attention to the infant should be in attendance at high-risk deliveries. Basic resuscitation equipment (a radiant warmer, resuscitation bags and masks, endotracheal tubes, laryngoscope, stethoscope, oxygen source and tubing) should be readily available for every delivery room. Those attending births at home should ensure that there is a means of keeping the baby warm, and that they carry resuscitation bag and masks, a stethoscope, and possibly an oxygen source. Because the need for resuscitation is not recognized prior to the birth of approximately half of all infants requiring resuscitation, the presence and proper working order of this equipment should be verified before each delivery.

While anaesthesia bags are likely to be required for the optimal resuscitation of severely asphyxiated infants, their hazards if used improperly (e.g. the application of dangerously high airway pressures) make them unsuitable for routine use by inexperienced caregivers. Likewise, the hazards of umbilical artery catheters and trochars for endotracheal tubes should preclude their use in delivery rooms except by highly experienced resuscitators.

4.1 Resuscitation

Artificial ventilation should be initiated promptly for infants with a heart rate less than 100 beats/minute after birth, and oxygen should be administered to any infant with generalized cyanosis. Regardless of heart rate or colour, artificial ventilation should also be commenced for infants with inadequate chest excursion and poor breath sounds, particularly small preterm infants likely to have surfactant deficiency.

Proper ventilation of the infant is the single most important aspect of neonatal resuscitation, and the heart rate is the most useful and easily measured criterion for its success. Caregivers who do not frequently intubate newborn infants should initiate resuscitation using a face mask, and consider intubation only when the heart rate does not increase promptly with properly performed bag-and-mask ventilation.

Before intubation is performed, attention should be given to the following points: proper positioning of the head ('sniffing position'), ensuring that the upper airway is clear, using sufficient pressure to produce adequate chest excursions, and administering an adequate inspired oxygen concentration. Observing distention of the throat as the resuscitation bag is squeezed indicates that a proper head position and clear airway (allowing delivery of gas to the level of the glottis) has been established. The careful use of an anaesthesia bag may be required to deliver more pressure, or a greater oxygen concentration, than can be delivered by self-inflating bags. Applying excessive pressure to the infant's head through the face mask may cause persistent bradycardia.

4.2 Oxygen

Supplemental oxygen (100 per cent concentration of the warm and humidified gas) is recommended for artificial ventilation of neonates who have not established effective spontaneous respiration by one minute of age. No comparative studies of the benefits and risks of various oxygen concentrations for the management of such infants have been reported.

Some concern has been expressed that blowing oxygen across the face of a newborn infant might result in bradycardia, but most concern has been focused on whether the risk of severe retinopathy would be appreciably increased in immature infants who experience short periods of exposure to high blood oxygen levels. However, there is no satisfactory evidence to suggest that the risk is any greater than that associated with the relatively high blood oxygen levels that occur at birth in all babies with the onset of air breathing.

4.3 Cardiac massage

Cardiac massage through the intact chest wall of the newborn baby can be life-saving when used for infants born with an absent heart beat. The procedure is not without hazard; it may cause rib fractures and trauma to the liver or lung.

Little information of the kind needed to recommend precise indications and methods is available. Current recommendations are to perform cardiac massage either by using both thumbs with the hands encircling the chest, or by using the tips of the middle finger and either the index or ring finger of one hand positioned directly above the chest. The sternum should be depressed a half to three-quarters of an inch (1.5 cm) 120 times per minute.

4.4 Naloxone

Naloxone hydrochloride is a narcotic antagonist believed to be virtually free of side-effects. It may be administered as an adjunctive measure *after* assisted ventilation has been established if depression is thought to be the result of a narcotic drug given to the mother before birth. It is probably wise not to give naloxone to the newborn of a narcotic-dependent mother for fear of precipitating withdrawal illness in the baby.

Apart from concern about the potential importance of endogenous opioid substances in newborn infants, fewer optimal maternal ratings of infant behaviour have been reported among naloxone-treated infants than among controls. Therefore administration of naloxone should be restricted to infants who have not only been exposed to narcotics during labour, but who also require active resuscitation in the immediate neonatal period.

4.5 Sodium bicarbonate

Randomized trials have failed to detect any benefit from either rapid or slow administration of sodium bicarbonate to asphyxiated neonates. Potential hazards include a transient rise in $PaCO_2$ and fall in PaO_2, a sudden expansion of blood volume, a reduction in cerebral blood flow, and an increased incidence of intracranial haemorrhage.

In the absence of any demonstrated benefits of giving sodium bicarbonate in the immediate postnatal period, its use cannot be recommended.

4.6 *Blood volume expanders*

The only clear-cut indication for the use of blood volume expanders in the early neonatal period is the combination of unmistakable signs of shock with evidence of acute blood loss, including fetomaternal haemorrhage. In this circumstance, shock may be treated with repeated infusions of blood volume expanders (usually 5–10 ml) and the infant's response assessed after each infusion. The volume expander may be 5 per cent albumin, Ringer's lactate, saline, or heparinized placental blood.

Volume expanders have also been used in the presence of hypotension unaccompanied by other signs of shock or blood loss. This practice is of far more dubious validity.

5 Indications for withholding or discontinuing resuscitation

The issue of when to withhold or discontinue resuscitation is the most difficult treatment decision to be made in the delivery room. Much of the information needed to define appropriate indications for using intensive care is lacking. Better data are required on the effect of intensive care on mortality and the quality of life of survivors, and the cost of care for severely impaired or malformed infants.

Ultimately, decisions to withhold or withdraw aggressive care involve value judgments about what is considered an acceptable outcome and an acceptable cost. Although much has been written to express the views of health care professionals, lawyers, and ethicists concerning aggressive care of extremely high-risk infants, little has been done to explore the views of the parents who, apart from the child, have most at stake in such decisions.

Given the limited amount of useful information which is available for reaching decisions about instituting or withholding aggressive neonatal care, a liberal policy of resuscitation must be recommended whenever doubt exists. This allows the physician time to gather important information about the infant, and the distressed parents time and opportunity to participate more effectively in joint decisions about subsequent treatment.

6 Conclusions

The vast majority of infants need only a vigilant caregiver, a clear airway, and a warm welcome. Moreover, most high-risk newborns can be successfully cared for by rather simple means.

Nasal and pharyngeal suction should be carried out on babies who have passed meconium *in utero*. Endotracheal suction should not be carried out on these babies unless they show clinical signs of depression (e.g. heart rate less than 100 beats/minute). There is no justification for routine suctioning of the stomach or for routine intubation of all very-low-birthweight infants in the delivery room.

The great majority of babies who are depressed at birth require only appropriate ventilation without the need to consider use of drugs, volume expanders, or other adjuncts. The most common serious error in neonatal resuscitation is the failure to recognize and correct hypoventilation, a problem that is preventable with sufficient training and experience. Each hospital must establish appropriate methods to facilitate the most effective care for asphyxiated or depressed neonates. Whenever feasible, the birth of high-risk infants should be attended by a caregiver experienced in neonatal resuscitation.

In the absence of further evidence, breastfed babies should receive supplemental vitamin K routinely to prevent haemorrhagic disease of the newborn. Although the evidence is not conclusive, it is probably best to administer vitamin K to formula-fed infants as well.

Where not required by law, observation for and prompt treatment of ophthalmia may be as effective as routine prophylaxis and may save many babies from unnecessary medication. If eye prophylaxis is required, erythromycin causes less chemical conjunctivitis than silver nitrate and is more effective against *Chlamydia* infection. Silver nitrate should not be used. There is no evidence to suggest that topical ophthalmic preparations must be given immediately after birth.

This chapter is derived from the chapters by Jon Tyson, William Silverman, and Joan Reisch (75), and by Howard Berger (83) in EFFECTIVE CARE IN PREGNANCY AND CHILDBIRTH, and from the chapter on immediate care of the newborn infant by Jon E. Tyson (3) in EFFECTIVE CARE OF THE NEWBORN INFANT.

References to primary sources and more complete data for statements made in this chapter can be found in the source chapters and/or in the following reviews from the *Cochrane pregnancy and childbirth database:*

Sinclair, J.C.
— Intubation and suction in vigorous meconium-stained babies. Review no. 05946.

Soll, R.F.
— Prophylactic administration of any surfactant. Review no. 05664
— Prophylactic administration of natural surfactant extract. Review no. 05207.
— Prophylactic administration of synthetic surfactant. Review no. 05253.
— Prophylactic surfactant vs treatment with surfactant. Review no. 05675.

— Prophylactic administration of modified surfactant extract. Review no. 07317.
— Prophylactic administration of Exosurf neonatal. Review no. 07319.

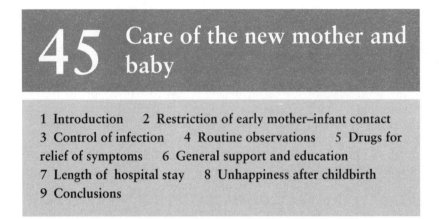

45 Care of the new mother and baby

1 Introduction   2 Restriction of early mother–infant contact
3 Control of infection   4 Routine observations   5 Drugs for relief of symptoms   6 General support and education
7 Length of hospital stay   8 Unhappiness after childbirth
9 Conclusions

1 Introduction

A new mother needs both emotional support and practical help in the days following childbirth. The form that this help takes and the way that it is given varies from culture to culture, and changes with the passage of time. The shift from home to hospital as the usual place of birth in this century has influenced the pattern of mother–baby interaction in at least two ways. First, women giving birth in unfamiliar surroundings and attended by caregivers whom they do not know may feel inhibited, and may behave differently toward their newborn child than they would have at home among familiar faces. Second, institutional rules and policies may obstruct spontaneous social interaction between newly delivered mothers and their babies.

The only justification for practices that restrict a woman's autonomy, her freedom of choice, and her access to her baby would be clear evidence that these restrictive practices do more good than harm.

2 Restriction of early mother–infant contact

Evidence concerning the possible adverse consequences of routine separation of mothers and their newborn infants in the early postnatal period accumulated during the thirty years or more during which this practice was widespread. Several controlled trials have compared the effects of institutional policies that restrict interaction between mothers and their newborn babies in the early hours after delivery with policies that encourage interaction at that time. Maternal affectionate behaviour was significantly less evident among mothers whose contact with their babies had been restricted than among mothers whose care encouraged liberal contact.

Trials have also compared standard hospital practices with attempts to encourage mother–infant interaction after the immediate postnatal period. The additional interaction in these trials ranged from a small amount of caretaking in the intensive care nursery by mothers of preterm infants to full rooming-in for mothers of normal infants. The few statistically significant differences that were observed suggest that restrictive policies were associated with less affectionate maternal behaviour and more frequent feelings of incompetence and lack of confidence. Perhaps most important of all, the results of one well conducted study suggest that, compared with a policy of rooming-in, the routine hospital policy of separating mothers from their babies led to an increase in the subsequent risk of child abuse and neglect among socially deprived first-time mothers.

Hospital policies can affect subsequent breastfeeding patterns among women who wish to breastfeed their babies. In all trials that have addressed this question, the proportion of women who had discontinued breastfeeding one to three months after delivery was substantially higher among women who had been subjected to more restrictive policies.

Maternal caretaking behaviour is so essential to the newborn baby that a variety of biological and social mechanisms must have evolved to promote it. The mutually reinforcing affectionate behaviour that occurs during early postnatal mother–infant contact is only one such mechanism. It is not the only mechanism. Most mothers who are constrained from early contact with their babies, whether through illness, misguided hospital policies, or personal preference, are likely to overcome any effects of this separation in the longer term.

3 Control of infection

Many of the regulations and routines in hospital postpartum care were instituted as attempts to prevent or contain cross-infection. As hospital nurseries began to fill up with members of the baby boom that followed the Second World War, staphylococcal skin disease among neonates became the major infectious problem. A variety of measures were taken in attempts to deal with this problem, including isolation, segregation, rules of dress and entry to the nursery, medicated bathing, and special treatment of the umbilical cord.

The inflexible use of central nurseries in which a number of babies were kept in close proximity to each other, but apart from their mothers, may actually have increased the risk of infection. A study performed almost fifty years ago showed lower rates of colonization and infection in babies who spent between eight and twelve hours a day with their mothers than in babies who were kept in a nursery and were rarely in contact with their mothers. The results of this controlled study had little impact on the then almost universal policy of routinely separating mothers and babies. Instead, the problem of cross-infection and nursery epidemics of staphylococcal infection was addressed with more technological approaches. The use of gowns, hats, and masks in nurseries became routine despite the lack of any evidence from controlled trials that they reduced infant colonization and infection rates.

As segregation and isolation failed to control neonatal infection, the mainstay of infection prophylaxis shifted to regimens of treatment of the umbilical cord and medicated bathing. The first controlled trial mounted to assess any of these new measures showed that staphylococcal colonization of the skin could be reduced by daily applications of triple dye (brilliant green, proflavin hemisulphate, and gentian violet) to the umbilical cord stump. This effect was confirmed in a number of studies conducted over the subsequent two decades.

Of the variety of methods used to treat the umbilical cord, only neomycin has been shown to reduce colonization rates more effectively than triple dye. Silver sulphadiazine offers no advantage over triple dye with respect to colonization rates, and there appears to be little if any difference between them in infection rates. Other preparations currently in use for treating the umbilical cord have simply not been adequately compared with the triple dye 'standard', and their introduction into clinical practice probably reflects commercial interests rather than scientific evidence.

A more complex approach involved the use of 3 per cent hexachlorophene in detergent to wash babies immediately after birth and daily until discharge from hospital. Compared with dry care or care with medicated soap, the hexachlorophene bathing reduced staphylococcal skin colonization and, more importantly, pyoderma. Unfortunately, no direct comparisons were made initially between the simpler regimen of painting the umbilical cord with triple dye, and the more elaborate policy of daily bathing with hexachlorophene (with its attendant risk of hypothermia). Evidence derived more than a decade later suggests that the two regimens have similar effects on staphylococcal skin colonization rates, but by this time it was also realized that hexachlorophene could be neurotoxic and too dangerous for routine use.

Therefore the fundamental question is not 'what medication should be put in the baby's bath water?', but 'is routine medicated bathing of babies justified at all in the light of the available evidence?'. The available evidence suggests that it is not.

Many restrictive and costly practices remain in force in hospital nurseries today. Gowning rules persist in almost three-quarters of newborn nurseries, despite the lack of evidence that this ritual has any beneficial effect. Many hospitals invoke concern about infection as a reason for restricting or forbidding siblings to visit, although studies using concurrent and historical control groups have been unable to detect any adverse effect of this on infant colonization rates.

It would be extremely dangerous to take a cavalier attitude to the undoubted reality of hospital-spread infection among new mothers and babies. Means of preventing or containing such infections must be kept under constant review. But maintaining or instituting restrictive measures without assessing whether or not they accomplish these ends is not an appropriate response.

4 Routine observations

Making and recording regular measurements of temperature, pulse, blood pressure, fundal height, and observations of lochia and the various wounds that a woman may sustain during delivery is still common practice in the days following childbirth. The intensity of this screening activity varies arbitrarily, and depends more on the hospital in which a mother happens to give birth, and on the length of time that she spends in it, than on her individual needs. While it is prudent to observe women in this way when they are known to be at increased risk of either infection or haemorrhage, it is difficult to justify this as a routine for all women.

5 Drugs for relief of symptoms

The postpartum period is often accompanied by a number of discomforts. Pain in the perineum and breasts (discussed in other chapters) is common. After-pains (uterine spasms) can also cause pain and distress in the first four days after birth. Women not infrequently count these symptoms among the most unpleasant memories that they have of childbirth.

In most hospitals, medication for such discomforts tends to be under the direct control and supervision of hospital personnel. This policy is valid in the case of dangerous drugs, drugs that might interact with other medications, or drugs that would normally be available only on a doctor's prescription. It does not make sense for drugs which, outside hospital, are readily available without a prescription. Because these medications are usually for the relief of symptoms, it would be more reasonable for women to use them when they feel the need to do so.

In many hospitals, non-prescription drugs intended for the relief of symptoms are still administered routinely and at set times. Professional belief in the value of early and regular bowel evacuation, for example, has led to some rather obsessional concerns on this matter. Laxatives, stool softeners, enemas, and anal ointments remain routine components of postpartum care in some hospitals. In the puerperium, as at other times, routinely prescribed laxatives and medicated enemas do result in earlier bowel movements, and bulk-forming laxatives are less likely to cause unpleasant cramps and diarrhoea than are irritant laxatives. However, this cannot be regarded as justification for administering any of these preparations routinely.

In an attempt to provide a more rational basis for care (including self-care), a number of hospitals have developed programmes to provide mothers with information about the drugs that they may use and have given them access to their own supply of medications. These descriptive studies all suggest that women are usually pleased with these arrangements and that staff time is used more efficiently after their introduction.

Routine medications for 'non-symptoms', such as oral ergometrine with the aim of hastening uterine involution, should not be used without evidence of a beneficial effect on substantive outcomes.

6 General support and education

Fragmentation of care is almost inevitable when the care of mothers is the responsibility of midwives or postpartum nurses and obstetricians, while care of the babies is carried out by nursery nurses and paediatricians. Consistent advice and continuity of care are prerequisites for effective support to mothers and their newborn babies. This is easier to achieve if mother and baby are accommodated together or, at the very least, if the same people are looking after mother and baby. The postpartum stay in hospital presents obvious opportunities for imparting information that may be of help to new mothers. Although there is little agreement on what should be the content of formal postpartum educational programmes, a number of studies, several of them controlled trials, have addressed the methods of postpartum education. These studies have dealt with educational programmes including subjects ranging from contraception, through feeding and immunization advice, to information about child safety. Although the demonstrable effects of teaching were rarely as dramatic as the investigators had hoped, this body of research shows that postpartum educational programmes can and do favourably affect parental behaviour and health outcomes.

7 Length of hospital stay

How long should a healthy woman and baby remain in hospital after childbirth? The generally accepted 'correct' length of postnatal stay has varied greatly from time to time and varies equally widely among institutions today. It appears to be determined more by fashion and the availability of beds than by any systematic assessment of the needs of recently delivered women and their new babies.

The period of 'lying in' in industrialized societies had its institutional foundations with the establishment of charitable lying-in hospitals in the mid-eighteenth century. In these first maternity hospitals, women were, quite literally, confined to bed for a period of 28 days following delivery! By the 1950s the usual hospital lying-in period was 12–14 days, but since then it has fallen in most countries, and in some cases is no more than one to two days.

Several formal early discharge programmes, often with strict inclusion criteria, have been instituted as a deliberate change from standard hospital policies. The most common impetus for initiating such programmes seems to have been the needs of the institution (because of a shortage of beds or personnel), although some programmes appear to have been inspired by the preferences of childbearing women.

Both randomized trials and observational studies have shown that few women or babies are readmitted to hospital after early discharge. These low readmission rates demonstrate that early discharge from hospital is feasible and that healthy mothers and babies who have help at home, and follow-up care and guidance available, may safely go home within a few hours of birth.

Providing a hospital telephone service to offer advice or information by postpartum nursing staff to new parents has been demonstrated to be of benefit.

Although no adverse consequences of early postnatal discharge from hospital have been demonstrated in terms of maternal and child health, there is also no evidence that such programmes save money or resources unless beds are closed as a result. If home support is provided, the costs of support given to women discharged early have been estimated to be about the same as the savings in hospital costs that result from earlier discharge. Also, when healthy women go home earlier, those remaining in hospital are, on average, more sick, require more intensive care, and involve additional work and strain for the staff. Early hospital discharge programmes inevitably put an additional load on the primary care sector and on the family or friends of the woman. The additional resources required to cope with this load must be assessed in any credible attempt to assess the costs of early discharge from hospital.

Demands for early discharge programmes have been made by a variety of commentators who believe themselves to be speaking on behalf of childbearing women. When such programmes have been instituted, they have not always been popular. The extent to which women who give birth in hospital want early discharge may have been overestimated, and may depend on the support available at home.

Women's physical, social, and psychological circumstances after delivery vary greatly. Professionals should be as flexible as possible in response to this variation; this may be easier in situations where there is continuity of care. Choice may well be the crucial issue. Many women feel that their hospital stay was too long, but some feel that it was too short. Perhaps a more extended escape from duties at home would have helped some women to recover from childbirth more effectively.

Although early discharge from hospital is feasible and safe, it is neither clearly beneficial (in health or economic terms) nor wanted by the majority of women. Firm decisions about when a woman should go home after childbirth should be delayed until after she has given birth and, as far as possible, be based on her individual needs and preferences rather than on any predetermined formula.

8 Unhappiness after childbirth

Lack of social and psychological support during the days and weeks following delivery is one of the main reasons that unhappiness after childbirth is such a common problem. This has traditionally, and wrongly, been seen as a medical subject, labelled postpartum depression. There is no persuasive evidence to support traditional explanations of so-called postpartum depression. No biochemical explanation of women's unhappiness after childbirth has been uncovered, and psychoanalytical explanations of postpartum depression cannot be validated empirically. Mothers of young children are often depressed and are no less likely to be so six months or a year after delivery than in the first few weeks or months following childbirth. Sociological and psychological studies have provided strong evidence of a relationship between some social conditions and postpartum depression. The social conditions linked with depression are only rarely 'out of the ordinary'; more often, they correspond to social expectations about what normal womanhood and normal motherhood must be.

Because many of the social factors that lead to postpartum unhappiness are rooted in society's expectations of new mothers, the solutions lie mainly in social change. However, there is considerable scope for professionals to reduce the difficulties and unhappiness experienced by women after childbirth. In particular, they should be more ready to listen to women, to learn about their social circumstances, and to provide them with information which will lead to more realistic predictions about the experience of pregnancy, childbirth, and early parenthood.

If, despite the best efforts being made to prevent the problem, women do become seriously depressed, then encouraging them to talk about their feelings to a non-judgemental person has been shown in controlled trials to increase their chances of early recovery.

9 Conclusions

The restriction of early postnatal mother–infant interaction, which has been such a common feature of the care of women giving birth in hospitals, has undesirable effects. Disruption of maternal–infant interaction in the immediate postnatal period may set some women on the road to breastfeeding difficulties, and possibly alter their subsequent behaviour towards their children. Nevertheless, for mothers who for whatever reason cannot have early contact with their babies, other mechanisms are likely to overcome any effects of this separation.

Most of the restrictive practices still perpetuated in some hospitals are ineffective and possibly harmful. Unless or until new evidence appears to the contrary, mothers should have unrestricted access to their babies. Caps and masks should be abolished and aprons and gowns used only by those who wish to protect their own clothing from the various kinds of messes that babies make. Bathing the baby should be seen not as a measure for preventing infection, but as an opportunity for a mother to interact with and gain confidence in handling her child.

Triple dye is an inexpensive and effective prophylactic for routine use to prevent infection of the umbilical cord. Neomycin powder is a somewhat more expensive and more effective alternative. There is currently no evidence to justify medicated bathing or the use of alcohol, powders containing hexachlorophene, or other medications outside the context of randomized controlled trials to assess whether they offer any advantages over triple dye or neomycin powder.

The standard hospital setting, with its orientation towards sickbed protocols and procedures and its division of care along the lines of the separate medical disciplines of obstetrics and paediatrics, is not conducive either to helping the new mother develop the skills and self confidence that she needs to care for herself and her new baby, or to enhancing her sense of personal worth and self-esteem. It is unlikely that any single scheme of care will prove to be right for all women. Treating the new mother as a responsible adult, giving her accurate and consistent information, letting her make her own decisions, and supporting her in those decisions is the essence of effective postpartum care.

This chapter is derived from the chapters by Molly Thomson and Ruta Westreich (77), and by Janet Rush, Iain Chalmers, and Murray Enkin (78) in EFFECTIVE CARE IN PREGNANCY AND CHILDBIRTH, and by Jill E. Baley and Avroy A. Fanaroff (20) in EFFECTIVE CARE OF THE NEWBORN INFANT.

References to primary sources and more complete data for statements made in this chapter can be found in the source chapters and/or in the following reviews from the *Cochrane pregnancy and childbirth database:*

Hodnett, E.D.
— Support from caregivers for socially disadvantaged mothers. Review no. 07674.
— Continuity of caregivers during pregnancy and childbirth. Review no. 07672.
— Support from caregivers for distressed postnatal women/couples. Review no. 07673.

Hay-Smith, J. and Renfrew, M.J.
— Education on contraceptive use by postpartum mothers. Review no. 07429.
— Provision of hospital telephone service for new parents. Review no. 07321.
— 'Anacal' ointment for puerperal haemorrhoids. Review no. 07428.
— Combination vs bulk-forming postpartum laxative. Review no. 07658.
— Irritant vs combination postpartum laxative. Review no. 07660.
— 'Senokot' vs 'Normax' laxatives postpartum. Review no. 07661.
— Irritant vs bulk-forming postpartum laxatives. Review no. 07659.

Renfrew, M.J.
— Early vs standard postnatal discharge from hospital. Review no. 02735.
— Oral ergometrine for four weeks postpartum. Review no. 04170.

Renfrew, M.J. and Hay-Smith, J.
— Medicated vs tapwater disposable postpartum enema. Review no. 03290.
— Postpartum laxatives. Review no. 03663.

1 Introduction

Breastfeeding is recognized to be important for both mother and baby. Women who choose to breastfeed do so because they feel that it is best for their babies, and also because they find it satisfying and enjoyable. However, many women encounter problems and stop breastfeeding before they wish to do so, particularly in the early days and weeks. As well as depriving the baby of the benefits of breastfeeding, this also causes distress for the mother and her family.

There have been attempts in several countries to increase the number of women who breastfeed their babies. Until women who wish to breastfeed are also supported in breastfeeding their babies for as long as they wish, promotion campaigns to encourage breastfeeding are likely to fall short of the objectives that are worth pursuing.

Many elements of care during pregnancy and childbirth can foster or jeopardize the successful establishment and maintenance of breastfeeding. Efforts to provide social and psychological support, for example, may increase the likelihood that mothers will breastfeed their babies successfully. In contrast, sedative and analgesic drugs given during labour can alter the behaviour of the newborn infant and compromise its crucial role in the initiation of lactation.

The establishment of lactation may be jeopardized at the time of delivery in other ways as well. Routine gastric suctioning and administration of silver nitrate eye drops in the immediate postnatal period can prejudice the infant's role in establishing lactation. Separating babies from their mothers, whether because of entrenched hospital routines or for necessary treatment of the baby, reduces the likelihood that breastfeeding will be established successfully.

Antenatal care practices, the time of the first feed, positioning, feeding frequency and duration, supplements for babies and mothers, and support for breastfeeding mothers can all affect the establishment and maintenance of breastfeeding.

2 Antenatal preparation

Most women who decide to breastfeed make this decision before or early in pregnancy. Those who choose to bottle feed tend to make up their minds later in pregnancy. This effect is not mediated by knowledge alone. Giving women well designed, well written, and well illustrated information about breastfeeding increases their knowledge of the subject, but has little effect on either their choice of feeding method or the duration of breastfeeding. It is likely that other influences, such as her own previous experiences, and the attitudes and experiences of her family and friends will have an important role in this decision. Once a woman has decided how she will feed her baby, she is unlikely to change her mind.

While information alone, at least in a written form, may not affect the *decision* to breastfeed, antenatal information given to women who have already decided to breastfeed may be beneficial. The available data suggest that antenatal classes may be effective in promoting breastfeeding, but more evidence is needed to find out which elements of the information and what kind of classes women find helpful.

Research studies to assess the efficacy of antenatal nipple 'conditioning' have not shown any significant differences, either objective or subjective, among the different methods of conditioning, the use of Massé cream, expression of colostrum, or no form of preparation. Two trials have assessed the effects of antenatal treatment for women with inverted or non-protractile nipples. Neither of the two treatments tested — Hoffman's nipple stretching exercises and breast shells — were shown to have any beneficial effect on the duration of breastfeeding.

3 Early versus later suckling

Early contact between mother and baby has beneficial effects on breastfeeding, in addition to other important benefits. It is difficult to separate the effects of early suckling *per se* from the effects of other early mother–baby interaction such as touching, gazing, and skin-to-skin contact. Feeding within the first two hours after birth increases the duration of breastfeeding when compared with a delay of four hours or more. No research has demonstrated a 'critical period' for the first feed in terms of breastfeeding success; that is, there is no evidence to suggest that her breastfeeding will suffer if a mother does not feed her baby immediately after birth. Therefore there are no research-based grounds for replacing old dogma ('no baby should breastfeed until four hours after delivery') with new dogma ('all babies should feed immediately after delivery') or for encouraging a mother to breastfeed her baby before she and the baby are ready. Babies have a wide range of behaviour following spontaneous birth, and are not all ready to feed at the same time. Unless or until more evidence is available, interventions aimed at either delaying or speeding up the time of the first feed should be avoided.

The first *feed* after delivery (as opposed to the immediate postdelivery nuzzle at the breast) should be given in privacy, at a time when the baby is receptive, and after the newly delivered mother and baby have been made comfortable. Skilled professional help would be useful at this time. If possible, it should be done while the father or someone else whom the mother has found supportive, is still present. The baby's behaviour and needs can be explained to the new parents. A brief explanation of the importance of correct positioning and the concept of supply and demand can be given before the mother positions the baby properly at the breast. This can be followed by a little more information on the importance of unrestricted feeding, potential problems, and how (and why) to summon help.

4 The importance of correct positioning

Correct positioning of the baby at the breast plays a crucial role in both the prevention of sore nipples and the successful establishment of breastfeeding. A woman's ability to position her baby correctly at her breast is a learned and predominantly physical skill, which the mother must acquire from observation and practice. On the whole, industrialized societies do not provide women with the opportunity to observe other breastfeeding women before they attempt breastfeeding themselves. This deficiency is compounded by the frequent lack of experienced breastfeeding mothers in the woman's immediate social sphere.

Professionals must understand the underlying mechanisms of suckling and acquire the skill and experience to help a mother to position her baby correctly before they can be of real value to the mother. The fragmentation of postnatal care that is so common today prevents many professionals from acquiring these skills.

When the baby is properly attached to the breast (Fig. 1), feeding should be pain free. The nipple, together with some of the surrounding

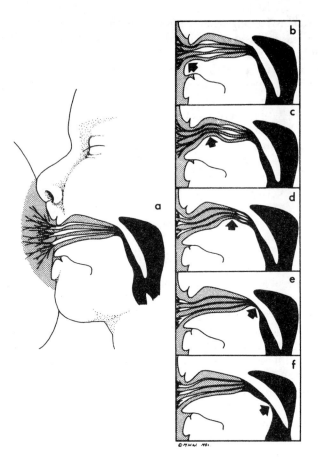

Fig. 1 (a-f): Diagrams of accurate positioning of the baby on the breast
(from Woolridge 1986a)

breast tissue, is drawn out into a teat by the suction created within the baby's mouth. Breaking this suction causes the nipple to recoil abruptly. The teat thus created extends as far back as the junction of the baby's hard and soft palate, with the nipple itself forming only about one third of the teat. At its base, the teat is held between the upper gum and the tongue, which covers the lower gum. It lies in a central trough formed by the raised edges of the baby's tongue, which directs the expressed milk backwards into the pharynx using a roller-like peristaltic movement. The peristaltic action begins as the front edge of the tongue curves upwards, closely followed by the raising of the lower jaw, which follows the tongue's movement with pressure from the lower gum. This wave of compression moves progressively backwards beyond the tip of the nipple, thus directing the milk into the pharynx and on into the oesophagus. Meanwhile, a fresh cycle of compression by the tongue has been initiated from its tip.

Thus the breast tissue opposed to the baby's lower jaw and tongue is the critical region in the transfer of milk; the tongue applies peristaltic force to the underside of the teat. The hard palate simply provides the necessary resistance to the tongue's action. Once sufficient breast tissue has been formed into the teat, there should be virtually no movement of this teat in and out of the baby's mouth. Friction from the tongue against the nipple should be minimal, and the gums should not come in contact with the nipple at all. If the baby is incorrectly positioned at the breast and is unable to form a teat out of the breast tissues as well as the nipple, then the nipple is likely to incur frictional damage as the teat is repeatedly drawn in and out of the mouth between the tongue and gums by the cyclical application of suction.

The mother needs to be taught how to elicit and use the two components of the baby's rooting reflex: the moving of the head towards the source of stimulation when the skin around the mouth is touched, and the accompanying gaping of the mouth preparatory to receiving the breast. She should be shown how to move the baby towards the breast and 'plant' the lower rim of the baby's mouth well below the nipple at the moment that the baby's mouth gapes widely. This should be followed by moving the baby close to the breast as it takes a good mouthful of breast tissue. The mother cannot rely, as her helper does, on seeing where the baby's lower lip and jaw are in relation to her nipple, for she has a poor view of the underside of the breast, the critical area of attachment. Observation of the baby's sucking pattern, as well as the sensations that the mother herself experiences (she should feel no pain) will serve to confirm that the baby is correctly positioned (Fig. 2).

Anything that interferes with the baby's ability to gape and grasp the breast can cause problems, and probably pain for the mother. Factors may include the baby being sleepy as a result of drugs used in labour, jaundice (and phototherapy), and tongue tie, which prevents the baby from using its tongue effectively to grasp and milk the breast. Early diagnosis of the problem and sensitive support for such mothers and babies is needed until the problem is resolved.

More widespread acquisition and use of the skills needed to achieve correct positioning of the baby on the breast would probably do more than anything else to reduce the frequency of the problems currently experienced by so many breastfeeding mothers.

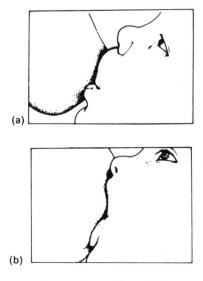

Fig. 2 (from Fisher 1981):
(a) Baby incorrectly positioned at breast.
(b) Baby correctly positioned at breast.

5 The importance of flexibility in breastfeeding practices

A baby needs to eat and sleep according to his or her own individual rhythms rather than those imposed by arbitrary regimens. Feeds are usually infrequent in the first day or so, but they become more frequent between the third and seventh day. The frequency then decreases more slowly over the next few days. Although a few babies are content to feed as infrequently as every 4 hours, most babies want to be fed more often than this. For the first few weeks of life at least, the interval between feeds is variable, ranging from 1 to 8 hours. Babies who are permitted to regulate the frequency of their feeds themselves gain weight more quickly and remain breastfed for longer than those who have external limitations imposed on them. There are no data providing any justification for the imposition of breastfeeding schedules.

Similar comments apply to the duration of feeds. The fact that limitation of suckling time is still being advocated in books addressed both to mothers and to professionals reflects the deeply ingrained belief that the nipples need to be 'toughened' to permit pain-free feeding. Mothers and professionals are often told that the baby should be permitted to feed for only 2 minutes at each breast for the first day, increasing the time by 2–3 minutes daily, so that at the end of the first week the baby has reached a maximum of 10 minutes on each side. These admonitions are based on the unwarranted belief that this practice will 'break the nipples in gradually', prevent their exposure to prolonged sucking, and thus prevent nipple soreness and cracks.

Controlled studies show that this is not the case. Comparison of restricted and unrestricted duration of feeds show no significant differences between the two groups in the proportion of women who develop sore or cracked nipples, but significantly more mothers in the regulated groups give up breastfeeding altogether by six weeks.

The still commonly given advice to limit sucking time to 10 minutes on each breast has repercussions beyond its failure to prevent nipple damage. The composition and the rate of flow of milk changes over time. The fat content increases and the flow rate decreases as the feed progresses. Thus at the start of a feed the baby takes a large volume of low calorie foremilk; this changes to a smaller volume of high calorie hindmilk at the end of the feed. Babies feed for differing lengths of time at the breast if left undisturbed. The length of the feed is probably determined by the effectiveness and rate of milk transfer between mother and baby. While many babies will terminate a feed spontaneously in under 10 minutes, those who have a slow rate of intake may well take longer than this. Even though the volume of milk that they consume after 10 minutes have elapsed may not be very great, it may be sufficiently high in calories to make a significant contribution in energy value. Thus the external imposition of a time limit for feeding will result in some babies having their calorie intake significantly curtailed.

Babies are driven to feed by the need to obtain calories, and thus take much larger volumes of low-calorie milk than they would of high calorie milk in an attempt to gain the required number of calories. Babies who are taken from the first breast before they spontaneously terminate the feed may take a much larger volume of milk from the second breast than they might otherwise have done in order to try to 'make up' their calories. For the same reason these babies may also require feeding much more frequently than they would have if they were allowed to finish feeding spontaneously.

In addition, interference with spontaneous feeding patterns may result in the baby being deprived of essential vitamins. Vitamin K, for example, is particularly concentrated in colostrum and hindmilk, and this may partly explain the increased incidence of haemorrhagic disease of the newborn in breastfed babies.

In some institutions there is still a need for more intensive policy discussions and in-service training to ensure that a policy that supports and encourages flexible breastfeeding is translated effectively into practice.

6 'Supplementing' the baby

There is no evidence to support the widespread practice of giving breastfed babies supplementary feeds of water, glucose, or formula. A healthy baby has no need for large volumes of fluid any earlier than these become available physiologically from the breast. Equally there is no evidence to support the widespread belief that giving additional fluids to breastfed babies prevents or helps to resolve physiological jaundice. In the only randomized controlled trial to have examined the question, there was no statistically significant reduction in mean plasma bilirubin levels associated with giving water supplements, nor any evidence that babies receiving extra fluids were any less likely to develop 'breast milk jaundice' or require phototherapy.

The practice of giving breastfed babies formula while lactation is becoming established is also misconceived. Women whose babies receive routine supplements are up to five times more likely to give up breastfeeding in the first week and twice as likely to abandon it during the second week as women who are not supplemented, but encouraged to feel that their own colostrum and milk are adequate.

Those hospitals that allow breastfeeding mothers to be given free samples of formula also prejudice the chances of successful establishment and maintenance of breastfeeding. This policy increases the chance that breastfeeding will have been abandoned within a few weeks of delivery.

7 'Supplementing' the mother

Advice given to breastfeeding women concerning their own fluid intake has been inconsistent and has led to confusion and misinformation. The results of controlled studies provide no evidence that increased fluid intake by breastfeeding mothers will result in improved lactation. Some women find it unpleasant to drink when they are not thirsty because it makes them feel 'turgid' and 'unwell'. Women with perineal and labial trauma may have their discomfort increased by the diuresis associated with greater fluid intake.

8 Combined oestrogen–progestogen contraceptives

Combined oestrogen–progestogen contraceptives may affect the composition of breast milk. Their effect on milk volume is not clear, probably because of limitations in the methods used to estimate milk volume. What is clear is that their use increases the incidence of both breastfeeding failure and supplementation with breast milk substitutes.

All the evidence from controlled trials suggests that combined oestrogen–progestogen contraceptives are unsuitable for women who wish to breastfeed their babies. This finding is particularly important in parts of the world in which breast milk substitutes pose a threat to infant life and health. Lactation itself, particularly if breastfeeding is unrestricted, has a contraceptive effect. Many women who breastfeed want a better guarantee of contraceptive effect than can be provided by lactation alone; they should use means other than combined oestrogen–progestogen preparations.

9 Supporting breastfeeding mothers and babies

Many women who wish to breastfeed begin to do so but discontinue long before their babies are four months old. Perhaps the most important factor in the efforts to increase the rate of breastfeeding is that those who attempt it should succeed. Many of the problems that confront women who are trying to breastfeed are avoidable. Fewer women would experience these problems if all breastfeeding women had access to accurate information, and appropriate and practical help and support when they need it. More pregnant women would then know of others who had breastfed successfully and be more confident that they themselves will succeed.

A number of controlled trials have assessed the effects of various forms of support for breastfeeding mothers. Although the results were not totally consistent, most show that the duration of breastfeeding can be increased by regular and frequent contact with the mother by the same caretaker, either in person or by visits followed by telephoning. Advice and support for mothers who wish to breastfeed can be important in helping them to achieve their objectives. If the advice given is flawed, it is unlikely to be helpful.

10 Nipple trauma

The most widely held explanation for the prevalence of nipple pain in industrialized cultures is the supposed thinness or sensitivity of the nipple epithelium. This probably explains the widespread belief, unsupported by the studies that have examined the question, that women with fair skin or red hair are more likely to experience problems.

A number of treatments, including ointments, tinctures, and sprays, have been used for the prevention or treatment of nipple damage. None that have been evaluated have been shown to be of benefit. The use of a nipple shield for any length of time, even by those who find its use acceptable, may add to a mother's problems by suppressing her milk production.

The only factor that has been shown to both prevent and treat nipple trauma is good positioning of the baby at the breast.

11 Problems with milk flow

If the milk is not removed as it is formed (as regulated by the baby's need to go to the breast) the volume of milk in the breast will exceed the capacity of the alveoli to store it comfortably. Overdistension of the alveoli with milk causes the milk secreting cells to become flattened, drawn out, and even to rupture. If severe, this will cause secondary vascular engorgement. Once the alveoli become distended, further milk production begins to be suppressed.

Engorgement results from limitations on feeding frequency and duration, and from problems with positioning the baby at the breast. A number of different treatments have been advocated. Some, such as the use of moist heat or ice packs, have not been evaluated. In the past, treatment of severe engorgement has included both administration of stilboestrol and binding of the breasts, two measures also advocated for suppression of lactation in women who do not wish to breastfeed.

Controlled evaluations of the effects of oxytocin failed to find any beneficial effect of oxytocin in relieving engorgement. Two early studies seem to indicate that manual expression, started antenatally and continued postnatally, will help to relieve engorgement and increase the duration of breastfeeding; however, both these studies were carried out under conditions where feeding was restricted and engorgement was, as a result, very common.

Placebo-controlled trials of oral proteolytic enzymes suggest that they may provide effective relief for women with breast pain, swelling, and tenderness. The evidence is not yet sufficiently strong to recommend this form of treatment.

Application of plant products directly onto the breast has also been advocated for the relief of engorgement. The plant remedy most commonly used is raw cabbage leaves. The only randomized controlled trial to evaluate the use of cabbage leaves found no difference in the relief of engorgement when women who applied cabbage leaves were compared with women who had routine care. However, the women who applied the cabbage leaves were more likely to be breastfeeding exclusively at six weeks and to breastfeed for longer. Both groups of women carried out a 'breast exercises' programme.

Allowing the baby unrestricted access to the breast still appears to be the most effective method of treating, as well as of preventing, breast engorgement.

The other common difficulty caused by milk flow problems is mastitis. Milk flow can be limited by restriction of feeding, by a badly positioned baby, or when some obstacle is placed in the way of milk draining from one section of the breast. This obstacle can result from such factors as blocked ducts, reduction in the normal frequency of feeding, compression from fingers holding the breast, bruising from trauma or rough handling, or a brassiere which is too small or too tight. In consequence, the milk collects in the alveoli and the pressure in the alveoli rises. The distension of the alveoli can often be felt as a tender lump in the breast tissue. If this distension is not relieved the pressure may force substances from the milk through the cell walls into the surrounding connective tissue, setting up an inflammatory reaction. The mother develops a swollen, red and painful area on her breast, a rise in her pulse and temperature, and an aching influenza-like feeling, often accompanied by shivering attacks and rigors. At this stage the process is not infectious and the problem can be resolved by relieving the obstruction. If this is not speedily accomplished bacterial infection may supervene, and may ultimately give rise to a breast abscess.

Perhaps understandably, the immediate response of the professional confronted with the symptoms of localized breast tenderness, redness, and fever in breastfeeding women is often the prescription of antibiotics. If not treated appropriately, what begins as a non-infectious inflammatory process may rapidly progress to an infectious process and delay in treating an infectious process will adversely affect the outcome. Nevertheless, a substantial proportion of women with mastitis do not have an infection.

For women with milk stasis, simply continuing to breastfeed, ensuring unlimited pain-free feeds, gives the best results. Expression of breast milk has not been shown to confer any advantage. The outcome for women with non-infectious mastitis is best with continuation of breastfeeding supplemented by breast milk expression. For women with infectious mastitis, antibiotics are necessary, and expression of breast milk improves the outcome for them.

12 Problems with milk supply

The most common reason given for discontinuing breastfeeding is insufficient milk. However, there is no information about the extent to which this insufficiency is inevitably physiological, as opposed to iatrogenic and thus preventable. Objective evidence of insufficient milk is hard to obtain, but it is likely that the high reported incidence reflects overdiagnosis of the problem. Observations in traditional societies suggest that less than 1–5 per cent of women would be physiologically incapable of producing an adequate milk supply.

It is important to be able to diagnose the occurrence and aetiology of insufficient milk accurately. Accurate measurement of milk production is possible with expensive and sophisticated research techniques, but the only clinically available method for this measurement is test weighing of the baby before and after feeds, and calculation of the breast milk intake by the difference in weights. Such test weighing, which is grossly inaccurate, has been used to estimate the baby's intake in cases of anxiety about milk supply, and as a routine practice in some hospitals.

The rationale for test weighing is to determine whether babies are taking 'too much' or 'too little' milk: if too little they can receive supplements; if too much the duration of breastfeeding can be curtailed. The hazards of these inappropriate responses to an inherently inaccurate test have already been discussed. In the one study mounted to examine this question directly, the effect of routine test weighing and supplementary feeding was compared with a policy of neither weighing nor supplementing. The total duration of breastfeeding was almost the same in both groups, but the mothers in the test weighed group were five times more likely to stop breastfeeding in the first week, and twice as likely to stop in the second week as those in the group whose babies were not test weighed.

The decision to give a healthy term breastfed baby supplementary feeds as a result of information gained by assessing milk intake is based on the unwarranted assumption that it is possible to know how much breast milk an individual baby needs. It would be more relevant to monitor the general condition of a baby (health, contentment/ behaviour, colour and consistency of stools, colour of urine, etc.) and note his or her progress, particularly the change in body weight.

The best means of preventing the occurrence of insufficient milk is unrestricted feeding by a well-positioned infant while giving good practical and emotional support to the breastfeeding mother. This is also the basis for the treatment of insufficient milk and is likely to solve the problem in a high proportion of, but not all, mothers.

Although the true incidence is unknown, professionals should remain alive to the possibility that milk insufficiency may persist. Babies may become seriously undernourished because of a dogged but mistaken belief that the problem will always be resolved by physiological means. When mothers and babies do not respond to the fundamental elements of good breastfeeding practice, other treatments should be considered.

In the past, when a baby's life depended on breast milk, many remedies were sought for those who seemed unable to produce enough milk. In addition to some rather bizarre prescriptions, a variety of herbal infusions such as the seeds of fennel (*Foeniculom vulgare*) and the flowers of goat's rue (*Galega officinalis*) were, and still are, recommended to increase milk production. We have been unable to identify any controlled evaluation of the effects of these preparations. However, four main types of drugs have been evaluated in attempted treatment of insufficient milk: dopamine antagonists, iodine, thyrotropin releasing hormone, and oxytocin. Since dopamine has been shown to have a critical role in the mechanisms which control prolactin production, several researchers have experimented with drugs that block dopamine receptors, including metoclopramide (Maxalon), sulpiride (Dolmatil), and domperidone (Motilium). There is some evidence that these drugs may be of use for women who are temporarily unable to feed their babies. Further research is required to clarify this.

As the let-down reflex, which is primed by oxytocin release from the posterior pituitary, is essential for successful breastfeeding, some investigators have reasoned that to give oxytocin may improve problems with milk supply. A variety of outcomes have been studied, but the most crucial information concerns weight changes in the baby. The trials conducted have produced conflicting results, and to date there is no strong evidence that oxytocin administration has a beneficial effect on milk supply.

13 Conclusions

Those who care for women during pregnancy and childbirth have a crucial role to play in enabling a woman to breastfeed successfully. Now that sound research-based information is readily available to them, the professional ignorance which may have been acceptable in the past is no longer tolerable. If the potential for helping women to breastfeed their babies is to be realized, professionals must reject many of the traditional practices in this field and pass on to women only those practices which have been demonstrated to be effective.

Those who are most likely to be closely involved with mothers at the time that breastfeeding is becoming established should have a clear understanding of how a baby breastfeeds. They should recognize that, although separation of babies from their mothers after delivery jeopardizes the successful establishment of lactation, there is no evidence to suggest that the timing of the first feed, in itself, is crucial to success. Interventions aimed at either delaying or speeding up the time of the first feed should be avoided.

Professionals should know how a mother can be helped to position her baby properly at the breast. They should impose no restrictions on the duration or frequency of feeds, and neither offer nor recommend additional fluids or formula for healthy breastfed babies. Free samples of formula given to women in hospital can be particularly detrimental to successful breastfeeding.

Normal lactating women with access to adequate fluid can depend on their thirst to regulate fluid intake effectively. Urging women to drink more than their thirst dictates has no justification.

Use of combined oestrogen–progestogen contraceptives compromises lactation. Women who wish to enhance the contraceptive effect of lactation should use means other than hormonal methods containing oestrogens and progestogens.

Women can be helped to establish and maintain breastfeeding in a number of ways, but the experimentally derived evidence suggests that continuity of personal support from an individual who is knowledgeable about breastfeeding is most effective.

The main reasons that women give for discontinuing breastfeeding are nipple trauma, breast engorgement, mastitis, and insufficient milk. The majority of these problems can be prevented by unrestricted breastfeeding by a baby who has been well positioned from the first feed on, and by giving mothers practical and emotional support.

If a woman does sustain nipple trauma, she should continue to breastfeed, express milk if necessary, and receive help with positioning. Discontinuing breastfeeding, and the application of any of a variety of preparations to the nipple, does not help. Indeed, some of these interventions have been shown to prejudice the success of breastfeeding.

Problems with milk flow can result in engorgement and possibly in mastitis. If mastitis does not resolve rapidly with good feeding and expression, then antibiotic treatment should be instituted. However, in all cases of engorgement and mastitis the key to successful treatment is good drainage of the breast. This is best achieved by unlimited feeds by a well-positioned baby.

Mothers and health professionals who suspect insufficient milk as a result of signs and symptoms in the baby or the mother face a challenging problem. Identifying the problem and its cause is always difficult and often impossible. Until diagnostic precision improves, the basis of the treatment offered when insufficient milk is suspected remains unrestricted breastfeeding by a well-positioned baby together with practical and emotional support for the mother. Only when mothers and babies do not respond to the fundamental elements of good breastfeeding practice should other treatments be considered.

This chapter is derived from the chapters by Sally Inch (21), Sally Inch and Sally Garforth (80), and Sally Inch and Mary Renfrew (81) in EFFECTIVE CARE IN PREGNANCY AND CHILDBIRTH.

References to primary sources and more complete data for statements made in this chapter can be found in the source chapters and/or in the following reviews from the *Cochrane pregnancy and childbirth database:*

Renfrew, M.J.
— Antenatal breastfeeding education. Review no. 04171.
— Antenatal expression of colostrum. Review no. 04028.
— Antenatal breastfeeding classes vs individual teaching. Review no. 07144.
— Postnatal support for breastfeeding mothers. Review no. 04173.
— Postnatal anticipatory guidance for mothers on infant feeding. Review no. 04177.
— Restricted schedule of breastfeeding. Review no. 04178.
— Provision of formula supplements to breastfed newborns. Review no. 04175.
— Provision of water supplements to breastfed newborns. Review no. 04174.
— Provision of free formula samples to breastfeeding mothers. Review no. 04172.
— Free bottles/water samples to breastfeeding mothers. Review no. 07135.
— Extra fluids for breastfeeding mothers. Review no. 04189.
— Combined oestrogen/progestogen contraceptive in breastfeeding mothers. Review no. 04376.
— Chlorhexidine/alcohol nipple spray. Review no. 04176.
— Oral sulpiride for women with poor lactation. Review no. 04187.

Renfrew, M.J. and Lang, S.
— Single daily bottle use in early postpartum period. Review no. 07908.

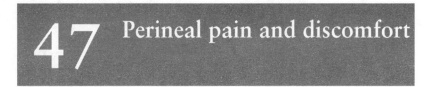

47 Perineal pain and discomfort

1 Introduction

Perineal pain constitutes a major problem for mothers in the early days after vaginal delivery, particularly if they have sustained perineal trauma. Avoidance of trauma when possible and proper repair when trauma occurs (see Chapter 36) are the primary approaches to avoiding or reducing these problems. In addition, a wide range of measures and active treatments are advocated for secondary prevention, or relief, of perineal pain.

2 Local applications

2.1 *Non-pharmacological applications*

Sprays, gels, creams, solutions, ice packs, baths, and douches are all commonly recommended for the relief of perineal discomfort after vaginal delivery, but they have received little, if any, formal evaluation.

Cooling with ice or sprays is often used in perineal care in the belief that pain and oedema are reduced. Ice-packs in the puerperium give immediate symptomatic relief by numbing the perineum, but this is usually shortlived and there is no evidence of any longer-term benefit. Sprays have been reported also to relieve perineal discomfort, probably by a cooling effect, although they occasionally cause a stinging discomfort.

There is evidence from a randomized trial that cold sitz baths are more effective than warm sitz baths in relieving perineal discomfort, but the differential effect is limited to the first half-hour after bathing and the cold baths are not popular with women.

The warmth of a hot bath may give some comfort in the immediate puerperium. In a recent survey, over 90 per cent of women reported that bathing had relieved perineal discomfort. However, this observation was uncontrolled and there is no knowing whether a similar proportion of women would have gained relief if they had not bathed at all, or had taken a shower instead. With the current trend towards showers and bidets, rather than baths, in modern maternity hospitals, the question of the difference in comfort resulting from these different approaches is important.

The growing popularity of locally applied herbal substances for relief of perineal pain probably reflects both a belief that 'time honoured natural medicines must be safe' and their increased availability from many retail outlets. No controlled studies of their use have been reported. Witch hazel soaked into gauze swabs or other pads and applied directly to the perineal tissues is commonly recommended for the relief of pain, but a randomized trial showed no evidence that it was any more effective than tap water.

Salt added to bath water is one of the oldest claimed 'remedies' for perineal and other trauma, and is still very popular. The salt is believed to soothe discomfort and to promote healing, although a precise mode of action is unclear. Claims that it has antiseptic or antibacterial properties have not been confirmed. There is no consensus as to the type of salt preparation or quantity which should be used. Recommendations about the quantity range from a heaped tablespoon in a small bath to 3 pounds in 30 gallons of water (about 10 grams/litre). In a large controlled trial, the addition of salt to the bath water had no detectable effect on either perineal pain or the patterns of perineal wound healing.

2.2 Local antiseptics

The addition of antiseptic solutions, particularly Savlon concentrate, to the bath water is also a common practice during the postnatal period. This too was studied in the large trial mentioned above, when women were asked to add Savlon to a daily bath for the first ten days after delivery. There was no evidence that this addition improved symptomatic relief from bathing or that it reduced perineal discomfort. Similarly, a study comparing antiseptic solution with unmedicated tap water for 'jug douching' also showed no differential effect on symptoms, healing, or infection rates.

The use of vaginal creams containing sulphonamides, once recommended for routine use in the postpartum period, has been evaluated in two controlled studies. The results suggested that minor benign cervical abnormalities, such as erosion or ectropion, were less common in the women who had used the sulphonamide creams and that these women reported less use of vaginal douches. Equal proportions of women in each group had resumed sexual intercourse by six weeks after delivery.

2.3 *Local anaesthetics*

Local anaesthetics are commonly applied as sprays, gels, creams, or foams. Double-blind comparisons of local anaesthetics show them to be clearly more effective for relief of perineal pain than placebo. Aqueous 5 per cent lignocaine spray or lignocaine gel appears to be the first choice of agent and formulation.

2.4 *Combinations of local anaesthetics and topical steroids*

Based on the assumption that much of the pain from perineal trauma arises from local oedema and inflammation, a local anaesthetic (pramoxine) has recently been combined with a steroid (hydrocortisone) as a single topical agent. Early uncontrolled studies gave very encouraging results, but two well-controlled studies produced conflicting findings. Whereas the first study reported less pain, better pain relief, and less use of oral analgesia in the pramoxine+hydrocortisone group, the other reported more oedema and a greater use of oral analgesia, particularly after the third day. Wound breakdown was also more common in the actively treated group in this study. As steroids are known to impair wound healing, the latter finding is biologically plausible. It would not seem sensible to use this combination except in the context of further properly controlled trials.

3 Local physiotherapies

3.1 *Relief of pressure on the perineum*

A variety of simple aids may be used during sitting or lying as a means of relieving pressure on the sore perineum. When a mother is resting in bed, a wedge or pillow may be used to support her on her side. These should be covered with a waterproof fabric so that they can be cleaned easily. Rubber or foam rubber rings have been widely advocated in the past, particularly for mothers needing to sit comfortably to feed their babies. Rubber rings have largely been withdrawn from use as they are believed to compress venous return, thereby increasing the risk of thrombosis in women already at higher risk postpartum. The fact that they are no longer supplied in hospital does not prevent many women from buying their own or substituting children's swimming rings. The popularity of this simple measure suggests that it gives relief to many women. Together with the fact that variations of this device are being developed, this suggests that further research would be worthwhile.

3.2 Ultrasound and pulsed electromagnetic energy

Developments in the physical treatment of soft-tissue injuries have recently led to the increased use of electrical therapies for the traumatized perineum. Two such treatments are currently gaining popularity: ultrasound and pulsed electromagnetic energy.

Evidence about the effectiveness of therapeutic ultrasound for other soft-tissue injuries is not wholly consistent, and the precise mode of action is not fully understood. Ultrasound therapy requires constant operator attendance during treatment and hence is costly in physiotherapist's time. The transducer is applied directly to the skin and must be moved during transmission as a safeguard against tissue damage; conduction is aided by a jelly or cream. This gentle movement of the transducer head over the injured tissues may alone provide some therapeutic or psychological benefit.

Similar benefits to those seen following ultrasound therapy have been claimed for pulsed electromagnetic energy on the basis of observational studies of treatment of soft-tissue injury. The interrupted transmission of the energy allows high intensity waves to be used while minimizing local heat. One possible advantage of pulsed electromagnetic energy is its ease of application. It can be transmitted through a sanitary towel, thus avoiding the need for constant operator attendance, although this is not always supported as good practice.

The effects of this therapy on perineal healing have been assessed in two published trials. Maternal reporting of perineal pain before and after treatment revealed no benefit from the active therapy.

A randomized controlled trial has compared ultrasound, pulsed electromagnetic energy, and placebo therapy given in the immediate postpartum period for the treatment of the 'severely traumatized' perineum. Preliminary analyses have not shown any clear differences in the major outcome measures among the groups. About 90 per cent of women in each group, including the placebo group, felt that the treatment had made the pain better, underlining both the necessity for well-designed comparative studies and the power of the placebo effect on the condition.

These preliminary analyses provide no basis for the widespread use of these two expensive modalities in the treatment of perineal problems. Further randomized trials are needed to investigate factors such as dosage and timing of their use. Any continued use of these technologies in the postpartum period should take place within the context of such a randomized trial, which would also provide an opportunity to assess the size of the placebo effect.

3.3 *Pelvic floor exercises*

The usual rationale for advising postnatal exercises of the pelvic floor muscles is the belief that they will reduce the risk of urinary stress incontinence and genital prolapse. In the only large controlled trial in which the effects of postnatal exercises on incontinence rates have been assessed, the rate of incontinence (three months after delivery) among women who had received intensive instruction and reinforcement for pelvic floor exercises was similar to that among women who had received the usual level of information and no special reinforcement. Although no beneficial effects of postnatal exercises on subsequent incontinence rates were detected, women who received the intensive postnatal exercise programme and reinforcement were significantly less likely to have perineal pain at three months, and they were also less likely to be depressed. This latter difference may have been mediated either by the greater attention paid to their progress postpartum or simply because they were in less pain.

4 Treatments taken by mouth

4.1 *Herbal preparations*

A range of herbal preparations for oral use, aimed at relieving perineal symptoms, are available. For example, arnica (leopard's bane), supplied as tablets, and comfrey, as a tablet or a tea, are claimed to reduce bruising. As far as we know, they have never been formally evaluated.

4.2 *Proteolytic enzymes*

Some of the pharmacologically active proteolytic enzymes occur naturally; for example, ananase is an extract of Hawaiian pineapple plants. Three such enzyme preparations (bromolain, chymotrypsin alone, and chymotrypsin + trypsin) have been evaluated in controlled trials. Although the trials individually gave somewhat conflicting results, an overview of the results shows significant decreases in oedema, pain on sitting, and pain on walking by the third day. These results suggest that oral proteolytic agents may have an important effect on perineal discomfort. Nevertheless, the trials vary in terms of quality, prevalence of outcomes, and estimates of treatment effect, and so this conclusion can be only tentative. Firm conclusions must await the results of better-controlled studies.

4.3 *Oral analgesics*

A bewildering choice of pharmacologically active preparations can be taken by mouth to relieve perineal pain. Despite the large number of randomized trials in which these agents have been evaluated, the experimental evidence is not helpful. There are two main reasons for this. First, most trials show that the active preparations are superior to placebo but fail to distinguish clinically important differences between alternative analgesics. Second, many of the drugs included in the trials are no longer commercially available.

A number of factors must be considered when making a choice of oral analgesic preparations. One is the severity of the pain being treated. Another is whether the formulation is likely to cause constipation, which is particularly important to avoid when treating perineal pain. Some oral preparations can cause stomach upset, and this too should be avoided if possible. Whether the drug or drugs are carried in breast milk and, if so, whether this has any potential danger for the baby is a further important consideration. In addition, some drugs have more serious, albeit rare, adverse effects. Finally, the relative costs of the alternative preparations should be taken into account.

On the basis of these criteria, paracetamol (acetaminophen) is probably the drug of choice for mild perineal pain. It has a useful analgesic effect and is largely free of unwanted side-effects. Aspirin is less satisfactory because it can cause gastric irritation, prolongs bleeding time, and poses a potential risk to the baby because of its carriage in breast milk. Of the other non-steroidal anti-inflammatory drugs, ibuprofen would seem to be the most appropriate if an alternative to paracetamol is required for treating perineal pain. Unlike some of the other non-steroidal anti-inflammatory drugs, it appears to be largely free of unwanted side-effects and very little is excreted in breast milk.

The choice of analgesics is less satisfactory when perineal pain is insufficiently relieved by paracetamol or ibuprofen. The opioid dextropropoxyphene may cause dependence and cannot be recommended. One option is to give paracetamol in combination with lower doses of codeine or dihydrocodeine than would be the case if the latter were being used on their own. Although it seems reasonable to combine the two types of analgesia, it is uncertain whether the analgesic effect is greater than with paracetamol on its own, and codeine predisposes to constipation.

If perineal pain does not respond adequately to paracetamol, it seems sensible to consider the additional use of local therapies such as heat and local anaesthetics. If the pain is likely to be associated with local inflammation, a non-steroidal anti-inflammatory agent such as ibuprofen may be helpful. If stronger analgesia is still required there is no obvious first choice. Individuals differ in their susceptibility to different analgesic formulations. Codeine derivatives are less suitable for perineal pain than for other types of pain because they predispose to constipation. For this reason the combinations of paracetamol (acetaminophen) with a stronger opioid analgesic may have a special place for the relief of perineal pain.

5 Conclusions

Cooling with crushed ice, witch hazel or tap water gives short-term symptomatic relief from perineal pain and discomfort. Locally applied anaesthetics such as aqueous 5 per cent lignocaine spray or lignocaine gel are also effective, and their effect may last longer. Adding a steroid to such local anaesthetics may do more harm than good. The addition of salt or antiseptic solution to bath water has no additional effect on perineal pain or healing.

The quality of personal care during the puerperium is likely to be a major determinant of postpartum perineal discomfort. On the basis of currently available evidence, the therapeutic effects noted from physiotherapies, such as therapeutic ultrasound, pulsed electromagnetic energy, and the teaching of postnatal exercises, may derive from the personal attention involved, rather than from the treatment modalities themselves. Further research is needed to investigate factors such as dosage and timing of these treatments.

Paracetamol (acetaminophen) is the oral analgesic of choice for mild perineal pain. If paracetamol in conjunction with the local therapies fails to control the pain, a non-steroidal anti-inflammatory agent such as ibuprofen is a useful alternative. The oral proteolytic enzymes may also be considered for relatively intractable perineal pain, although their effectiveness has still not been clearly established. There is no obvious oral analgesic for more severe pain which is inadequately controlled by paracetamol. The tendency for codeine derivatives to cause constipation makes these drugs less suitable for perineal pain than for pain in other sites.

Until recently, the prevention and treatment of perineal pain following childbirth using approaches other than systemic analgesia have been the subject of little formal evaluative research. Yet, postpartum perineal pain is so common that alternative strategies can be readily compared in randomized controlled trials. Such trials are needed if more effective treatments for this common problem are to be developed.

This chapter is derived from the chapter by Adrian Grant and Jennifer Sleep (39) in EFFECTIVE CARE IN PREGNANCY AND CHILDBIRTH.

References to primary sources and more complete data for statements made in this chapter can be found in the source chapter and/or in the following reviews from the *Cochrane pregnancy and childbirth database*:

Kaufman, K.
— Adding Savlon concentrate to bath water for perineal trauma. Review no. 03796.
— Adding salt to bath water for perineal trauma. Review no. 03691.
— Intensive postnatal pelvic floor exercises. Review no. 05563.

— Warm vs cold sitz baths. Review no. 05564.
— Local heat vs local cold for perineal injury. Review no. 05565.
— Ultrasound treatment of perineal pain. Review no. 05566.
— Pulsed electromagnetic energy for perineal pain. Review no. 05567.
— Electromagnetic energy vs ultrasound for perineal pain. Review no. 05568.
— Pramoxine/hydrocortisone for perineal pain. Review no. 05569.
— Pramoxine/hydrocortisone vs local anaesthetic for perineal pain. Review no. 05570.
— Pramoxine/hydrocortisone vs ice for perineal pain. Review no. 05571.
— Pramoxine/hydrocortisone vs witch hazel for perineal pain. Review no. 05572.
— Witch hazel vs ice for perineal pain. Review no. 05573.
— Local anaesthetic for perineal pain. Review no. 05575.
— Alcoholic vs aqueous lignocaine for perineal pain. Review no. 05577.
— Lignocaine vs cinchocaine for perineal pain. Review no. 05576.
— Oral proteolytic enzymes for perineal trauma. Review no. 03204.

48 Breast symptoms in women who are not breastfeeding

1 Introduction   2 Non-pharmacological approaches
3 Pharmacological approaches   3.1 *Sex hormones*
3.2 *Bromoergocriptine (bromocriptine)*   3.3 *Cabergoline*
3.4 *Other drugs*   4 Conclusions

1 Introduction

Women may not breastfeed their babies after childbirth for a variety of reasons, ranging from personal choice to stillbirth. The decision not to breastfeed results in considerable breast pain and engorgement during the days after childbirth until lactation becomes spontaneously suppressed. A number of approaches have been adopted in attempts to hasten the suppression of lactation and reduce the symptoms that accompany it.

2 Non-pharmacological approaches

Until forty years ago tight binding of the breasts and fluid restriction were the most common approaches to the suppression of lactation. They appear to be equally effective and are still the most frequently adopted of the non-pharmacological methods. Almost no formal investigation of these methods has been undertaken, although the results of one small randomized trial showed that breast pain was less frequent in women who also restricted their fluid intake than in those just wearing brassieres.

Non-pharmacological methods of inhibiting lactation have been implicitly compared with pharmacological methods in trials of different drug agents. In one study, in which breast binders were compared with bromoergocriptine, the drug was shown to be more effective in controlling symptoms during the first week postpartum, by the second week there was no difference, and by the third week symptoms were more frequent in women who had received bromoergocriptine. Therefore, it is probable that there may be short-term disadvantages, but longer-term benefits, of non-pharmacological approaches to suppress lactation.

3 Pharmacological approaches

3.1 *Sex hormones*

The use of stilboestrol reduces the incidence of continuing lactation, breast pain, and engorgement during the first week postpartum, but there are long-term costs associated with these short-term benefits. More women given stilboestrol than women given placebo required additional treatment after discharge from hospital, and four times as many in the stilboestrol-treated group reported abnormal vaginal bleeding after the end of treatment.

Trials comparing the effects of different oestrogens show that stilboestrol suppresses lactation and pain more effectively than quinoestrol. Chlorotrianisene, another stilboestrol analogue, has also been shown to reduce lactation, breast pain, and engorgement in placebo-controlled trials.

Various combinations of an oestrogen and testosterone have been shown to have dramatic short-term effects on lactation, breast pain, and breast engorgement. The only trial in which long-term effects have been reported shows a recurrence of pain and engorgement by the end of the second week.

The risk of thromboembolic complications is increased with oestrogen use, although the absolute level of risk is low. Withdrawal bleeding after hormonal treatment is reported by about 15 per cent of women, irrespective of the type of drug used.

3.2 *Bromoergocriptine (bromocriptine)*

The short-term beneficial effects of bromoergocriptine, compared with placebo, have been well established. The results of well controlled trials show that it greatly decreases lactation, breast pain, and engorgement in the first postpartum week. Only limited data are available on effects during the second week, but the beneficial effects by this time are much less dramatic. Rebound lactation is common. Comparisons show that bromoergocriptine suppresses lactation and breast engorgement in the first postpartum week more effectively than stilboestrol.

No major adverse effect has been reported in women treated with bromoergocriptine for suppression of lactation. The reported frequency of nausea is less than 5 per cent.

3.3 *Cabergoline*

Cabergoline has recently been compared with a placebo and with bromocriptine in three trials. The results indicate that a single dose of cabergoline is as effective and results in fewer side-effects and less rebound lactation than a twice daily dose of bromocriptine for 14 days. Cabergoline should now be the drug of choice for the pharmacological suppression of lactation, although more infirmation is needed about the best dose and timing of administration.

3.4 *Other drugs*

Pyridoxine has been compared with placebo in three studies; the few data available show little effect on continued lactation.

In the early 1960s, the effects on lactation and breast symptoms of spraying synthetic oxytocin intranasally was studied in at least three trials, one of which is unpublished. None of these studies provided any evidence that the treatment was effective.

4 Conclusions

The available evidence suggests that physical methods of lactation suppression, like breast binding, are associated with more pain in the first week after delivery than pharmacological methods, but they appear to be at least, if not more, effective in the longer term. Fluid restriction may further reduce symptoms among women who do not wish to breastfeed. Women should be informed of these relative advantages and disadvantages when a method to suppress lactation is chosen.

If they decide to use one of the pharmacological approaches, the available evidence suggests that cabergoline should be the drug of choice.

Cabergoline and newer drugs that may be developed should be compared formally with physical methods of suppressing lactation in controlled trials with adequate sample sizes and duration of follow-up. Women's views of the relative merits and disadvantages of the alternative methods should be an essential element in the evaluation, and more serious attention should be given to documenting the frequency of short- and long-term adverse reactions.

This chapter is derived from the chapter by Fabio Parazzini, Flavia Zanaboni, Alessandro Liberati, and Gianni Tognoni (82) in EFFECTIVE CARE IN PREGNANCY AND CHILDBIRTH.

References to primary sources and more complete data for statements made in this chapter can be found in the source chapter and/or in the following reviews from the *Cochrane pregnancy and childbirth database*:

Renfrew, M.J.
— Breast binder vs fluid limitation for lactation suppression. Review no. 03878.
— Brassiere vs fluid limitation for lactation suppression. Review no. 03879.
— Breast binder vs brassiere for lactation suppression. Review no. 03880.
— Stilboestrol for lactation suppression. Review no. 03379.
— Quinoestrol for lactation suppression. Review no. 03380.
— Stilboestrol vs quinoestrol for lactation suppression. Review no. 03386.
— Chlorotrianisene for lactation suppression. Review no. 03382.
— Oestrogen/testosterone combination for lactation suppression. Review no. 03381.
— Quinestrol vs oestrogen/testosterone for lactation suppression. Review no. 06473.
— Bromocriptine for lactation suppression. Review No. 03384.
— Bromocriptine vs breast binders for lactation suppression. Review no. 03887.
— Bromocriptine vs stilboestrol for lactation suppression. Review no. 03387.
— Bromocriptine vs antioestrogens for lactation suppression. Review no. 05727.
— Bromocriptine vs prostaglandin E_2 for lactation suppression. Review no. 05729.
— Bromocriptine vs jasmine flowers for lactation suppression. Review no. 05730.
— Bromocriptine vs chlorotrianisene for lactation suppression. Review no. 05717.

— Bromocriptine vs oestrogen/testosterone combinations for lactation suppression. Review no. 05718

— Bromocriptine vs other dopaminergics for lactation suppression. Review no. 05726.

— Pyridoxine for lactation suppression. Review no. 03385.

— Intranasal oxytocin for lactation suppression. Review no. 03888.

— Cabergoline (ergot derivative) for lactation suppression. Review no. 05723.

— Cabergoline vs bromocriptine for lactation suppression. Review no. 06471

— Antioestrogen vs breast binder and brassiere for lactation suppression. Review no. 05719.

— Antioestrogens for lactation suppression. Review no. 05728.

— Chlormezanone for lactation suppression. Review no. 05720.

— Prostaglandin E_2 for lactation suppression. Review no. 05721.

— Chlormezanone vs stilboestrol for lactation suppression. Review no. 05722.

— Proteolytic enzymes for lactation suppression. Review no. 05724.

— Antihistamines for lactation suppression. Review no. 05725.

— Pyridoxine vs bromocriptine for lactation suppression. Review no. 06474.

— Lisuride $300\mu g$ vs $600\mu g$ for lactation suppression. Review no. 06475.

49 Loss and grief in the perinatal period

1 Introduction

There have been great changes in attitudes towards neonatal illness, childhood impairment, and perinatal death over the past few decades. Improvements in perinatal and infant mortality rates have been accompanied by ever-increasing expectations by parents that their children will be born safely and survive. When things do go wrong it comes as a great shock. Parents may suffer much more than the sense of the loss of the healthy baby that they had anticipated. They may also experience a loss of their faith in modern medicine and doctors, and of belief in their own ability to produce a normal baby. In a similar way it is often shattering for the caregivers to witness the apparent failure of their skills.

To compound the difficulties further, in industrialized societies we have lost our day-to-day familiarity with death and bereavement, and with the mourning rituals that used to play an important part in meeting the psychological needs of the bereaved. Thus we have become poorly equipped to cope with this tragic situation.

Grieving will follow the birth of an ill or impaired baby, as well as after a baby's death. The two components of normal grief are the acute symptoms (episodes of restlessness, angry pining, and anxiety) set against a background disturbance consisting of chronic low mood, loss of purpose in life, social withdrawal, impaired memory and concentration, and disturbances of appetite and sleep. These symptoms occur as bereaved people go through the process of coming to terms with the reality of their loss, and of psychologically withdrawing from their relationships with the impaired or dead child. This process, which is necessary to let them continue with their own lives in a positive manner, may take months or years to complete. A successful outcome depends on the personalities and life experiences of the people concerned, on the circumstances of the impairment or death, and on the effectiveness of the supportive network surrounding them.

2 Perinatal loss

2.1 *Illness and impairment*

The reaction of parents to a newborn baby who is gravely ill or impaired is a form of grief reaction to the loss of the healthy child that they had expected. The initial phase is marked by shock and panic ('I can't look after a handicapped child'), denial ('He's not my baby'), grief, guilt, and anger. This is followed by a phase of bargaining ('I will look after him if he can be taught to be clean and dry') and finally acceptance, when parents cope with the reality of the situation. Some parents fail to adapt and remain in a state of chronic sorrow. It is important for caregivers to form an effective alliance with the parents, on which plans for care can be based.

In addition to a grief reaction, most parents of ill or impaired babies suffer high levels of anxiety which appear to be increased by contact with the baby. This does not mean that separation of mother and baby is to be recommended. On the contrary, close contact between parents and the baby, with support from caregivers, will allow the parents to form a relationship with a real child. If the mother planned to breast-feed, she can be encouraged to initiate lactation by expressing her milk. Providing this milk for her baby may help in the development of her relationship with the child and demonstrate her own unique role in caring for the baby.

Apart from the emotional stress of the situation, parents have the physical and financial stress of visiting their baby in hospital, particularly if the baby has recurrent medical crises and needs care over a long period of time. Some parents withdraw emotionally and physically from their baby before the medical staff have given up hope of the baby's survival. This is termed 'anticipatory mourning'. It can be precipitated by giving an excessively gloomy prognosis, or even by a casual remark indicating a possible bad prognosis. It carries with it a risk of rejection if the baby does eventually survive.

2.2 *Perinatal death*

Grief after perinatal death is no different from that following the death of any loved person. However, there are some special features to be considered.

Bereaved parents may feel anxious and angry and direct blame at their caregivers, other members of the family, or themselves. This may be due in part to the suddenness of the death, and is probably compounded when there is no 'scientific' explanation for what has gone wrong. Parents desperately seek for a cause of the baby's death. It is easier for those who have an explanation, such as malformations or extreme immaturity. 'Empty arms' is another common and distressing symptom after the phase of numbness has passed. Mothers are frequently tormented by hearing their dead baby cry. Some bereaved parents experience negative feelings towards other babies and are fearful of losing control, while others long to hold a baby — any baby — however painful this might be. Many mothers do not expect to lactate once the baby has died and find the fact that they do upsetting. Most mothers experience a great loss of self-esteem and a sense of being failures, both as women and as wives.

Most parents will not have been bereaved before, and have difficulty coping with the complicated registration and funeral procedures. Many are unprepared for the emotional turmoil of their grief reaction, and may feel they should be 'over it' after a few weeks. This view may be reinforced by well-meaning friends, relatives, and even medical practitioners, who may advise the couple to go ahead with another pregnancy long before they have recovered sufficiently from their loss. There is evidence that fathers recover from their grief more quickly than mothers. This in itself may lead to problems with their relationship, particularly if the couple are not used to sharing their feelings, or if one of them is blaming the other for the baby's death. Sexual and marital difficulties are common.

Another difficult area is the reaction of other young children in the family to the loss of the baby. They may be confused about what has happened to the baby, and even feel responsible for the disappearance. Behavioural changes are common; they may take the form of overactivity, naughtiness, regression, and problems at school, as well as other emotional problems. These reactions are usually fairly short lived (a few weeks or months) unless the emotional state of the parents is such that there is an absence of normal warmth in family relationships for an extended period, or if serious difficulties develop in the relationships between the mother and her living children.

It may be particularly difficult to work through one's grief when a baby is stillborn. There is no real object to mourn. The baby has never lived outside the womb and there are no memories to help. The problems are accentuated if the stillborn baby is rapidly removed from the delivery room before the parents have a chance to see or hold him or her, and if the hospital, for whatever reason, takes over the funeral arrangements without involving the parents.

Long-term follow up studies show that a significant proportion, up to a fifth of women interviewed, still suffer from serious psychological symptoms for years after losing a baby. Although it is not possible to identify with great confidence those most at risk of developing problems, the most frequently reported markers are not seeing or holding the baby, having an unsupportive partner or social network, and embarking immediately on another pregnancy.

There may be problems with parental relationships when babies have been conceived too quickly after a loss. If the dead child has been incompletely mourned before the start of a new pregnancy, mourning may be postponed until after the birth of the baby, when it can reappear as 'postpartum depression'. The new baby's identity can become confused with that of the idealized baby, causing great emotional problems. The new child may never be able to live up to the parents' expectations, and may become the focus of any unresolved anger that the parents have as a result of their loss. The survivor of a twin pregnancy may be involved in similar problems if the dead twin is not properly mourned at the time.

3 Care by hospital staff

The maternity unit staff have a vital role to play in the care of bereaved parents. A programme of care should encourage the parents to see, hold, and name their baby, and to hold a funeral. Arrangements should be made for them to see senior obstetrical, midwifery, and paediatric staff to discuss what went wrong, to obtain genetic and obstetrical counselling, and to receive the autopsy results. Providing informed compassionate care will help the recovery process after a perinatal death.

Effective care for most families can and should be provided by the maternity unit staff. These professionals and the family doctor are in the ideal position to help bereaved families by facilitating the establishment of normal grieving from the start, thus preventing abnormal reactions. Special bereavement counselling services are not often required.

3.1 Communication

Good care hinges upon good communication. Parents frequently comment on communication failures when describing their experiences. Caregivers must give bereaved parents opportunities to talk about the loss of their baby and, even more importantly, listen sympathetically to their expressions of grief. The senior obstetrical, midwifery, or paediatric staff must help parents with their search for a cause of death and create opportunities for discussing this with them.

Seeing both parents together helps to strengthen their relationship as they share the experience of their baby's loss, and prevents misunderstandings and inconsistencies in explanation. Arranging for the same caregivers to attend regularly to the parents also helps this. Any information given in the first few days of the loss will probably need to be repeated later, as the initial shock passes. A follow-up interview a few weeks later seems to be the best way of overcoming this difficulty.

Good communication between professionals about the loss of the baby is necessary to prevent painful situations, such as a member of staff being unaware that the baby has died and asking the mother about the baby. The primary care team should be informed about the baby's loss immediately, so that they can make contact with the family as soon as, or even before, the mother is discharged. Parents may want the support of their own religious adviser, and the hospital should check on this and contact him or her if required.

3.2 Immediate and early care when the baby is dead or dying

For mourning to begin, parents must be enabled to face their fear of death and dying so that they can experience the painful reality of their loss. This involves encouraging them to have as much contact as possible with their baby, both before and after death. It is particularly important for parents of a stillborn baby to see, hold, and name their baby.

When an intrauterine death is suspected, the fears for the baby's condition should not be denied, but shared with the parents, together if at all possible. If the mother is at the clinic, efforts should be made to contact her partner or a friend or family member so that she is not left to travel home alone and unsupported. The technicians in the ultrasound scanning room have an important role to play when the confirmatory scan is done. They need to be sympathetic to the situation and to allow the mother to be accompanied by anyone she chooses (see Chapter 27).

Most women are frightened at the prospect of delivering a dead baby, as well as shocked by their loss. It helps if caregivers take time to explain carefully what will happen, that adequate pain relief will be available, and what the baby will look like at delivery. This is usually successful in overcoming any reluctance that the parents may have about seeing or holding their baby. It may help if a malformed or macerated baby is wrapped up before being first shown to the parents.

A few parents will not be able to cope with seeing and holding the baby at the time of delivery. A photograph should be taken and kept in the medical notes for possible use later, and further opportunities for seeing the baby should be offered to parents over the next few days, as they often change their minds. Photographs and other mementos of the baby such as a lock of hair, a piece of the umbilical cord, or a print of the baby's hand or foot are important, as they provide tangible evidence of the reality of the baby's existence and loss. They should be available for parents as keepsakes, if they wish.

When the baby lives long enough to be transferred to a neonatal unit, it is again important for caregivers to keep parents as fully informed as possible about the baby's condition and to encourage them to share in the care. Photographs of the baby are helpful, particularly for fathers to keep at home or if the mother is too unwell to visit the unit. In a randomized trial of the use of routine polaroid photographs of sick neonates in the first week of life, there was a significant increase in visiting by the parents of the photographed babies compared with the non-photographed group.

When the baby's condition is known to be terminal, it is important to involve the parents in the decision to cease life support, and then to let them take their dying baby in their arms, if at all possible, free of all equipment that has been necessary until then. In describing this, authors quote parents saying such things as, 'It was all I could do for her to hold her in my arms as she died'. Some parents may wish to take the baby home to die; they should be supported in this decision. Feelings of guilt about removing the baby from the life-support system have not been reported.

Many parents like to help with the laying out of the baby's body, and this should be encouraged. Often they have selected special clothes or toys to be placed in the coffin with the baby. Supporting the parents' contact with the reality of the death of their baby in these ways will facilitate their grief reaction. They will need privacy to express their grief, and this should be provided, however busy the unit happens to be.

The choice of site for the aftercare of the mother is important, as mothers differ in their requirements at this time. Some want to be on their own, far away from the sound of babies crying; others long to return to familiar faces on the ward. It is helpful if as much flexibility as possible is offered to them and if, at least for the first night, partners are allowed to remain with them. Ideally, the hospital should provide a couch in the mother's room so that the parents can share their grief together. Lactation and help with its suppression is an important issue for the mother whose baby has died. If the mother is physically fit to return home immediately, and wishes to do so, it is important to ensure that she has a supportive network of family, friends, and professionals before discharging her.

3.3 *Autopsy*

Consent for an autopsy and chromosome studies should always be requested after a perinatal death. These investigations may provide information about the cause of death, help parents with their grief, and assist the planning of future pregnancies. Most parents agree to an autopsy, although it is often a painful decision for them. Having consented, parents cherish great hopes that the results will provide answers to their questions about why the baby died. It is important that they receive the results in a form that makes sense to them. The best person to do this would be a senior member of staff who can interpret the pathological findings.

3.4 Death registration and funeral arrangements

Knowledge of the legal procedures required when a baby dies or is stillborn is fundamental to good care. It is necessary to be familiar with the registration and funeral arrangements operating in one's own locality, as these are often complicated and baffling for parents still suffering from the shock of their baby's death. Religious practices vary greatly as well, and an awareness of these and sensitivity to the wishes of individual parents is crucial. The funeral may involve considerable expense; helping those in financial difficulties, besides encouraging them to attend, are therapeutic aspects of care.

Many units have prepared leaflets outlining their own procedures and giving helpful advice for parents.

3.5 Follow-up

Most mothers will be discharged home within a few days of their baby's death, still too shocked by it to grasp properly what has happened and why. Careful and supportive follow-up is extremely important. Parents should be able to contact the staff who cared for them by telephone after they leave the hospital. Some units are able to offer home visits by their social worker. An appointment should be made for both parents to see a senior member of staff two to six weeks later as soon as the chromosome studies and autopsy results are available, and some form of perinatal mortality conference has taken place. Caregivers should be aware that returning to the hospital is likely to be traumatic for the parents.

The next pregnancy will inevitably be an extremely anxious time, and the mother will need extra support during it and in the first few months after the birth.

4 Care in the community

4.1 Health professionals

The general practitioner, health visitor, community midwife, and other primary health care workers will form the professional supportive network once the mother has been discharged home. These professionals can help by continuing to support parents in the expression of their grief and putting them in touch with local support groups for parents who have lost a child.

The general practitioner or midwife can watch for signs of pathological grief reactions and refer the parents for specialist help if necessary. These pathological reactions can take the form of an inhibited reaction, with no sign of any sense of loss, or a prolonged reaction, with unremitting symptoms of depression, severe anxiety, or the appearance of psychosomatic illness. There may be drug or alcohol abuse.

Unremitting anger is another feature of a pathological grief reaction. General practitioners and midwives may need to deal with anger focused on the maternity unit. To do so, they must ensure that the parents have good relationships with the obstetrical, midwifery, and paediatric staff, and are fully informed about the course of events which led to the baby's loss. The parents may blame the general practitioner or midwife as well as the maternity unit. When this happens, it is essential that he or she meets with the family as soon as possible so that they can ventilate their feelings, and, hopefully, re-establish their relationship. Many parents remain angry simply because they were denied any compassionate response to their situation: no one said 'I'm so sorry your baby died.'

The general practitioner or health visitor will probably be the person to whom the family will turn to for help in coping with the reactions of their other children to the baby's death. Parents may need help to allow their children to express their feelings about so painful a subject. It must be remembered that young children will use play as a vehicle for doing so. Explaining death to children under five years of age is difficult because they are not yet able to grasp the concept. Even simple statements like 'The baby's gone' will be interpreted literally and lead to questions about where the baby has gone, and when a visit can be made. The parents will need to add more information as the child's capacity for understanding develops.

4.2 Self-help groups

Self-help can be effective in providing the right kind of support for parents facing many different kinds of problems, and perinatal bereavement is no exception. However, it is important that the people running the group have recovered sufficiently from their own loss to be able to help others, and that they have access to professionals for help and advice as and when necessary. Parents can benefit from sharing their experiences together, from discovering that they are not alone in their suffering, and from learning that time does help to heal the wounds. Not everyone can cope with group support, and it is unwise to rely on a local self-help group to meet the needs of all bereaved families. While it is invaluable to give parents the telephone number or address of a local contact, this should not replace follow-up by the hospital staff, general practitioner, and health visitor.

5 The role of specialist counsellors

Prevention of prolonged grief reactions through appropriate care is not always successful. About one in five families will show reactions that are detrimental to their health and are likely to be accompanied by problems in family relationships. The help of specialist counsellors trained in grief work will be needed in these situations, either to advise other colleagues giving care or to take over responsibility for care themselves.

The treatment required is often protracted, and antidepressant drugs and psychiatric surveillance may be necessary for severe depressive symptoms. Child and family psychiatrists may be particularly helpful in dealing with the relationship problems within families. Specialist counsellors can also try to promote normal grieving in parents most at risk of pathological reactions.

Specialist counsellors can also be useful in supporting the staff of the unit (through regular staff meetings, case discussions, or training sessions) and can offer help and advice to self-help groups. The training of caregivers in the care of families who lose their baby, or who are faced with a baby with a severe impairment, deserves as much emphasis as the development of their technical expertise.

6 Conclusions

Much can be done to help a bereaved family cope with their loss and recover from their grief. The extent to which this is accomplished will depend on the importance that is attached to training in this area and on the attitudes of individual professionals, both in the maternity unit and in the community. Parents need the opportunity to have contact with their ill child and support for the mother to lactate if she wishes. Parents of stillborn or dying babies should similarly to encouraged to touch and hold their baby. Photographs of their baby will provide tangible evidence of the reality of the baby's existence and loss. Giving photographs to parents of sick babies has also been shown to increase their visits to their babies in the first week of life.

The practical aspects of death registration and funeral arrangements for babies should receive careful attention. Time must be spent, listening as well as talking, with parents whose baby has died or is impaired. The senior members of staff need to play a central role in caring for the parents, sharing their experience and expertise with more junior caregivers. The primary health care team must accept the role of monitoring and supporting the parents during the ongoing bereavement process.

The provision of adequate support will almost certainly lead to improved rapport with grieving families. It will help professionals to cope better with their own grief because they feel more able to help. Most importantly, it will help families to emerge from their grief able to resume normal functioning.

This chapter is derived from the chapter by Gillian Forrest (85) in EFFECTIVE CARE IN PREGNANCY AND CHILDBIRTH.

References to primary sources and more complete data for statements made in this chapter can be found in the source chapter and /or in the following review from the *Cochrane pregnancy and childbirth database*:

Hodnett, E.D.
— Support from caregivers for distressed postnatal women/couples. Review no. 07673.

SYNOPSIS

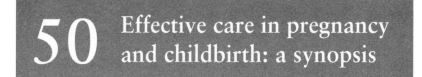

The underlying thesis of this book is that evidence from well controlled comparisons provides the best basis for choosing among alternative forms of care in pregnancy and childbirth. This evidence should encourage the adoption of useful measures and the abandonment of those that are useless or harmful.

In this final chapter we have tried to summarize the main conclusions reached in earlier chapters. This summary takes the form of six tables which list the following:

(1)   beneficial forms of care;

(2)   forms of care that are likely to be beneficial;

(3)   forms of care with a trade-off between beneficial and adverse effects;

(4)   forms of care of unknown effectiveness;

(5)   forms of care that are unlikely to be beneficial;

(6)   forms of care that are likely to be ineffective or harmful.

Tables 1 and 6 are based on clear evidence from systematic reviews of randomized trials. Tables 2 and 5 are based on information from reviews of controlled trials or good observational evidence, but for which the conclusions cannot be as firmly based as those for Tables 1 and 6. Table 3 lists forms of care with both beneficial and adverse effects, which women and caregivers should weigh according to their individual circumstances and priorities; and Table 4 lists forms of care for which there are insufficient data or data of inadequate quality on which to base a recommendation.

We have been explicit about our criteria for choosing which table to use for each intervention, but there is inevitably some subjectivity in our choice. We worked from two basic principles: firstly, that the only justification for practices that restrict a woman's autonomy, her freedom of choice, and her access to her baby, would be clear evidence that these restrictive practices do more good than harm, and secondly, that any interference with the natural process of pregnancy and childbirth should also be shown to do more good than harm. We believe that the onus of proof rests on those who advocate any intervention that interferes with either of these principles.

A tabulated summary such as this is necessarily selective. Nuances discussed in the chapters cannot find full expression in summary tables. Nevertheless, we hope that the explicit form in which these conclusions have been stated will be useful, and that the advantages of this summary approach will outweigh its drawbacks.

The inclusion of a particular form of care in Tables 1 or 2 does not necessarily imply that it should always be adopted in practice. Research based on the study of groups may not always apply to individuals, although it should be relevant to guide broad policies of care. Forms of care listed in Tables 5 and 6 may still be useful in particular circumstances, although, once again, they should be discouraged as a matter of policy. Practices listed in Table 3 will require careful consideration by the individuals concerned, while those in Table 4 should usually be avoided except in the context of trials to better evaluate their effects.

Some of the conclusions that we have reached will be controversial, but they must be judged in the light of the methods used by our colleagues and ourselves to assemble and review the evidence on which they are based. While we have made great efforts to ensure that the data presented are comprehensive and accurate, it is possible that errors and misinterpretations have crept in. We conclude by reiterating the invitation extended to readers in our first edition, to bring omissions and mistakes to our attention for inclusion and correction in The *Cochrane database of systematic reviews* and in later editions of this book.

Table 1   Beneficial forms of care

Effectiveness demonstrated by clear evidence from controlled trials

	Chapter

BASIC CARE

Support for socially disadvantaged mothers to improve child care	3
Women carrying their case notes during pregnancy to enhance their feeling of being in control	3
Pre-and periconceptional folic acid supplementation to prevent recurrent neural tube defects	5,6
Folic acid supplementation (or high folate diet) for all women contemplating pregnancy	5,6
Programmes (particularly behavioural strategies) to assist stopping smoking during pregnancy	5
Balanced energy and protein supplementation of diet when supplementation is required	6
Vitamin D supplementation for women with inadequate exposure to sunlight	6
Iodine supplementation in populations with a high incidence of endemic cretinism	6

SCREENING

Doppler ultrasound in pregnancies at high risk of fetal compromise	12

PREGNANCY PROBLEMS

Antihistamines for nausea and vomiting of pregnancy if simple measures are ineffective	13
Local imidazoles for vaginal candida infection (thrush)	13
Local imidazoles instead of nystatin for vaginal candida infection (thrush)	13
Postpartum adminstration of anti-D immunoglobulin to rhesus-negative women with a rhesus-positive fetus	18
Administration of anti-D immunoglobulin to rhesus-negative women at 28 weeks of pregnancy	18
Antibiotic treatment of asymptomatic bacteriuria	19
Antibiotics during labour for women colonized with group B streptococcus	19
Tight as opposed to too strict or moderate control of blood sugar levels in diabetic women	20

Table 1 *continued*

Beneficial forms of care

CHILDBIRTH

PROBLEMS DURING CHILDBIRTH

TECHNIQUES OF INDUCTION AND OPERATIVE DELIVERY

CARE AFTER CHILDBIRTH

Table 1 | 393

Table 1 *continued*

Beneficial forms of care

	Chapter
Consistent support for breastfeeding mothers	46
Personal support from a knowledgeable individual for breastfeeding mothers	46
Unrestricted breastfeeding	46
Local anaesthetic sprays for relief of perineal pain postpartum	47
Cabergoline instead of bromocriptine for relief of breast symptoms in non-breastfeeding mothers	48

Table 2 Forms of care likely to be beneficial

The evidence in favour of these forms of care is not as firmly established as for those in Table 1

	Chapter

BASIC CARE

Adequate access to care for all childbearing women	3
Social support for childbearing women	3
Financial support for childbearing women in need	3
Legislation on paid leave and income maintenance during maternity or parental leave	3
Midwifery care for women with no serious risk factors	3
Continuity of care for childbearing women	3
Antenatal classes for women and their partners who want them	4
Advice to avoid excessive alcohol consumption during pregnancy	5
Avoidance of heavy physical work during pregnancy	5

SCREENING

Selective use of ultrasound to answer specific questions about fetal size, structure, or position	8
Selective use of ultrasound to assess amniotic fluid volume	8
Selective use of ultrasound to estimate gestational age in first and early second trimester	8,9
Ultrasound to facilitate intrauterine interventions	8,9
Ultrasound to determine whether the embryo is alive in threatened miscarriage	8,14
Ultrasound to confirm suspected multiple pregnancy	8,17
Ultrasound for placental location in suspected placenta praevia	8,21
Early second trimester amniocentesis to identify chromosomal abnormalities in pregnancies at risk	9
Genetic counselling before prenatal diagnosis	9
Transabdominal instead of transcervical chorion villus sampling	9
Regular monitoring of blood pressure during pregnancy	10
Testing for proteinuria during pregnancy	10
Uric acid levels for following the course of pre-eclampsia	10
Fundal height measurements during pregnancy	12

Table 2 | 395

Table 2 *continued*
Forms of care likely to be beneficial

	Chapter

PREGNANCY PROBLEMS

Antacids for heartburn of pregnancy if simple measures are ineffective	13
Bulking agents for constipation if simple measures are ineffective	13
Local metronidazole for symptomatic trichomonal vaginitis after first trimester	13
Antihypertensive agents to control moderate to severe hypertension in pregnancy	15
Antithrombotic and antiplatelet agents to prevent pre-eclampsia	15
Anticonvulsant agents for eclampsia	15
Screening all pregnant women for blood group isoimmunization	18
Anti-D immunoglobulin to rhesus-negative women after any bleeding episode during pregnancy	18,21
Anti-D immunoglobulin to rhesus-negative women after any intrauterine procedure	18
Anti-D immunoglobulin to rhesus-negative women sustaining abdominal trauma	18,21
Intrauterine transfusion for a severely affected isoimmunized fetus	18
Routine screening for and treatment of syphilis in pregnancy	19
Rubella vaccination of seronegative women postpartum	19
Screening for and treatment of *Chlamydia* in high prevalence populations	19
Caesarean section for active herpes (with visible lesion) in labour with intact membranes	19
Prepregnancy counselling for women with diabetes	20
Specialist care for pregnant women with diabetes	20
Home instead of hospital glucose monitoring for pregnant women with diabetes	20
Ultrasound surveillance of fetal growth for pregnant women with diabetes	20
Allowing pregnancy to continue to term in otherwise uncomplicated diabetic pregnancies	20
Careful attention to insulin requirements postpartum	20

Table 2 *continued*

Forms of care likely to be beneficial

CHILDBIRTH

Table 2 | 397

Table 2 *continued*

Forms of care likely to be beneficial

	Chapter
Presence of a companion on admission to hospital	29
Giving women as much information as they desire	29
Change of mother's position for fetal distress in labour	30
Intravenous betamimetics for fetal distress in labour to 'buy time'	30
Woman's choice of position for the second stage of labour or giving birth	32
Oxytocics to treat postpartum haemorrhage	33
Intramyometrial prostaglandins for severe postpartum haemorrhage	33

PROBLEMS DURING CHILDBIRTH

	Chapter
Regular top-ups of epidural analgesia instead of top-ups on maternal demand	34
Maternal movement and position changes to relieve pain in labour	34
Counter-pressure to relieve pain in labour	34
Superficial heat or cold to relieve pain in labour	34
Touch and massage to relieve pain in labour	34
Attention focusing and distraction to relieve pain in labour	34
Music and audio-analgesia to relieve pain in labour	34
Epidural instead of narcotic analgesia for preterm labour and birth	34,37
Amniotomy to augment slow or prolonged labour	35
Continuous subcuticular suture for perineal skin repair	36
Primary repair of episiotomy breakdown	36
Delivery of a very preterm baby in a centre with adequate facilities to care for immature babies	37,44
Presence of a paediatrician at a very preterm birth	37,44
Trial of labour after previous lower-segment caesarean section	38
Trial of labour after more than one previous lower-segment caesarean section	38
Use of oxytocin when indicated after previous caesarean section	38
Use of epidural analgesia in labour when needed after previous caesarean section	38

Table 2 *continued*

Forms of care likely to be beneficial

Table 2 | 399

Table 2 *continued*

Forms of care likely to be beneficial

	Chapter
Encouraging parental contact with a dying or dead baby	49
Providing parents with prompt accurate information about a severely ill baby	49
Encouraging autopsy for a dead baby and imparting results to parents	49
Help with funeral arrangements for a dead baby	49
Self-help groups for bereaved parents	49
Specialist counsellors for parents with prolonged grief reactions	49

Table 3 Forms of care with a trade-off between beneficial and adverse effects

Women and caregivers should weigh these effects according to individual circumstances and priorities

	Chapter

BASIC CARE

Continuity of caregiver for childbearing women	3
Legislation restricting type of employment for childbearing women	3

SCREENING

Formal systems of risk scoring	7
Routine early ultrasound	8
Chorion villus sampling versus amniocentesis for diagnosis of chromosomal abnormalities	9
Serum alpha-fetoprotein screening for neural tube defects	9
Routine fetal movement counting to improve perinatal outcome	12

PREGNANCY PROBLEMS

Screening for toxoplasmosis during pregnancy	19
Corticosteroids to promote fetal maturation before preterm delivery in diabetic women	20,25
Induction of labour for prelabour rupture of membranes at term	23
Betamimetic drugs to delay preterm delivery for implementation of effective measures	24
Oral betamimetics to maintain labour inhibition	24
Cervical cerclage for women at risk of preterm birth	24
Betamimetic drugs to stop preterm labour	24
Expectant care versus induction of labour after fetal death	27

CHILDBIRTH

Continuous EFM plus scalp sampling versus intermittent auscultation during labour	30
Mid-line versus mediolateral episiotomy, when episiotomy is necessary	32
Prophylactic oxytocics in the third stage of labour	33
Active versus expectant management of third stage of labour	33

Table 3 | 401

Table 3 *continued*

Forms of care with a trade-off between beneficial and adverse effects

Table 4 Forms of care of unknown effectiveness

There are insufficient or inadequate quality data upon which to base a recommendation for practice

	Chapter
BASIC CARE	
Social support for high-risk women to prevent preterm birth	3,24
Formal preconceptional care for all women	5
Fish oil supplementation to improve pregnancy outcome	6,15
Prostaglandin precursors to improve pregnancy outcome	6,15
Changes in salt intake to prevent pre-eclampsia	6,15
Calcium supplementation to improve pregnancy outcome	6,15
Magnesium supplementation to improve pregnancy outcome	6,15
Zinc supplementation to improve pregnancy outcome	6
Antigen avoidance diets to reduce risk of an atopic child	6
SCREENING	
Placental grading by ultrasound to improve perinatal outcome	8,12
Fetal biophysical profile for fetal surveillance	12
PREGNANCY PROBLEMS	
Acupressure for nausea and vomiting of pregnancy if simple measures are ineffective	13
Vitamin B_6 for nausea and vomiting of pregnancy if simple measures are ineffective	13
Ginger for nausea and vomiting of pregnancy	13
Prostigmine for heartburn of pregnancy if simple measures are ineffective	13
Dilute acid or lemon juice for heartburn of pregnancy if antacids do not provide relief	13
Increased salt intake for leg cramps	13
Progestogens for threatened miscarriage with a live fetus	14
Human chorionic gonadotrophin for threatened miscarriage with a live fetus	14
Immunotherapy for recurrent miscarriage	14
Bed-rest for women with pre-eclampsia	15
Plasma volume expansion for pre-eclampsia	15

Table 4 | 403

Table 4 *continued*

Forms of care of unknown effectiveness

	Chapter
Choice among magnesium sulphate, benzodiazepines, and phenytoin for eclampsia	15
Hospitalization and bed-rest for impaired fetal growth	16
Abdominal decompression for impaired fetal growth	16
Betamimetics for impaired fetal growth	16
Oxygen therapy for impaired fetal growth	16
Hormone therapy for impaired fetal growth	16
Calcium-channel blockers for impaired fetal growth	16
Plasma volume expanders for impaired fetal growth	16
Prophylactic betamimetics for multiple pregnancy	17
Hospitalization and bed-rest for triplet and higher-order pregnancy	17
Treatment of group B streptococcus colonization during pregnancy	19
Antiviral agents for women with a history of recurrent genital herpes	19
Routine elective caesarean for breech presentation	22
Postural techniques for cephalic version of breech presentation	22
Prophylactic antibiotics for prelabour rupture of membranes at term or preterm	23
Postpartum prophylactic antibiotics after prelabour rupture of membranes	23
Home uterine activity monitoring for prevention of preterm birth	24
Bed-rest to prevent preterm birth	24
Magnesium supplementation to prevent preterm birth	24
Calcium supplementation to prevent preterm birth	24
Progestogens to prevent preterm birth	24
Magnesium sulphate to stop preterm labour	24
Calcium antagonists to stop preterm labour	24
Routine cervical assessment for prevention of preterm birth	24
Antibiotic therapy in preterm labour	24
Oxytocin antagonists to stop preterm labour	24
Adding thyrotropin-releasing hormone to corticosteroids to promote fetal maturation	25
Sweeping of membranes to prevent post-term pregnancy	26,40
Nipple stimulation to prevent post-term pregnancy	26
Induction instead of surveillance for pregnancy at 41+ weeks gestation	26

Table 4 *continued*

Forms of care of unknown effectiveness

	Chapter

CHILDBIRTH

Routine amnioscopy to detect meconium in labour	30
Routine artificial rupture of membranes to detect meconium in labour	30
Short periods of electronic fetal monitoring as an admission screening test in labour	30
Fetal stimulation tests for fetal assessment in labour	30
Maternal oxygen administration for fetal distress in labour	30
Routinely repeated blood pressure measurements in labour	31
Guarding the perineum versus watchful waiting during birth	32
Prophylactic ergometrine + oxytocin versus oxytocin alone in third stage of labour	33
Early versus late clamping of the umbilical cord	33
Controlled cord traction in third stage of labour	33
Intraumbilical vein oxytocin for retained placenta	33

PROBLEMS DURING CHILDBIRTH

Abdominal decompression to relieve pain in labour	34
Immersion in water to relieve pain in labour	34
Acupuncture to relieve pain in labour	34
Acupressure to relieve pain in labour	34
Transcutaneous electrical nerve stimulation to relieve pain in labour	34
Intradermal injection of sterile water to relieve pain in labour	34
Aromatherapy to relieve pain in labour	34
Hypnosis to relieve pain in labour	34
Continuous infusion versus intermittent top-ups for epidural analgesia	34
Early use of oxytocin to augment slow or prolonged labour	35
'Active management' of labour	35
Cervical vibration for slow or prolonged labour	35
Histoacryl tissue adhesive for perineal skin repair	36
Phenobarbitone to the mother to prevent intraventricular haemorrhage in the very preterm infant	37
Vitamin K to the mother to prevent intraventricular haemorrhage in the very preterm infant	37

Table 4 | 405

Table 4 *continued*

Forms of care of unknown effectiveness

TECHNIQUES OF INDUCTION AND OPERATIVE DELIVERY

CARE AFTER CHILDBIRTH

Table 5   Forms of care unlikely to be beneficial

The evidence against these forms of care is not as firmly
established as for those in Table 6

	Chapter

BASIC CARE

	Chapter
Reliance on expert opinion instead of on good evidence for decisions about care	2
Routinely involving doctors in the care of all women during pregnancy and childbirth	3
Routinely involving obstetricians in the care of all women during pregnancy and childbirth	3
Not involving obstetricians in the care of women with serious risk factors	3
Fragmentation of care during pregnancy and childbirth	3
Advice to restrict sexual activity during pregnancy	5
Prohibition of all alcohol intake during pregnancy	5
Imposing dietary restrictions during pregnancy	6
Routine vitamin supplementation in late pregnancy in well-nourished populations	6
Routine haematinic supplementation in pregnancy in well-nourished populations	6
High-protein dietary supplementation	6

SCREENING

	Chapter
Routine use of ultrasound for fetal anthropometry in late pregnancy	8,12
Using oedema to screen for pre-eclampsia	10
Cold pressor test to screen for pre-eclampsia	10
Roll-over test to screen for pre-eclampsia	10
Isometric exercise test to screen for pre-eclampsia	10
Measuring uric acid as a diagnostic test for pre-eclampsia	10
Screening for 'gestational diabetes'	11
Routine glucose challenge test during pregnancy	11
Routine measurement of blood glucose during pregnancy	11
Insulin plus diet therapy for 'gestational diabetes'	11
Diet therapy for 'gestational diabetes'	11
Routine use of Doppler ultrasound screening in all pregnancies	12
Measurement of placental proteins or hormones (including oestriol and human placental lactogen)	12

Table 5 | 407

Table 5 *continued*

Forms of care unlikely to be beneficial

	Chapter

PREGNANCY PROBLEMS

Table 5 *continued*

Forms of care unlikely to be beneficial

Table 5 | 409

Table 5 *continued*

Forms of care unlikely to be beneficial

Table 6   Forms of care likely to be ineffective or harmful

Ineffectiveness or harm demonstrated by clear evidence

Table 6 | 411

Table 6 *continued*

Forms of care likely to be ineffective or harmful

Index